Y0-AAC-179

STATISTICAL IMAGE PROCESSING AND GRAPHICS

STATISTICS: Textbooks and Monographs

A SERIES EDITED BY

D. B. OWEN, Coordinating Editor

Department of Statistics
Southern Methodist University
Dallas, Texas

Vol. 1: The Generalized Jackknife Statistic, *H. L. Gray and W. R. Schucany*
Vol. 2: Multivariate Analysis, *Anant M. Kshirsagar*
Vol. 3: Statistics and Society, *Walter T. Federer*
Vol. 4: Multivariate Analysis: A Selected and Abstracted Bibliography, 1957-1972, *Kocherlakota Subrahmaniam and Kathleen Subrahmaniam* (out of print)
Vol. 5: Design of Experiments: A Realistic Approach, *Virgil L. Anderson and Robert A. McLean*
Vol. 6: Statistical and Mathematical Aspects of Pollution Problems, *John W. Pratt*
Vol. 7: Introduction to Probability and Statistics (in two parts), Part I: Probability; Part II: Statistics, *Narayan C. Giri*
Vol. 8: Statistical Theory of the Analysis of Experimental Designs, *J. Ogawa*
Vol. 9: Statistical Techniques in Simulation (in two parts), *Jack P. C. Kleijnen*
Vol. 10: Data Quality Control and Editing, *Joseph I. Naus* (out of print)
Vol. 11: Cost of Living Index Numbers: Practice, Precision, and Theory, *Kali S. Banerjee*
Vol. 12: Weighing Designs: For Chemistry, Medicine, Economics, Operations Research, Statistics, *Kali S. Banerjee*
Vol. 13: The Search for Oil: Some Statistical Methods and Techniques, *edited by D. B. Owen*
Vol. 14: Sample Size Choice: Charts for Experiments with Linear Models, *Robert E. Odeh and Martin Fox*
Vol. 15: Statistical Methods for Engineers and Scientists, *Robert M. Bethea, Benjamin S. Duran, and Thomas L. Boullion*
Vol. 16: Statistical Quality Control Methods, *Irving W. Burr*
Vol. 17: On the History of Statistics and Probability, *edited by D. B. Owen*
Vol. 18: Econometrics, *Peter Schmidt*
Vol. 19: Sufficient Statistics: Selected Contributions, *Vasant S. Huzurbazar (edited by Anant M. Kshirsagar)*
Vol. 20: Handbook of Statistical Distributions, *Jagdish K. Patel, C. H. Kapadia, and D. B. Owen*
Vol. 21: Case Studies in Sample Design, *A. C. Rosander*
Vol. 22: Pocket Book of Statistical Tables, *compiled by R. E. Odeh, D. B. Owen, Z. W. Birnbaum, and L. Fisher*
Vol. 23: The Information in Contingency Tables, *D. V. Gokhale and Solomon Kullback*
Vol. 24: Statistical Analysis of Reliability and Life-Testing Models: Theory and Methods, *Lee J. Bain*
Vol. 25: Elementary Statistical Quality Control, *Irving W. Burr*
Vol. 26: An Introduction to Probability and Statistics Using BASIC, *Richard A. Groeneveld*
Vol. 27: Basic Applied Statistics, *B. L. Raktoe and J. J. Hubert*
Vol. 28: A Primer in Probability, *Kathleen Subrahmaniam*
Vol. 29: Random Processes: A First Look, *R. Syski*

Vol. 30: Regression Methods: A Tool for Data Analysis, *Rudolf J. Freund and Paul D. Minton*
Vol. 31: Randomization Tests, *Eugene S. Edgington*
Vol. 32: Tables for Normal Tolerance Limits, Sampling Plans, and Screening, *Robert E. Odeh and D. B. Owen*
Vol. 33: Statistical Computing, *William J. Kennedy, Jr. and James E. Gentle*
Vol. 34: Regression Analysis and Its Application: A Data-Oriented Approach, *Richard F. Gunst and Robert L. Mason*
Vol. 35: Scientific Strategies to Save Your Life, *I. D. J. Bross*
Vol. 36: Statistics in the Pharmaceutical Industry, *edited by C. Ralph Buncher and Jia-Yeong Tsay*
Vol. 37: Sampling from a Finite Population, *J. Hajek*
Vol. 38: Statistical Modeling Techniques, *S. S. Shapiro*
Vol. 39: Statistical Theory and Inference in Research, *T. A. Bancroft and C.-P. Han*
Vol. 40: Handbook of the Normal Distribution, *Jagdish K. Patel and Campbell B. Read*
Vol. 41: Recent Advances in Regression Methods, *Hrishikesh D. Vinod and Aman Ullah*
Vol. 42: Acceptance Sampling in Quality Control, *Edward G. Schilling*
Vol. 43: The Randomized Clinical Trial and Therapeutic Decisions, *edited by Niels Tygstrup, John M. Lachin, and Erik Juhl*
Vol. 44: Regression Analysis of Survival Data in Cancer Chemotherapy, *Walter H. Carter, Jr., Galen L. Wampler, and Donald M. Stablein*
Vol. 45: A Course in Linear Models, *Anant M. Kshirsagar*
Vol. 46: Clinical Trials: Issues and Approaches, *edited by Stanley H. Shapiro and Thomas H. Louis*
Vol. 47: Statistical Analysis of DNA Sequence Data, *edited by B. S. Weir*
Vol. 48: Nonlinear Regression Modeling: A Unified Practical Approach, *David A. Ratkowsky*
Vol. 49: Attribute Sampling Plans, Tables of Tests and Confidence Limits for Proportions, *Robert E. Odeh and D. B. Owen*
Vol. 50: Experimental Design, Statistical Models, and Genetic Statistics, *edited by Klaus Hinkelmann*
Vol. 51: Statistical Methods for Cancer Studies, *edited by Richard G. Cornell*
Vol. 52: Practical Statistical Sampling for Auditors, *Arthur J. Wilburn*
Vol. 53: Statistical Signal Processing, *edited by Edward J. Wegman and James G. Smith*
Vol. 54: Self-Organizing Methods in Modeling: GMDH Type Algorithms, *edited by Stanley J. Farlow*
Vol. 55: Applied Factorial and Fractional Designs, *Robert A. McLean and Virgil L. Anderson*
Vol. 56: Design of Experiments: Ranking and Selection, *edited by Thomas J. Santner and Ajit C. Tamhane*
Vol. 57: Statistical Methods for Engineers and Scientists. Second Edition, Revised and Expanded, *Robert M. Bethea, Benjamin S. Duran, and Thomas L. Boullion*
Vol. 58: Ensemble Modeling: Inference from Small-Scale Properties to Large-Scale Systems, *Alan E. Gelfand and Crayton C. Walker*
Vol. 59: Computer Modeling for Business and Industry, *Bruce L. Bowerman and Richard T. O'Connell*
Vol. 60: Bayesian Analysis of Linear Models, *Lyle D. Broemeling*
Vol. 61: Methodological Issues for Health Care Surveys, *Brenda Cox and Steven Cohen*
Vol. 62: Applied Regression Analysis and Experimental Design, *Richard J. Brook and Gregory C. Arnold*

Vol. 63: Statpal: A Statistical Package for Microcomputers – PC-DOS Version for the IBM PC and Compatibles, *Bruce J. Chalmer and David G. Whitmore*
Vol. 64: Statpal: A Statistical Package for Microcomputers – Apple Version for the II, II+, and IIe, *David G. Whitmore and Bruce J. Chalmer*
Vol. 65: Nonparametric Statistical Inference, Second Edition, Revised and Expanded, *Jean Dickinson Gibbons*
Vol. 66: Design and Analysis of Experiments, *Roger G. Petersen*
Vol. 67: Statistical Methods for Pharmaceutical Research Planning, *Sten W. Bergman and John C. Gittins*
Vol. 68: Goodness-of-Fit Techniques, *edited by Ralph B. D'Agostino and Michael A. Stephens*
Vol. 69: Statistical Methods in Discrimination Litigation, *edited by D. H. Kaye and Mikel Aickin*
Vol. 70: Truncated and Censored Samples from Normal Populations, *Helmut Schneider*
Vol. 71: Robust Inference, *M. L. Tiku, W. Y. Tan, and N. Balakrishnan*
Vol. 72: Statistical Image Processing and Graphics, *edited by Edward J. Wegman and Douglas J. DePriest*
Vol. 73: Assignment Methods in Combinatorial Data Analysis, *Lawrence Hubert*
Vol. 74: Understanding Statistics, *Bruce J. Chalmer*
Vol. 75: Econometrics and Structural Change, *Lyle D. Broemeling and Hiroki Tsurumi*
Vol. 76: Statistical Tools for Simulation Practitioners, *Jack P. C. Kleijnen*
Vol. 77: Randomization Tests, 2nd Edition, *Rusty Edgington*

OTHER VOLUMES IN PREPARATION

STATISTICAL IMAGE PROCESSING AND GRAPHICS

edited by

Edward J. Wegman

Center for Computational Statistics and Probability
George Mason University
Fairfax, Virginia

Douglas J. DePriest

Lighting Business Group
General Electric Company
Cleveland, Ohio

MARCEL DEKKER, INC. New York and Basel

Library of Congress Cataloging-in-Publication Data

Statistical image processing and graphics.

(Statistics, textbooks and monographs ; v. 72)
Includes index.
1. Image processing--Statistical methods.
2. Computer graphics. I. Wegman, Edward J.,
[date]. II. DePriest, Douglas J., [date].
III. Series.
TA1632.S75 1986 621.36'7 86-11683
ISBN 0-8247-7600-3

MARCEL DEKKER, INC.
270 Madison Avenue, New York, New York, 10016

Current printing (last digit):
10 9 8 7 6 5 4 3 2 1

PRINTED IN THE UNITED STATES OF AMERICA

Dedicated to the memory of

Warren Joseph Wegman
August 16, 1947 – July 27, 1984

and

Howard Elliott
October 11, 1952 – November 18, 1984

Viewing Instructions for Color Anaglyphs

A section of red-green anaglyphs and other color art, Plates 1 to 18, is inserted between pages 240 and 241.

Filter eyeglasses are provided at the back of the book for use in viewing the two-color anaglyphs. Hold the eyeglasses with the red lens over the right eye and the green lens over the left. It may take a few seconds for the three-dimensional effect to appear.

Shifting your head from left to right may help in achieving stereo perception. However, some people with strong eye dominance will not find it possible to fuse the stereo images. Faint ghosts can be eliminated by moving further away from the plot or using a less bright light source.

Preface

Image processing techniques and computer graphics techniques are converging. This is made evident by the common use of similar hardware (for example, high resolution raster scan display devices), the common use of similar mathematical and statistical methods (for example, cluster analysis, discriminant analysis, and random field theory), and, finally, the pervasive attempts to create ever more realistic looking graphics using methods such as random tesselations and fractals. This convergence suggested to us that an interesting juxtaposition could be created by inviting members of each community to a thematic workshop. This was done and an Office of Naval Research (ONR) sponsored workshop was held in Luray, Virginia, from 24 through 27 May 1983.

The central themes of the workshop were statistical image processing and statistical graphics, both reflecting research programs at ONR. The former area is frequently reckoned to be in the domain of the electrical engineering community. Indeed, much of the ONR program is centered in this community although the tools of mathematics and statistics have received increasing emphasis. The papers of Elliott et al., Janowitz, Alexander, Lee, Eshera et al., Friedlander, and Gibson represent a variety of sophisticated mathematical and statistical approaches to various problems of image processing. The papers by Yang and Chang, Brown, and Reader represent a stronger computer science orientation addressing several software and hardware issues related to image processing. These ten papers constitute the section of this book that is devoted to statistical image processing.

On the other hand, the statistical graphics area is represented by a variety of researchers primarily based in the mathematics and statistics community. This is an area of research exhibiting tremendous vigor, particularly since 1980. The paper of Banchoff describes the early experiences of a pure mathematician trying to visualize hyperdimensional phenomena, now a common theme of the statistical graphics community. The papers by Donoho et al., Carr et al., McDonald, and Gabriel and Odoroff describe a variety of graphical tools for exploring or visualizing data in three or more dimensions. The paper by Carr et al. is a particularly noteworthy expository treatment. Exploratory analysis of hyperdimensional data is often coupled with curve, surface, or hypersurface fitting. The papers by Wahba, Scott, and Sager describe innovative approaches to such fitting techniques. Finally, the paper by Getty and Higgins explores some of the basic psychology of perception of three dimensions. We felt compelled to include this material since a fundamental tool of hyperdimensional analysis is the use of computer graphics to visualize three-dimensional projections.

This book itself is something of an experiment in the use of color and three-dimensional presentations in a printed format. So far as we know, this is the first attempt to include three-dimensional statistical graphics in a printed format. It has been an interesting if somewhat time-consuming challenge to accomplish these presentations, and we hope that the readers of this volume will enjoy them. We would particularly like to think that Warren Wegman and Howard Elliott, to whom this book is dedicated, would have enjoyed these innovations, for both had a fine sense of delight with the unusual. It is for this bittersweet reason that we have chosen to dedicate this volume to their memory. They were both young men whose full potential was yet to be realized. We miss them both.

During preparation on this volume, we learned of the death of Professor K. S. Fu on 29 April 1985. We are likewise saddened by his passing.

Edward J. Wegman
Douglas J. DePriest

Contributors

Winser E. Alexander Department of Electrical and Computer Engineering, North Carolina State University, Raleigh, North Carolina

Thomas F. Banchoff Department of Mathematics, Brown University, Providence, Rhode Island

Christopher M. Brown Department of Computer Science, University of Rochester, Rochester, New York

D. B. Carr Pacific Northwest Laboratory, Richland, Washington

Shi-Kuo Chang Department of Electrical and Computer Engineering, Illinois Institute of Technology, Chicago, Illinois

R. Cristi* Department of Electrical and Computer Engineering, University of Massachusetts, Amherst, Massachusetts

H. Derin Department of Electrical and Computer Engineering, University of Massachusetts, Amherst, Massachusetts

H. S. Don School of Electrical Engineering, Purdue University, West Lafayette, Indiana

David L. Donoho† Department of Statistics, Harvard University, Cambridge, Massachusetts

H. Elliott‡ Department of Electrical and Computer Engineering, University of Massachusetts, Amherst, Massachusetts

---

**Current affiliation*: Department of Electrical Engineering, University of Michigan, Dearborn, Michigan
†*Current affiliation*: Department of Statistics, University of California, Berkeley, California
‡Deceased

M. A. Eshera School of Electrical Engineering, Purdue University, West Lafayette, Indiana

Benjamin Friedlander Systems Control Technology, Inc., Palo Alto, California

K. S. Fu* School of Electrical Engineering, Purdue University, West Lafayette, Indiana

K. Ruben Gabriel Department of Statistics, University of Rochester, Rochester, New York

D. Geman Department of Mathematics and Statistics, University of Massachusetts, Amherst, Massachusetts

David J. Getty BBN Laboratories, Inc., Cambridge, Massachusetts

Jerry D. Gibson Department of Electrical Engineering, Texas A&M University, College Station, Texas

D. L. Hall† Pacific Northwest Laboratory, Richland, Washington

Peter J. Huber Department of Statistics, Harvard University, Cambridge, Massachusetts

A. W. F. Huggins BBN Laboratories, Inc., Cambridge, Massachusetts

M. F. Janowitz Department of Mathematics and Statistics, University of Massachusetts, Amherst, Massachusetts

Jong-Sen Lee Digital Image Processing Laboratory, Naval Research Laboratory, Washington, D.C.

R. J. Littlefield Pacific Northwest Laboratory, Richland, Washington

John Alan McDonald‡ Department of Statistics, Stanford University, Stanford, California

K. Matsumoto National Aerospace Laboratory, Science and Technology Agency, Tokyo, Japan

W. L. Nicholson Pacific Northwest Laboratory, Richland, Washington

Charles L. Odoroff Division of Biostatistics, University of Rochester, Rochester, New York

Ernesto Ramos Department of Statistics, Harvard University, Cambridge, Massachusetts

---

*Deceased
†*Current affiliation*: Boeing Computer Services Company, Seattle, Washington
‡*Current affiliation*: Department of Statistics, University of Washington, Seattle, Washington

Cliff Reader International Imaging Systems, Inc., Milpitas, California

Thomas W. Sager Department of General Business, University of Texas, Austin, Texas

David W. Scott Department of Mathematical Sciences, Rice University, Houston, Texas

Hans-Mathis Thoma Department of Statistics, Harvard University, Cambridge, Massachusetts

Grace Wahba Department of Statistics, University of Wisconsin, Madison, Wisconsin

Chung-Chun Yang Digital Image Processing Laboratory, Naval Research Laboratory, Washington, D.C.

Contents

Part I
STATISTICAL IMAGE PROCESSING

Application of the Gibbs Distribution to Image Segmentation[1]

H. Elliott, H. Derin, and R. Cristi[2]

Department of Electrical and Computer Engineering
University of Massachusetts
Amherst, MA

D. Geman

Department of Mathematics and Statistics
University of Massachusetts
Amherst, MA

ABSTRACT

This paper describes a new statistical approach to image segmentation. Making use of Gibbs distribution models of Markov random fields a dynamic programming based segmentation algorithm is developed. A number of examples are presented which give an indication of the potential of this approach.

1 INTRODUCTION

This report presents a new statistical approach to the image segmentation problem. By modelling image data as a Markov random field characterized by a Gibbs distribution, a dynamic programming algorithm is developed. The primary contribution of the paper is this new near optimal method for processing scenes described by the noncausal Gibbs model.

Image segmentation, the process of grouping image data into regions with similar features is a component process in image understanding systems and also serves as a tool for image enhancement. As such it has received considerable attention in the literature. Many techniques work well on noise free images with slow spatial variation in intensity. However, when the data is noisy or textured, these algorithms become less reliable. In this case it can be advantageous to statistically model the noise and any texture which is random in

---

[1] Research supported by ONR contract N00014-83-0059 at the University of Massachusetts, Amherst, MA 01003.

[2] *Current affiliation*: Department of Electrical Engineering, University of Michigan, Dearborn, Michigan.

nature. Furthermore, one must also take advantage of two-dimensional spatial ergodicity to average the effects of noise. If a region is spatially ergodic then a pixel and its neighbors will have similar statistical properties. In its simpler forms, the Gibbs model can be used to exploit this type of spatial continuity, and this is its primary role in the segmentation algorithm.

Use of the Gibbs distribution dates back to the work of Ising [1] in 1925 who modelled molecular interaction in ferromagnetic materials, and it has received considerable attention in both the statistical mechanics and statistics literature [2]. However, only recently have attempts been made to apply it to problems in image processing. In [4], the autobinomial form of the Gibbs distribution was used to model texture. The algorithm in [5] segments textured images hierarchically operating on successively smaller blocks and uses Gibbs distributions to model texture. To our knowledge, the work in [6] represents the first application of the Gibbs model to image segmentation. The algorithm in [6] is highly parallel in nature with the flavor of a "relaxation" algorithm and requires a number of iterative passes on the image data. The algorithm we propose processes the data in a raster scan fashion, and only requires a single scan of the data. It will be important to further study the trade-offs in the various algorithms.

The report is organized as follows. Section II defines the segmentation problem in a statistical framework, introduces some notation, and presents some background on Markov random fields. Section III then presents the dynamic programming algorithm in detail for the case of segmenting images consisting of uniform intensity regions in high levels of additive white Gaussian noise. Section IV presents results of applying the algorithms to some experimentally generated images consistent with this model as well as some synthetic aperture radar images which are clearly inconsistent with the assumed model. These results clearly demonstrate the applicability of the technique to realistic data as well as the robustness of the algorithm with respect to modelling assumptions. In Section V, some comments and concluding remarks are given, and extensions to this work which are in progress are briefly outlined.

2 PROBLEM FORMULATION AND MATHEMATICAL BACKGROUND

Preliminary Definitions

Let a class of scenes be characterized by a discrete finite random field $X = [X_{ij}]$ of size $(N_1 \times N_2)$, and let a realization of this field or a specific scene be represented by the matrix $x = [x_{ij}]$. It will be assumed that each pixel (i, j) can belong to one of M distinct region types and that $X_{ij} = m$ if pixel (i, j) is a member of region m, $m \epsilon [1, 2, \ldots, M]$. Associated with a specific scene is a set of K, $(N_1 \times N_2)$ observation matrices $\bar{y} = \{y^k\}_1^K$, $y^k = [y_{ij}^k]$. For simplicity of exposition, we will assume $K = 1$ and simply define y^1 to be $y = [y_{ij}]$. However, it should be pointed out that the algorithms presented below extend trivially to the case of multiple observations such as with Landsat data. Since regions can be textured or contain observation noise the range space of y_{ij} is larger than that of X_{ij}. Thus y will be assumed to be a realization of a real valued random field $Y = [Y_{ij}]$. The general model which can we will employ is

$$Y_{ij} = F_{ij}(X_{ij}) + W_{ij}. \tag{1}$$

The field W_{ij} is a random noise field, and the mapping F_{ij} can be used to characterize texture models. In the case which will be described in most detail, a region will be characterized by constant intensity so that

$$F_{ij}(X_{ij}) = r_m \text{ if } X_{ij} = m \tag{2}$$

Furthermore, N_{ij} will be assumed to be a white Gaussian field with zero mean and variance σ^2, i.e.,

$$W_{ij} \sim N\ (0, \sigma^2). \tag{3}$$

MAP Segmentation

The segmentation problem can now be simply stated as follows. Given the observation matrix $y = [y_{ij}]$ find an estimate $\tilde{x} = [\tilde{x}_{ij}]$ of the scene realization $x = [x_{ij}]$. The algorithm presented below attempts to maximize the posterior probability or likelihood of $\tilde{x}$ given y. In particular if $P(\cdot)$ is an appropriate probability measure, then one would like to find the estimate $\tilde{x}$ which maximizes $P(X = \tilde{x} \mid Y = y)$. Using Bayes rule

$$P(X = \tilde{x} \mid Y = y) = \frac{P(Y = y \mid X = \tilde{x})\ P(X = \tilde{x})}{P(Y = y)}. \tag{4}$$

Since $P(Y = y)$ is independent of the estimate $\tilde{x}$, we can equivalently maximize

$$P(X = \tilde{x},\ Y = y) = P(Y = y \mid X = \tilde{x})\ P\ (X = \tilde{x}) \tag{5}$$

or

$$ln\ P\ (X = \tilde{x},\ Y = y) = ln\ P(Y = y \mid X = \tilde{x}) + ln\ P(X = \tilde{x}). \tag{6}$$

The dynamic programming algorithm presented in the next section is an approximation to one which guarantees finding the $\tilde{x}$ which maximizes (6). It should be pointed out that the difficulty in maximizing (6) is that it is a joint log-likelihood for all the image data. It does not simply describe the likelihood of a single pixel. In particular the maximizing $\tilde{x}$, is one of $M^{N_1 N_2}$ possibilities.

Markov Random Fields and the Gibbs Distribution

Obviously, any processing algorithm for maximization of (6) will depend critically on the form of $P(X = x)$ and $P(Y = y \mid X = x)$. In this subsection, these measures are defined in the case where X is a Markov random field characterized by a Gibbs distribution, and Y is given by (1) — (3).

To begin it will be helpful to introduce some additional notation. Let L be defined as the $N_1 \times N_2$ lattice characterizing all pixel locations in the scene, i.e.

$$L = \{(i,j)\colon 1 \leqslant i \leqslant N_1,\ 1 \leqslant j \leqslant N_2\}. \tag{7}$$

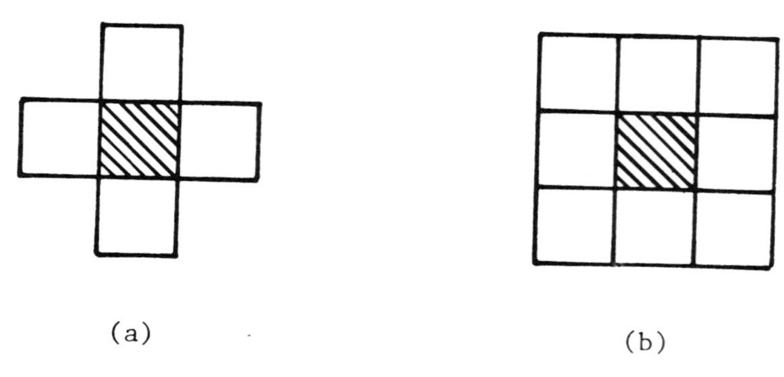

Figure 1 Neighborhoods η^1_{ij} (a) and η^2_{ij} (b)

Next let η_{ij} define a set of neighbor sites for the pixel (i,j) but excluding (i,j) itself. Two simple neighborhoods are depicted in Figure 1. These are:

$$\eta^1_{ij} = \{(l,m)\colon 0 < (i-l)^2 + (j-m)^2 \leqslant 1\} \tag{8}$$

$$\eta^2_{ij} = \{(l,m)\colon 0 < (i-l)^2 + (j-m)^2 \leqslant 2\} \tag{9}$$

Finally, define η_L to be the collection of all neighbors in L, or a neighborhood system. Then η^1_L and η^2_L characterize the collection of all neighborhoods η^1_{ij} and η^2_{ij}, i.e.

$$\eta^1_L = \{\eta^1_{ij} : (i, j) \in L\} \tag{10}$$

$$\eta^2_L = \{\eta^2_{ij} : (i, j) \in L\}. \tag{11}$$

Given this notation, a Markov random field can then be defined as:

Definition 1: A random field on a lattice L is a Markov random field with respect to a neighborhood system η_L if and only if

$$\begin{aligned} P(X_{ij} = x_{ij} \mid X_{lm} = x_{lm}, (l, m) \in L, (l, m) \neq (i, j)) \\ = P(X_{ij} = x_{ij} \mid X_{lm} = x_{lm}, (l, m) \in \eta_{ij}). \end{aligned} \tag{12}$$

If in addition to (12) $P(X = x) > 0$ for all realizations x, then X can be characterized by a Gibbs distribution defined on the neighborhood system η_L [2], [3]. In order to define the Gibbs distribution it is necessary to introduce the notion of the cliques of a neighborhood system. Simply, a clique is any set of pixel locations for which any two are neighbors of each other. Figure 2 show the types of cliques found for the η^1_L and η^2_L neighborhood systems. Let $C(\eta_L)$ be the collections of cliques for the neighborhood system η_L. The Gibbs distribution can now be defined as follows:

Definition 2: Gibbs probability distribution has the form

$$P(X = x) = \frac{1}{Z} e^{U(x)} \tag{13}$$

$$U(x) = \sum_{c \in C(\eta_L)} V_c(x) \tag{14}$$

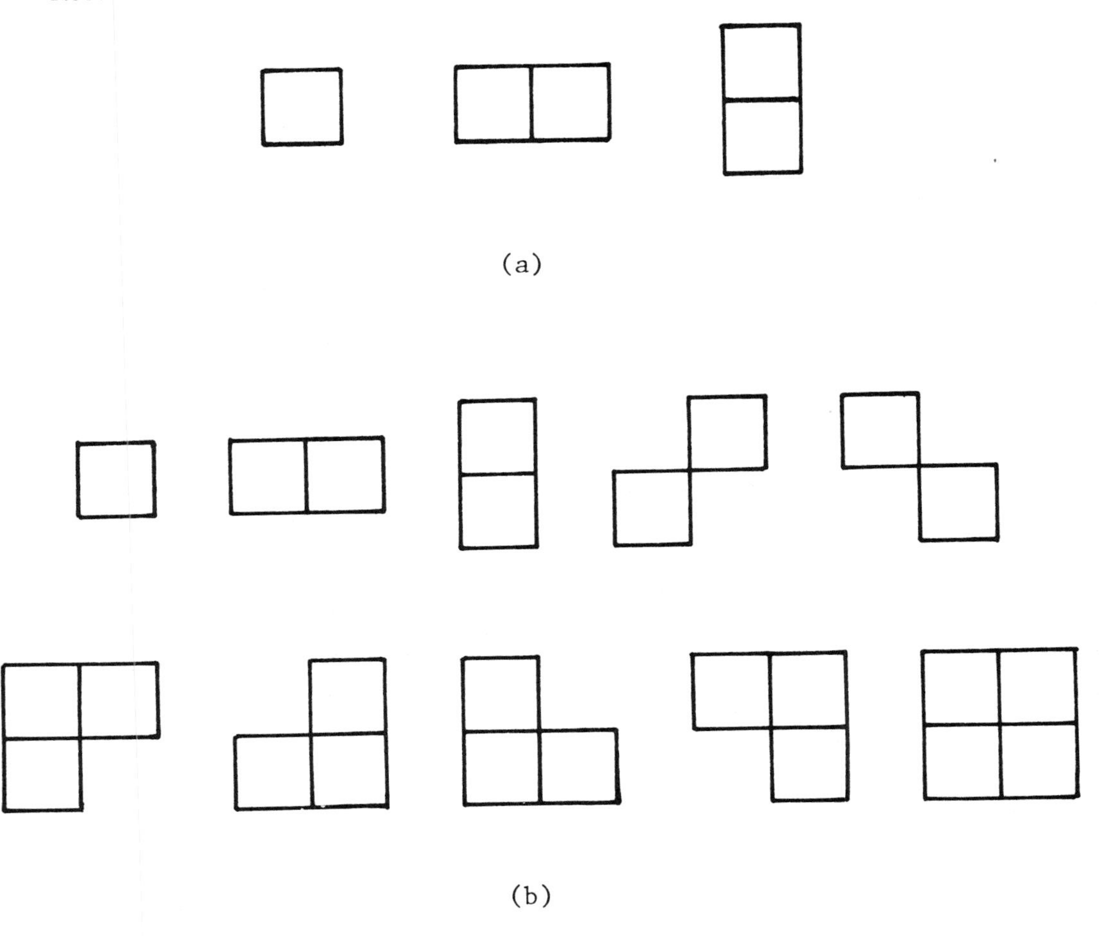

Figure 2 Cliques for neighborhood systems η_L^1 (a) and η_L^2 (b)

$\triangleq$ energy function

$V_c(x) \triangleq$ potential associated with clique c

$$Z = \sum_x e^{U(x)} \tag{15}$$

$\triangleq$ partition function

As can be seen from the definition Z is simply a normalizing constant so that the sum of the probabilities of all realizations, x, add to one. Thus the key functions in determining the properties of the distribution are the potential functions $V_c(x)$. The only limitation on $V_c(x)$ is that it only depends on the values of the pixels in clique c. We will consider homogeneous fields where the form of $V_c(x)$ is fixed by the structure of the clique c and not its location in the lattice. For the segmentation algorithm discussed in the next section,

Table 1 Potential Functions for Segmentation Algorithm

No.		Clique Structure	Potential
1	*	$[(i,j)]$	$V_c(x) = \alpha_m$ if $x_{ij} = m$
2	**	$[(i,j),\ (i,\ j+1)]$	$V_c(x) = \beta_1$ if $x_{ij} = x_{i\ j+1}$ $= -\beta_1$ otherwise
3	* *	$[(i,j),\ (i-1,j)]$	$V_c(x) = \beta_2$ if $x_{ij} = x_{i+1,\ j}$ $= -\beta_2$ otherwise
4	* *	$[(i,j),\ (i-1,\ j+1)]$	$V_c(x) = \beta_3$ if $x_{ij} = x_{i-1,j+1}$ $= -\beta_3$ otherwise
5	** **	$[(i,j),\ (i+1,\ j+1)]$	$V_c(x) = \beta_4$ if $x_{ij} = x_{i+1,\ j+1}$ $= -\beta_4$ otherwise
6	** *	$[(i,j),\ (i+1,\ j),$ $(i,\ j+1)]$	$V_c(x) = \gamma_1$ if all x_{ij} in c are equal $= -\gamma_1$ otherwise
7	* **	$[(i,j),(i-1,j),\ (i,j+1)]$	$V_c(x) = \gamma_2$ if all x_{ij} in c are equal $= -\gamma_2$ otherwise
8	** *	$[(i,j),(i,j+1),(i+1,j+1)]$	$V_c(x) = \gamma_3$ if all x_{ij} in c are equal $= -\gamma_3$ otherwise
9	* **	$[(i,j),(i,j+1),(i-1,j+1)]$	$V_c(x) = \gamma_4$ if all x_{ij} in c are equal $= -\gamma_4$ otherwise
10	** **	$[(i,j),(i-1,j),(i,j+1),$ $(i-1,\ j+1)]$	$V_c(x) = \sigma_1$ is all x_{ij} in c are equal $= -\sigma_1$ otherwise

the potential functions were chosen to exploit spatial continuity, and for simplicity of calculation. We have modelled the region clustering as being characterized by a Markov random field over the neighborhood system η_L^2. The most general form we use for the potential functions $V_c(x)$ are given in Table 1.

The parameters α_i, β_i, γ_i, σ_i need to be estimated for certain classes of scenes. Although this is a difficult problem, there are methods in the literature for estimating these parameters, see [3] — [5]. However, we are not using the Gibbs distribution to model textures or detailed shape. We just want to model the fact that regions are clusters of pixels, i.e., spatial continuity within regions. For the examples in section IV, we use only cliques 1—5, and in many cases just cliques 2 and 3, e.g., $\gamma_i = \sigma_i = 0$. The nonzero parameters were chosen by trial and error. Although for a given signal to noise ratio the algorithm was relatively insensitive to the choice of these values, as the signal to noise ratio changed we found it necessary to modify the values. We are presently studying this phenomena more carefully as well as determining new schemes for estimating these parameters for the context in which they are being used.

Finally, since a particular realization of the Gibbs field, x, assigns each pixel to one of the M region types, using the Gaussian noise model an appropriate form for $P(Y=y \mid X=x)$ is

$$P(Y=y \mid X=x) = \prod_{m=1}^{M} \prod_{(i,j)\in S_m} \frac{1}{(2\pi\sigma^2)^{1/2}} \exp\left(\frac{-1}{2\sigma^2} (y_{ij}-r_m)^2\right) \tag{16}$$

$$S_m = \{ (i,j): X_{ij} = m \} \tag{17}$$

3 NEAR OPTIMAL MAP SEGMENTATION

In this section, an optimal algorithm is posed for processing images consisting of a few rows. The complete near-optimal algorithm is then obtained by applying this optimal processor on overlapping strips of the larger $N_1 \times N_2$ image. This algorithm will be near optimal when correlation between the random variables in X_{ij} drops rapidly as their vertical distance increases. This assumption appears reasonable in the sense that it can often be shown for one dimensional Markov chains. However, calculation of correlations in a Markov random field is extremely difficult even for the simplest case (Ising Model [2]), and is an unresolved problem in the statistics literature.

First consider the problem of segmenting a $D \times N_2$ image, $D << N_1$. Let $l(.)$ denote a log-likelihood function. Using the Gibbs likelihood and the conditional data likelihoods derived in the last subsection of Section II, we can write the following expression for the joint log-likelihood of the image data y and a realization x of the Gibbs field X:

$$l(Y=y,\, X=x) = l(Y=y \mid X=x) + l(X=x) \tag{18}$$

$$l(X=x) = -\ln Z + \sum_{c\in C_L(\eta^2)} V_c\,(x) \tag{19}$$

$$l(Y=y \mid X=x) = \frac{-DN_2}{2} \ln(2\pi\sigma^2) - \sum_{m=1}^{M} \sum_{(i,j)\in S_m} (y_{ij}-r_m)^2 \tag{20}$$

Observe that we can calculate (18) recursively as follows

$$l_{N_1} = l(Y=y,\, X=x)$$

$$l_0 = \frac{-DN_1}{2} \ln(2\pi\sigma^2) - \ln\, Z$$

$$l_k = l_{k-1} + \sum_{c\in C^{k-1,k}} V_c(x) - \sum_{m=1}^{M} \sum_{(i,j)\in S_m^k} \frac{1}{2\sigma^2} (y_{ij} - r_m)^2$$

$C^{k-1,\,k} = \{c \in C(\eta_L^2)$: c contains only pixels in column k or only pixels in column $k-1$ and $k\,\}$

$$S_m^k = \{ (i,j): x_{ij} = m,\; j = k\} \quad 1 \leqslant i \leqslant W$$

This recursion in conjunction with the principle of optimality allows formulation of a forward dynamic programming algorithm [7] for finding $\tilde{x}$. The state space associated with the dynamic programming algorithm has dimension M^D since there are M^D possible segmentations of each column of the $D \times N_2$ scene. This implies that the algorithm would have N_2 iterations with on the order of M^{2D} calculations during each iteration. Thus this algorithm may only be computationally tractable for small values of M and D, e.g., $2 \leqslant M, D \leqslant 4$. Although using the technique described below, full size images can be processed with reasonable computation speeds, we do only recommend the algorithm be implemented for segmenting images into at most M=4 region types. We are presently working on a method which allows this algorithm to be applied iteratively to segment images for which $M > 4$. More will be said regarding this in Section V. Since this is a standard dynamic programming application, details will not be given. However, two important remarks are in order.

Remark 1: Observe that the value of l_0 is independent of any segmentation $\tilde{x}$, and hence the algorithm can be initialized by setting $l_0 = 0$. In particular, there is no need to undertake the difficult task of calculating the partition function Z.

Remark 2: In order to calculate potentials for all cliques in column one and row one of the $(D \times N_2)$ image, it is necessary to assume a segmentation for a fictitious column zero and row zero. This corresponds to a boundary condition for the Markov random field X. An appropriate choice for these boundary conditions will be discussed below.

In order to use the dynamic programming algorithm described above which is capable of optimally processing a $D \times N_2$ strip of an $N_1 \times N_2$ image $N_1 >> D$, we will assume the random variables X_{ij} and $X_{i+D,\,j}$ for all (i,j) to have negligible correlation or covariance. Thus, the segmentation for row i should be negligibly impacted by the segmentation of row $i+D$. In view of this, the complete segmentation procedure is described by the following algorithm.

Segmentation Algorithm

Step (0) Choose a value for D, $2 \leqslant D \leqslant 4$,
Step (1) Set $I = 1$
Step (2) Apply Dynamic Programming Algorithm to rows I through $I+D-I$
Step (3) Store the segmentation for row I
Step (4) Set $I = I+1$
Step (5) If $I \leqslant N_1-D+1$, go to 2
Step (6) $D = D-1$
Step (7) If $D \geqslant 1$, go to 2
Step (8) Stop

To summarize, the dynamic programming algorithm is applied to overlapping image strips of width D, but only the segmentation of the first row of that strip is used. For example, the processing of rows 1 through D yields a segmentation for row 1, and the processing of rows 2 through $D+1$ yields a segmentation of row 2. Under the correlation assumption above, this algorithm is near optimal since the data in row $I + D$ will have little impact on the segmentation of row I.

As pointed out in Remark 2 above, the strip dynamic programming algorithm requires a fixed segmentation or boundary condition for row $I - 1$ and column 0. For $I = 1$, (i.e., row 0) and column zero, we arbitrarily assume all pixels to be background pixels, i.e., $\tilde{x}_{i0} = \tilde{x}_{0i} = M_b$ where $M_b \epsilon [1,2,\ldots,M]$ and M_b is the assumed background intensity. Although this is somewhat arbitrary, if the correlation assumption holds both vertically and horizontally, it will have negligible impact on the segmentation of the scene for $i > D$ and $j > D$. For $I > 1$ we use the fixed segmentation of the previous row to initialize the strip processor.

4 EXAMPLES

In this section, some examples are presented which are representative of the performance of the algorithm and which highlight some of its properties. To begin, consider Figs. 3 and 4. They show the results of applying the algorithm to (64 × 64) test images consisting of an object either an ellipse or a diamond, on a background. Thus, M=2. For these examples, the images were corrupted by additive white zero mean Gaussian noise fields with variances such that the signal to noise ratio S/N=2 where we define

$$S/N = \frac{|r_1 - r_2|}{\sigma}$$

The values of r_1, r_2, and σ were assumed known and the algorithm was applied with D=4 and all Gibbs parameters set to zero except the β_i. For the ellipse of Fig. 3, $\beta_i = 0.2$ while for the diamond of Fig. 4, $\beta_i = 0.15$, i=1,2,3,4. The top row of each figure from left to right show the original and noise corrupted images, while the bottom row from left to right show the Gibbs algorithm segmentation, and the result of filtering the segmented image through a (3 × 3)

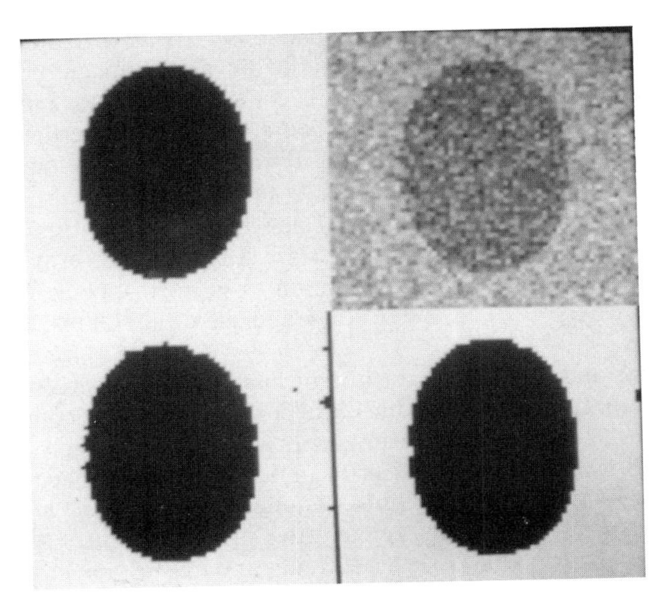

Figure 3 Segmentation of an ellipse with S/N = 2

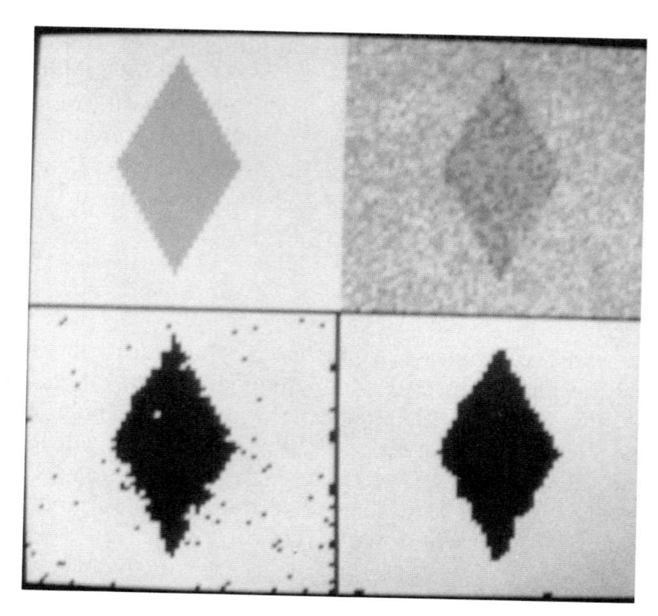

Figure 4 Segmentation of a diamond with S/N = 2

median filter. As can be seen by comparison of the two figures, the algorithm performs well at this signal to noise ratio but as the magnitude of the Gibbs parameters decreases, the algorithm becomes subject to more single pixel errors. This is consistent with expectation since in this case the Gibbs distribution is weighted less relative to the data terms in the likelihood function. Thus there is less emphasis on spatial continuity in the likelihood.

Figures 5 — 7 show the result of applying the algorithm to an ellipse in noise such that $S/N = 1$ using $D=2,3,4$. Although performance is best for $D=4$ there is surprisingly little degradation in performance as D decreases. In all three figures $\beta_i = 0.175$, $i = 1,2,3,4$ and all other Gibbs parameters were zero.

Figures 8 and 9 show the results of applying the algorithm to a diamond in noise such that $S/N=1$. In both Figs. $D=4$, however by comparison of Figs. 8 and 9, one can see the improvement in performance which can be obtained by assuming some additional prior knowledge of shape. For Fig. 8 $\beta_i=0.2$ $i=1,2,3,4$ and all other Gibbs parameters were zero. For Fig. 9, realizing that the object had diagonal edges, the diagonal cliques were emphasized relative to the horizontal and vertical cliques, i.e., $\beta_1=\beta_2=0.15$ and $\beta_3=\beta_4=0.25$ while all other parameters were zero. For the latter case, the shape of the diamond was more obvious in the segmentation.

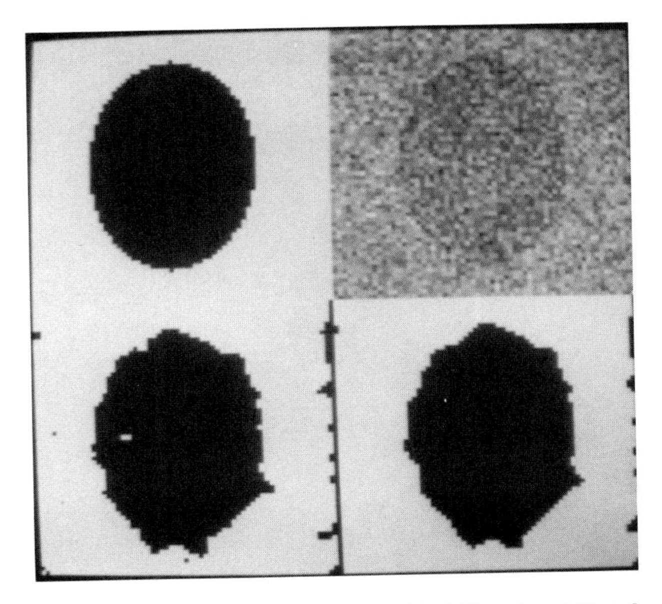

Figure 5 Segmentation of an ellipse with S/N = 1 and D = 2

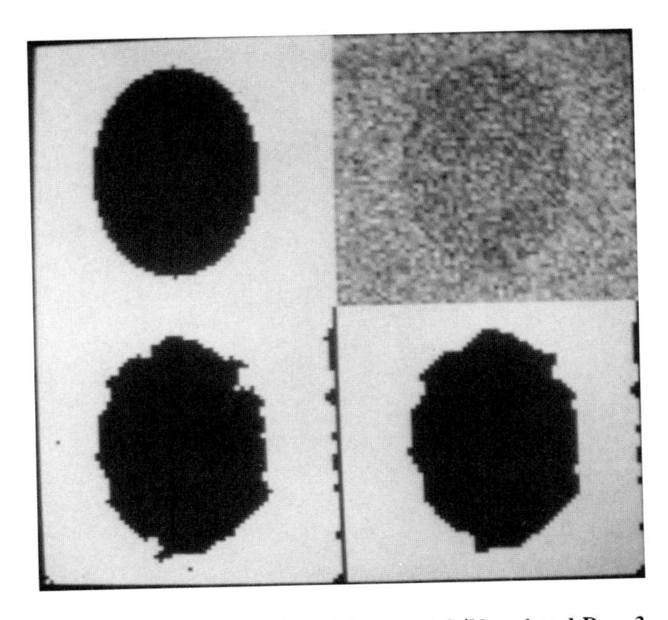

Figure 6 Segmentation of an ellipse with S/N = 1 and D = 3

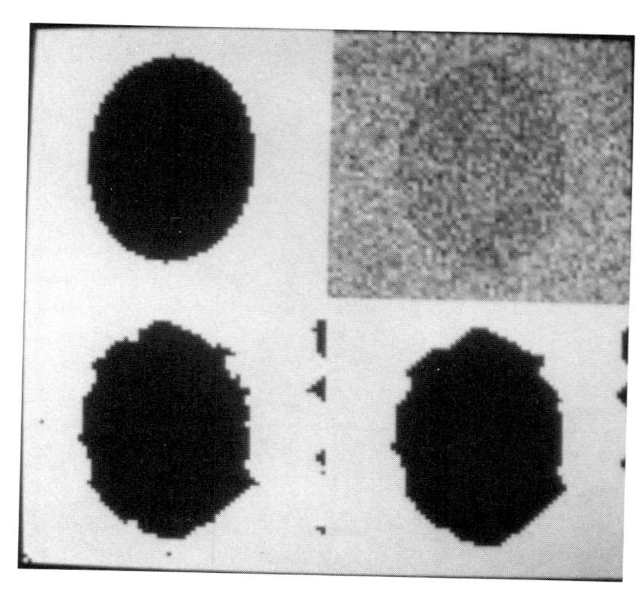

Figure 7 Segmentation of an ellipse with S/N = 1 and D = 4

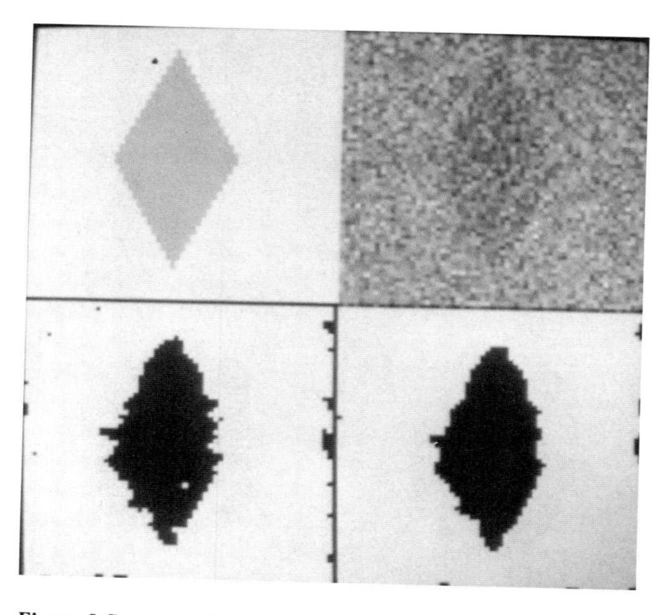

Figure 8 Segmentation of a diamond with S/N = 1 all β_i equal

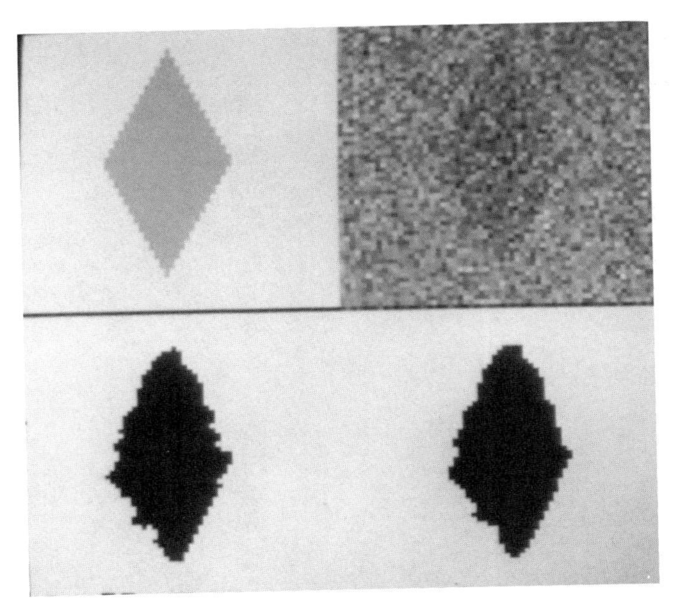

Figure 9 Segmentation of a diamond with S/N = 1 and β_i for diagonal cliques emphasized

For Figs. 10 — 17, the algorithm was applied to two test images, each containing $M=4$ region types. For Figs. 10 — 15, $D=2$ while for Figs. 16 and 17, $D=3$. We used $D=2$ to keep computation times down, however as can be seen from Figs. 12 — 15, while performance at $S/N=1.5$ is reasonable, performance at $S/N=1$ is not. For $M=4$, we define the signal to noise ratio as

$$S/N = \min_{i \neq j} \frac{|r_i - r_j|}{\sigma} .$$

Increasing to $D=3$ did considerably improve performance at $S/N=1$, however CPU time for this case on a VAX 11/780 was 60 minutes as compared to 4 minutes for $M=4$ and $D=2$ and 7 minutes for $M=2$ and $D=4$. This is our motivation for going to the hierarchical scheme for handling multiregion images ($M \geqslant 3$) which will be briefly outlined in Section IV below. We also point out that we are rewriting our code to make use of look-up tables and anticipate a factor of 10 or better improvement in computation speed.

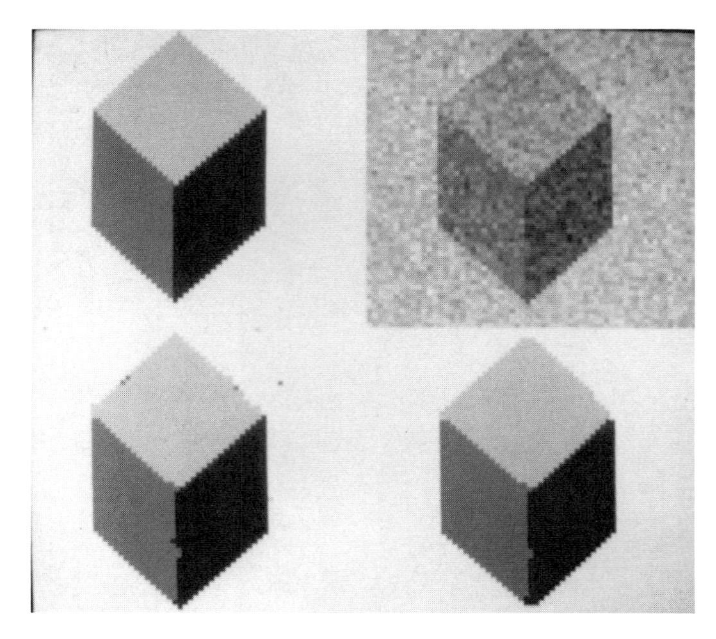

Figure 10 Segmentation of a four region diamond test image with S/N = 2 and D = 2

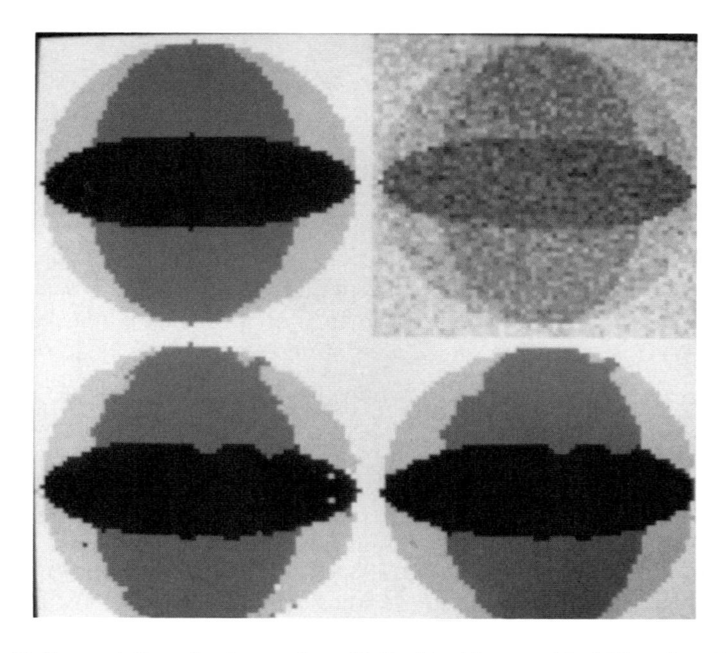

Figure 11 Segmentation of a four region elliptical test image with S/N = 2 and D = 2

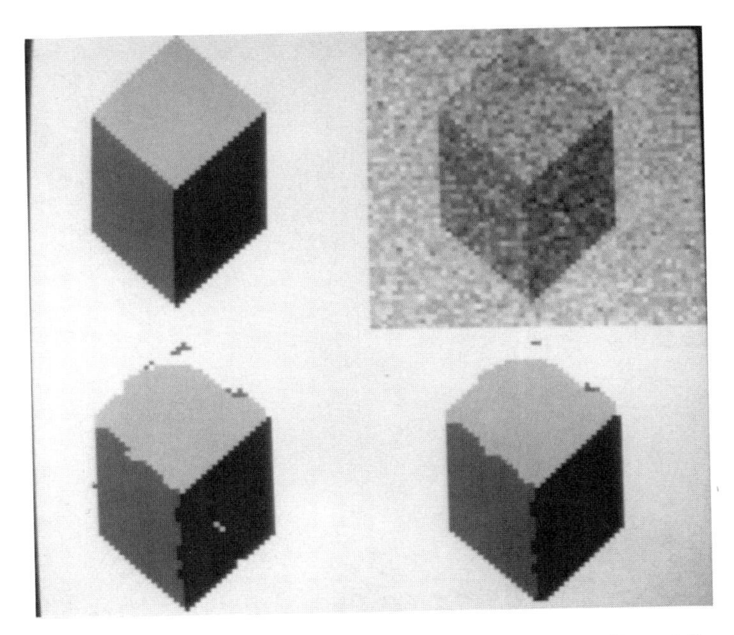

Figure 12 Segmentation of a four region diamond test image with S/N = 1.5 and D = 2

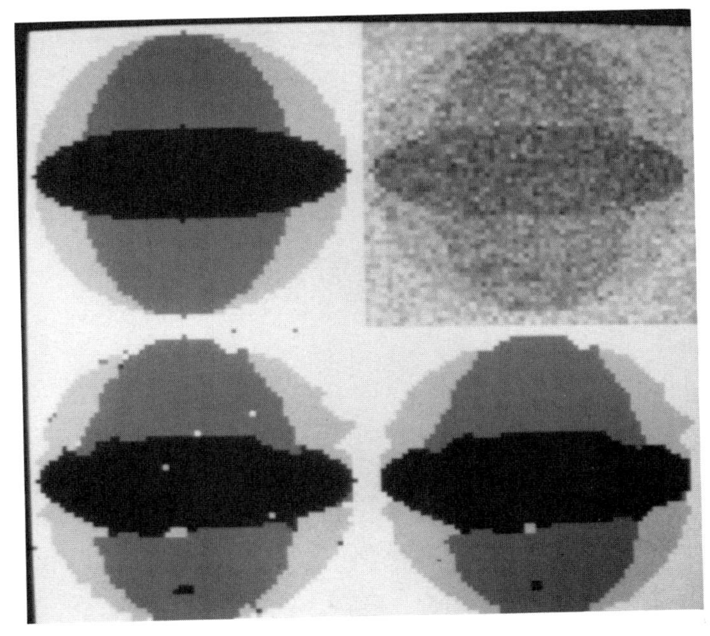

Figure 13 Segmentation of a four region elliptical test image with S/N = 1.5 and D = 2

Figure 14 Segmentation of a four region diamond test image with S/N = 1 and D = 2

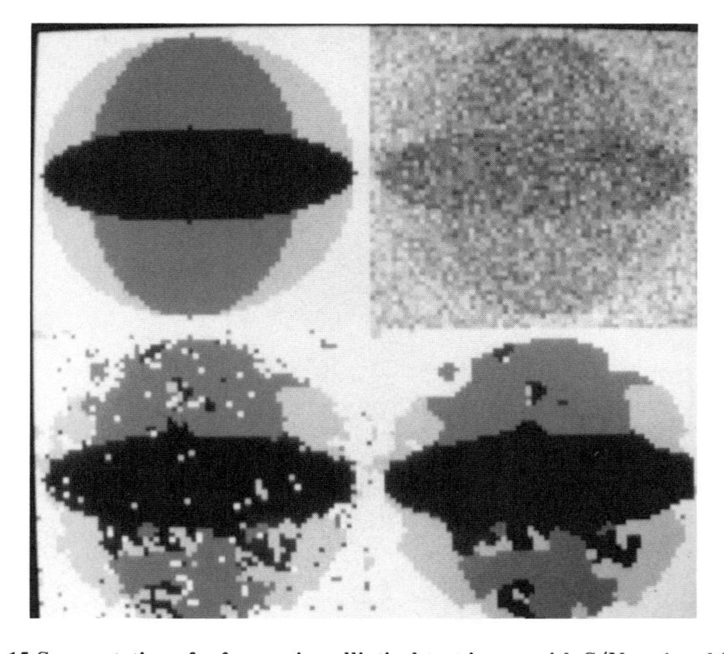

Figure 15 Segmentation of a four region elliptical test image with S/N = 1 and D = 2

Figure 16 Segmentation of a four region diamond test image with S/N = 1 and D = 3

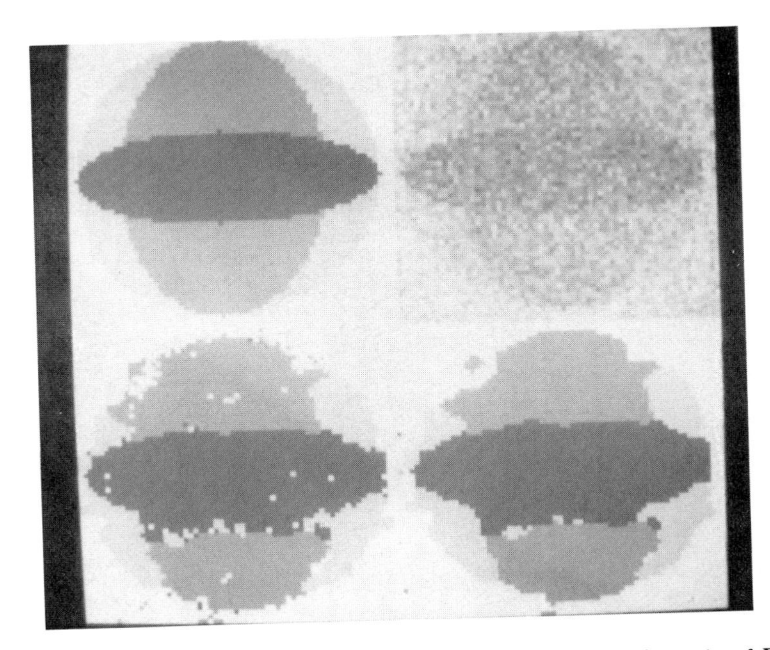

Figure 17 Segmentation of a four region elliptical test image with S/N = 1 and D = 3

Finally, for Figs. 19 — 21, the algorithm was applied to (64 × 64) sections of the synthetic aperture radar image shown in Fig. 18. Figure 19 shows the result of applying algorithm to a small bay, while in Fig. 20, there is a river with some bridges, and in Fig. 21 there is a boat. In all cases, we assumed $M=2$, and obtained r_1, r_2 and σ by applying the method of moments under the assumption that $(64)^2$ pixels in the scene where samples of a random variable characterized by a distribution which was a mixture of two Gaussian distributions. Each picture shows the original scene and three segmentations corresponding to different Gibbs parameters. For Fig. 19, proceeding clockwise around the picture we had $\alpha_1 = -.05$ (white), $\alpha_2=.05$ (black) and all $\beta_i=0.125$, $\alpha_1=\alpha_2=0$ and all $\beta_i=0.125$, $\alpha_1=\alpha_2$ and all $\beta_i=0.15$. For Fig. 20, we had, $\alpha_1=.05$ (white), $\alpha_2=-.05$ (black), and all $\beta_i=0.1$, $\alpha_1=.05$, $\alpha_2=-.05$, and all $\beta_i=0.125$, $\alpha_1=\alpha_2=0$ and all $\beta_i=0$. For Fig. 21 we had, $\alpha_1=\alpha_2=0$ and all $\beta_i=0.2$, $\alpha_1=\alpha_2=0$ and all $\beta_i=0.35$, $\alpha_1=0.1$, $\alpha_2=-0.1$ and all $\beta_i=0.35$.

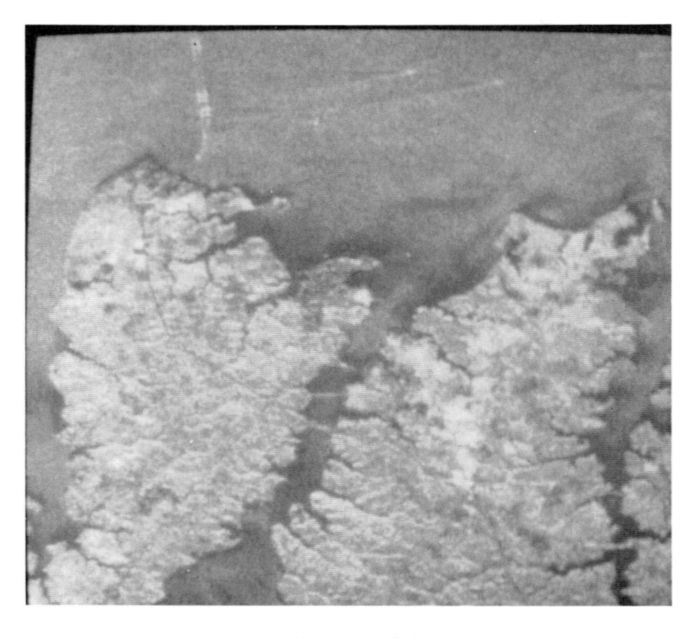

Figure 18 (512 × 512) SAR Image

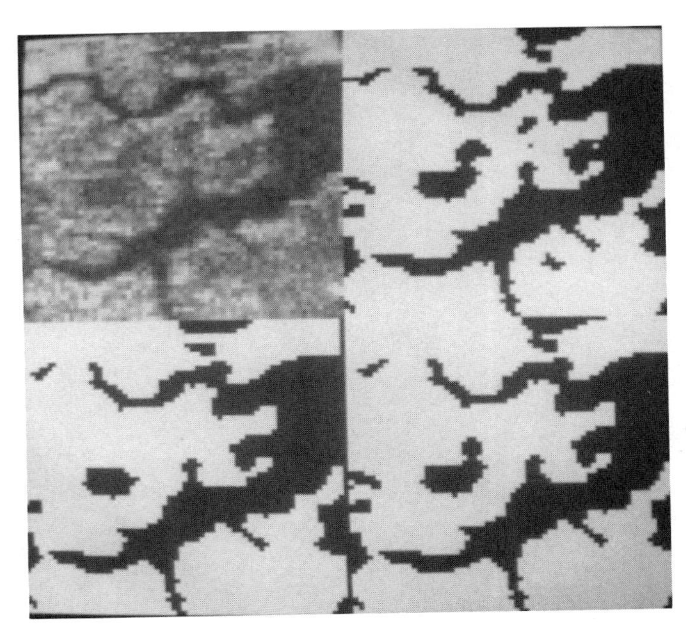

Figure 19 Three segmentations of a (64 x 64) section of SAR image containing a small bay

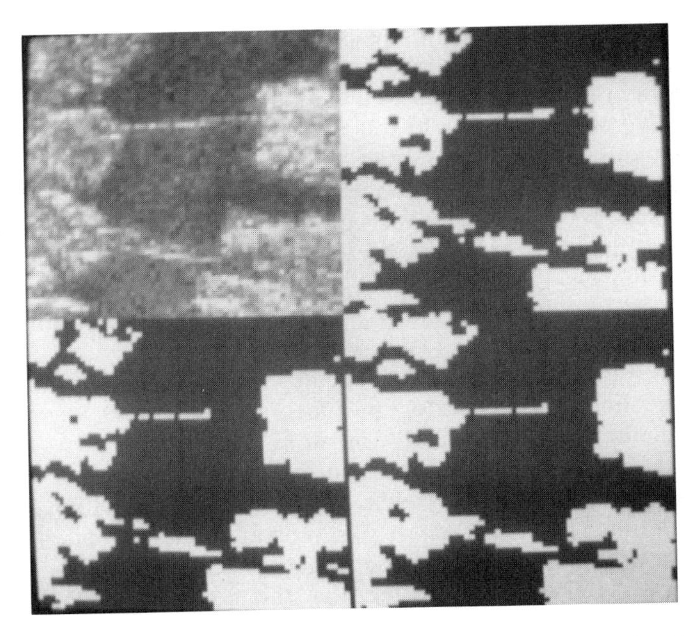

Figure 20 Three segmentations of a (64 x 64) section of SAR image containing river with bridges

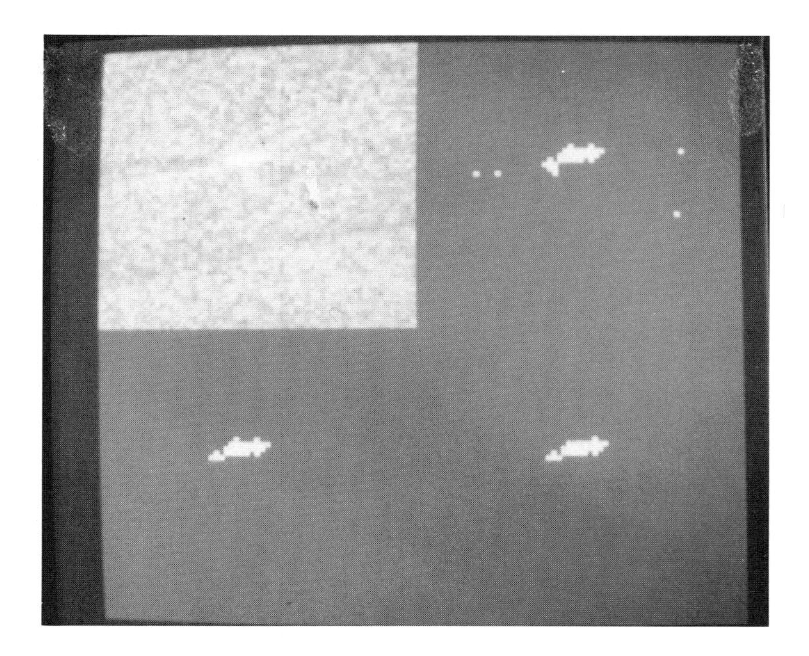

Figure 21 Three segmentations of a (64 × 64) section of SAR image containing a boat

5 CONCLUDING REMARKS

This paper has presented a new approach to segmentation of noisy images. It uses the Gibbs distribution to model spatial continuity or clustering properties of regions. The algorithm is recursive in nature, requires a single pass on the data, and works well at low signal to noise ratios.

Presently, two extensions to this algorithm are being developed. The first allows computationally efficient segmentation of images with more than two region types. If there are M region types, the two region version of the algorithm is applied $M-1$ times. Thus computation time grows linearly with the number of regions. To see how this can be done consider the case where $M=4$ and for convenience, the regions have been labelled so that $r_1 < r_2 < r_3 < r_4$. The two region algorithm can first be applied using region means of r_2 and r_3. Since $r_1 < r_2$ and $r_4 > r_3$ the result will be a segmentation grouping regions 1 and 2 together, and regions 3 and 4 together. Next the two region algorithm is applied to the image using region means of r_1 and r_2. However, only pixels classified as being in the region with mean r_2 during the first pass are classified in the second pass. Those classified as being in the region with mean r_3 during the first pass are ignored. This yields a three class segmentation into regions 1, 2 and the combined regions 3 and 4. To separate regions 3 and 4 a third pass is made using region means r_3 and r_4 but ignoring those pixels assigned to regions 1 and 2 during the first pass. We feel that this hierarchical algorithm will have little effect on the near optimality of the overall approach, however we have no examples to show at this time.

The second extension is to textured images. In this case, instead of modelling regions as being of constant intensity but imbedded in uncorrelated noise, we assume textured regions are realizations of a second Gibbs distribution. In this case, we are developing algorithms similar to the one given above, but with the Gaussian (quadratic) data term in the log-likelihood replaced by terms associated with the second Gibbs distribution characterizing texture. In this context, we are also developing new methods for Gibbs parameter estimation in texture images. The main difference between our work on texture modelling and that in [4], [5] is that we are experimenting with different types of potential functions $V_c(x)$.

In conclusion, we feel that the applications of Gibbs models to problems in image processing and image understanding are just beginning to emerge. It is potentially a very powerful tool but there are still many problems to be resolved.

REFERENCES

[1] E. Ising, Zeitschrift Physik, Vol. 31, p. 253, 1925.

[2] R. Kinderman, J. L. Snell, *Contempory Mathematics: Markov Fields and their Applications*, Vol. 1, American Mathematical Society Providence, R.I., 1980.

[3] J. Besag, "Spatial Interaction and the Statistical Analysis of Lattice Systems," J. Royal Statistical Society, Series B, Vol. 36, pp 192 - 236, 1974.

[4] G. R. Cross, and A. K. Jain, "Markov Random Field Texture Models," *IEEE Trans on PAMI*, Vol. PAMI-5, No. 1, pp. 25-39, 1983.

[5] F. Cohen, *Parallel Adaptive Heirarchical Segmentation of Textured Images Using Noncausal Markovian Fields*, Ph. D. Dissertation, Brown University, August 1983.

[6] D. Geman and S. Geman, "Stochastic Relaxation, Gibbs Distributions, and the Bayesian Restoration of Images," August 1983 (In preparation).

[7] R. Bellman, *Dynamic Programming*, Princeton University Press, Princeton N.J., 1957.

[8] K. Abend, T. J. Harley, and L. N. Kanal, "Classification of Binary Random Patterns," *IEEE Trans. Inform. Theo.*, Vol. IT-21, pp 538 - 544, October 1965.

[9] L. N. Kanal, "Markov Mesh Models," *Image Modeling*, Academic Press, 1980.

[10] D. K. Pickard, "A Curious Binary Lattice Process," *J. Appl. Prob.*, Vol. 14, pp. 717 - 731, 1977.

[11] D. B. Cooper, H. Elliott, F. Cohen, L. Reiss and P. Symosek, "Stochastic boundary estimation and object recognition," *Computer Graphics and Image Processing*, Vol. 10, pp. 326 - 355, April 1980.

[12] F. R. Hansen and H. Elliott, "Image Segmentation Using Simple Markov Random Field Models," *Computer Graphics and Image Processing*, Vol. 20, pp. 101 - 132, 1982.

A Model for Ordinal Filtering of Digital Images[1]

M.F. Janowitz

Deparment of Mathematics and Statistics
University of Massachusetts
Amherst, MA

ABSTRACT

Properties of spatially defined filters for digital images are investigated. If one wants to be able to use a piece-wise linear mapping to change the "lookup" table for the display of a monochromatic image, and if one wants to be able to reasonably interpret its effect on the filtered version of the image, it is shown that a certain class of ordinal filter must be used. Properties of these filters are examined, and in the presence of certain types of errors, the filter is explicitly given which has the highest probability of correctly classifying pixels. These "best" filters are the "median" filters for interiors of regions, and modifications of median filters near boundaries between regions.

1 INTRODUCTION

A monochromatic digital image may be thought of as a finite rectangular array of numbers. Each number in the array represents some sort of average value for a specified type of radiation from a fixed geographic or geometric region, with adjacent numbers corresponding to adjacent regions. The digital image is subsequently transformed to an actual picture by means of some sort of video display device. As will soon be seen, however, a given digital image can produce more than one picture, so one must examine the meaning of the display of this type of data. This question will be examined within the context of a model that allows one to classify various filter techniques, and to compare com-

[1] This research was supported by the Office of Naval Research under contract N00014-79-C-0629.

peting techniques so as to decide which is better. In that the underlying model is order theoretic in nature, it is appropriate at this point to present some needed background material from the theory of partially ordered sets. Though the material can be presented in a more abstract fashion, we choose to stay within the image processing framework in which it will be applied.

In what follows, X will denote a finite rectangular array of points, with X having m rows and n columns. The point in the ith row and j^{th} column is denoted $x(i,j)$ or simply x if it is not important to specify the exact coordinates. A monochromatic digital image may then be thought of as a mapping $d{:}X \rightarrow L$ where L denotes a finite chain. Though the exact nature of L may vary from computer to computer, a typical representation may be obtained by taking $L = \{0,1.\ldots,255\}$ where $i < j$ in L has its usual meaning in the real number system. For M a nonempty subset of L, we agree to let $\vee M$ and $\wedge M$ denote respectively the largest and smallest elements of M. The symbol $\mathscr{P}'$ or $\mathscr{P}'(X)$ will be used to denote the collection of nonempty subsets of X, partially ordered by set inclusion. In connection with this, the symbols $\subset, \subseteq, \cup, \cap$ will have their usual set theoretic meanings.

Now let $d{:}\ X \rightarrow L$ be any mapping. Notice that d may be uniquely extended to a join homomorphism $D^*{:}\,\mathscr{P}'(X) \rightarrow L$ by the rule

$$D^*(Y) = \vee\{d(y) : y \in Y\}.$$

One can also associate with d a mapping

$$D{:}\ L\ [h_d,255] \rightarrow \mathscr{P}'$$

where

$$h_d = \vee\ \{d(x){:}x \in X\}$$

by the rule

$$D(h) = \{x \in X{:}d(x) \leqslant h\}.$$

The pair (D*, D) of mappings has a number of interesting features in that

($R1$) $D^*$is isotone

($R2$) D is isotone

($R3$) $Y \subseteq D[D^*(Y)]$ for all $Y \in \mathscr{P}'(X)$

($R4$) $h \geqslant D^*[D(h)]$ for all $h \geqslant h_d$.

This says that $D^*{:}P' \rightarrow [h_d,255]$ is residuated in the sense of [1], p. 11. The reader is referred to that source for a development of the properties of residuated and residual mappings. What we shall need is the fact that we have established one-one correspondences between three classes of objects:

(A) mappings $d{:}X \rightarrow L$

(B) residuated mappings $D^*{:}\,\mathscr{P}'(X) \rightarrow [h_d,255]$

where

$$h_d = \wedge\ \{d(x){:}x \in X\}$$

(*C*) residual mappings $D:[h_d,255] \to \mathscr{P}'(X)$.

The link from (B) to (A) is accomplished by observing that d is the restriction of D^* to singleton subsets of X with the remaining links arising from the theory of residuated mappings [1].

Suppose now that $\theta:L \to L$ is isotone. If $M = [\theta(0),255] = \{h: h \geqslant \theta(0)\}$, it is easy to see that if θ is regarded as a mapping from L into M, then θ is residuated with $\theta^+: M \to L$ given by

$$\theta^+(k) = \vee\{h: \theta(h) \leqslant k\}.$$

This may be illustrated by defining θ on {0,1,2,3,4,5,6,7,} by means of the table

Table 1

h	0	1	2	3	4	5	6	7
$\theta(h)$	3	3	4	5	6	6	6	6

Then $M = \{3,4,5,6,7\}$ and θ^+ is given by

Table 2

h	3	4	5	6	7
$\theta^+(h)$	1	2	3	7	7

There is one final needed item. If $d: X \to L$ and $\theta: L \to L$ is isotone, then $\theta d:X \to L$ with $\theta \circ D^*$ its associated residuated mapping. As in the proof of [2], Lemma 7.1, p. 68,

$$(\theta \circ D^*)^+ = D \circ \theta^+.$$

In summary then, a picture may either be viewed in the form (A), (B) or (C) as previously described. Somewhat closer contact may be made with the model described in [2] by working with the lattice of all subsets of $X, \mathscr{P}(X)$. In a manner similar to that described earlier in the section, one establishes a bijection between the following classes of objects:

($\overline{A}$) mappings $d:X \to L$

3 FILTERS

It is often desirable to take a picture $d: X \to L$ and enhance it to produce a picture $F(d): X \to L$. This is done by applying some sort of spatial or frequency domain filter to d. To achieve the most generality, let us agree to call any transformation of a picture $d : X \to L$ to a picture $F(d) : X \to L$ a *filter*. It is understood, however, that for a filter to be useful, it must somehow or other aid in the understanding of the original picture. At any rate, applying a filter to d produces F(d). This may be illustrated schematically by

$$\boxed{d} \xrightarrow{F} \boxed{F(d)}$$

We now apply an isotone mapping θ to the lookup table. This transforms both the original and filtered versions of the picture as indicated below:

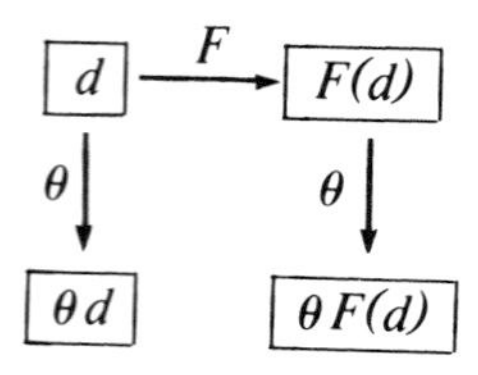

Suppose now that an object is observed in $\theta F(d)$ that was not observed in any of the other images. Is it real or is it an artifact of F and/or θ? Unless one understands the relations between θd and $\theta F(d)$, this is difficult to answer. An ideal relationship would be if

$$F(\theta d) = \theta F(d),$$

since this says that the process that produced F(d) from d also produced $\theta\ F(d)$ form $\theta\ (d)$. This now leads to a natural question. For a given class of isotone mappings $\theta : L \rightarrow L$, what filters F have the property that

$$F(\theta d) = \theta F(d)$$

for all digital images $d : X \rightarrow L$? In connection with this, it will often be useful to view a digital image $d : X \rightarrow L$ as a residual map $D : M \rightarrow \mathcal{P}'(X)$ where $M = [h_d, 255]$ and $h_d = \vee\{d(x) : x \in X\}$. The filter

$$d \rightarrow F(d)$$

then takes the form

$$D \rightarrow \bar{F}(D)$$

where $\bar{F}(D)$ is the residual mapping associated with F(d). Notice then that by [2], Lemma 7.1, p. 60,

$$F(\theta d) = \theta F(d) \Longleftrightarrow \bar{F}(D \circ \theta^{+}) = \bar{F}(D) \circ \theta^{+}.$$

DEFINITION

(i) Let $M \subseteq L$ and $\theta : L \rightarrow L$ be isotone. The filter F is said to be (θ, M) *-compatible* in case

$$F(\theta d) = \theta F(d)$$

for all images d with range contained in M.

(ii) The filter F is said to be *monotone equivariant* if it is (θ, M)-compatible for all θ such that $\theta|_M$ is injective.

Our first goal will be an examination of the meaning of monotone equivariance for filters F. It should be mentioned that though this has essentially already been done in [3], it will still be useful to repeat the proofs here.

$(\overline{B})$ residuated mappings $D^*:p(X) \rightarrow [h_d,255]$

where h_d is some fixed lower band for $\{d(x;x \epsilon X)\}$

$(\overline{C})$ residual mappings $D:[h_d,255] \rightarrow \mathscr{P}(X)$

The situation will be examined in more detail in the next section.

2 DISPLAY OF DIGITAL IMAGES

Let $d:X \rightarrow L$ be a monochromatic digital image, where L might be $\{0,1,\ldots,255\}$. The digital image d is converted to an actual picture by means of some sort of visual display device. The display device either produces a black and white picture by associating with each integer i in L an intensity of grey $C[i]$, or it produces a color picture by associating with i a color $C[i]$. In the latter situation, the display is controlled by a color lookup table of the form

0	1	2	....	255
C [0]	C [1]	C [2]	....	C [255]

where the 256 colors are chosen from a palette consisting of 2^{3k} possible colors with k representing the number of bits assigned to represent a given color. The colors in the table are then linearly ordered by the rule

$$C[i] < C[j] \text{ if } i < j \text{ in } L.$$

For purposes of this discussion, it will be assumed that the colors or grey shades C[0], $C[1],\ldots,C[255]$ are fixed. Many image processing systems come equipped with a means of dynamically modifying the lookup table so as to enhance certain portions of the picture. This deserves a careful consideration.

We are given a digital picture d where

$$d: X \rightarrow L.$$

If θ is an isotone mapping on L, a new picture may be created by looking at

$$\theta d: X \rightarrow L.$$

But this involves applying θ to each and every pixel $x \epsilon X$ to change the color representation from

$$x \rightarrow d(x) \rightarrow C[d(x)]$$

to

$$x \rightarrow \theta d(x) \rightarrow C[\theta d(x)].$$

The same effect may be achieved by changing the lookup table so that it becomes

0	1	2	...	255
$C[\theta\ (0)]$	$C[\theta\ (1)]$	$C[\theta\ (2)]$	...	$C[\theta\ (255)]$.

Since this only involves changing 256 entries, it is the preferred approach. This, however, has dramatic implications. When a digital image is entered into the computer, one can obtain not just a single picture but a whole class of pictures that arise by means of lookup table modifications. This should cause one to examine the significance of the actual data values in a given picture d.

An alternate view of a digital picture involves associating with each color C[h] that portion of the picture that is painted with the color C[h]. Thus one would look at

$$C[h] \to h \to \{x \in X: d(x) = h\}$$

instead of

$$x \to d(x) \to C[d(x)].$$

What happens to the alternate representation when an isotone mapping θ is applied to d? As long as the restriction of θ to the range of d is injective, this is easy to answer, since one simply looks at

$$C[\theta(h)] \to h \to \{x: d(x) = h\}.$$

What happens though if θ merges distinct levels of the picture? One can certainly still consider

$$C[\theta(h)] \to \{x: \theta d(x) = \theta(h)\},$$

but it turns out to be more useful to do a second alternate representation by means of

$$C[h] \to \{x: d(x) \leqslant h\}.$$

Since for $h \geqslant h_d = \vee\{d(x): x \in X\}$ it is true that

$$\theta d(x) \leqslant h <==> d(x) \leqslant \theta^+(h),$$

we then have

$$C[h] \to \theta^+(h) \to \{x:d(x) \leqslant \theta^+(h)\}.$$

The second alternate representation associates with each color C[h] those pixels x that are painted with C[h] or some color that preceded it in the lookup table.

Lemma 1. If F is monotone equivariant, then range $F(d) \subseteq$ range d, for every $d : X \to L$.

Proof. If $h \in$ range d, one can define $\theta : L \to L$ so that $\theta(k) = k$ for all $k \in$ range d, but $\theta(h) \neq h$. Then $d = \theta d$, whence

$$F(d) = F(\theta d) = \theta F(d).$$

If $k = (Fd)(x)$, then

$$k = (Fd)\ (x) = \theta[(Fd)(x)] = \theta(k).$$

In that $h \neq \theta(h)$ we conclude that $h \notin$ range $F(d)$.

Theorem 2. If F is monotone equivariant and if d has range $h_0 < h_1 < \ldots < h_t$, then F(d) depends only upon the sequence

$$D(h_0) \subset D(h_1) \ (\ldots \subset D \ h_t)$$

and is independent of the actual levels $h_0 < h_1 < \ldots < h_t$.

Proof. Let $d : X \rightarrow L$ be as above, and let $Y_i = D(h_i)(i = 0, I, \ldots, t)$. In view of Lemma 1, the output is determined by the sequence

$$(\bar{F}D) \ (h_0) \subseteq (\bar{F}D)(h_1) \subseteq \ldots \subseteq (\bar{F}D)(h_t)$$

since $h_i \leqslant h < h_{i+1}$ implies $(\bar{F}D)(h) = (\bar{F}D)(h_i)$. To show that the output is independent of the actual levels h_i, choose a sequence

$$k_0 < k_1 < \cdots < k_t$$

of elements of L and define

$$c : X \rightarrow L$$

so that $Y_i = C(k_i)$. If $\theta : L \rightarrow L$ is defined so that $\theta(h_i) = k_i$ one then has

$$\theta d = c,$$

whence

$$F(c) = F(\theta d) = \theta F(d).$$

Then

$$\bar{F}(C) = (\bar{F}D) \circ \theta^+$$

and

$$(\bar{F}C)(k_i) = (\bar{F}D)(\theta^+ \ (k_i)) = (\bar{F}D)(h_i).$$

This shows a monotone equivariant filter to be ordinal in nature. It transforms a strictly increasing sequence

$$A_0 \subset A_1 \subset \cdots \subset A_t = X,$$

of subsets of X to an increasing sequence

$$B_0 \subseteq B_1 \subseteq \cdots \subseteq B_t = X,$$

where B_i depends on the entire input sequence.

Remark 3. The converse of Theorem 2 is also true. To see this assume that d has range $h_0 < h_1 < \cdots < h_t$, that range F (d) $\subseteq$ range d, and that D (d) depends only on the sequence

$$D(h_0) \subset D(h_1) \subset \cdots \subset D(h_t) = X$$

of subsets of X. If $\theta|_{\text{range d}}$ is injective, we can let $k_i = \theta(h_i)$ and $c = \theta(d)$. Then

$$d(x) \leqslant h_i \Longleftrightarrow c(x) \leqslant k_i$$

so that $C(k_i) = D(h_i)$ for $i = 0, 1, \ldots, t$. But then

$$(\bar{F}C)(k_i) = (\bar{F}D)(h_i),$$

so

$$(Fc)(x) \leqslant k_i \Longleftrightarrow (Fd)(x) \leqslant h_i$$

and consequently

$$(Fc)(x) = k_i \Longleftrightarrow (Fd)(x) = h_i.$$

It is immediate that

$$F(\theta d) = \theta F(d)$$

as desired.

In the most general situation, a change of lookup table involves the application of a linear or piecewise linear mapping θ to L. In the linear case, one is given non-negative integers a, b such that

$$0 \leqslant a < b \leqslant 255$$

and $\theta_{a,b}$ is defined by the rule

$$\theta_{a,b}(h) = \begin{cases} 0 & h \leqslant a \\ [\frac{h-a}{b-a} \circ 255] & a \leqslant h \leqslant b \\ 255 & h > b \end{cases}$$

where [t] denotes the result of rounding the real number t to the nearest integer. When $b = a + 1$, we may denote $\theta_{a,b}$ as θ_a and observe that

$$\theta_a(h) = \begin{cases} 0 & h \leqslant a \\ 255 & h > a \end{cases}$$

It turns out that (θ,M)—compatibility for all θ_a is in itself a powerful condition on a filter F.

Definition A filter F is said to be flat if there is an isotone mapping $\gamma : \mathscr{P}(X) \rightarrow \mathscr{P}'(X)$ such that

$$\bar{F}(D) = \gamma \circ D$$

for all $d : X \rightarrow L$.

Theorem 4 Let F be (θ_a,M) – compatible for all θ_a $(0 \leqslant a < 255)$ and all $M \subseteq L$. Then F is flat.

Proof. Let c, d : X → L be digital images, and let a,b, ∊ L be such that $C(a) = D(b)$. We claim that $(\bar{F}C)(a) = (\bar{F}D)(b)$. This is a direct consequence of the fact that

$$(C \circ \theta_a^+(h) = \begin{cases} C(a) & h \neq 255 \\ X & h = 255 \end{cases}$$

$$(D \circ \theta_b^+)(h) = \begin{cases} D(b) & h \neq 255 \\ X & h = 255 \end{cases}.$$

Thus $\theta_a\ c = \theta_b d$, so

$$[\bar{F}(C)] \circ \theta_a^+ = [\bar{F}(D)] \circ \theta_b^+$$

and

$$[\bar{F}(C)](a) = [\bar{F}(D)](b).$$

We may now define $\gamma : \mathscr{P}(X) \rightarrow \mathscr{P}'(X)$ by

$$\gamma(Y) = [\bar{F}(D)](b)$$

$$\gamma(\phi) = \phi,$$

where $D(b) = Y$ and note that γ is well defined. It is then clear that

$$[\bar{F}(D)](h) = (\gamma \circ D)(h)$$

so F is flat.

Corollary 5. If F is (θ_a, M)—compatible for every θ_a and every $M \subseteq L$, then F is (θ, M)—compatible for all isotone mappings $\theta : L \rightarrow L$ and all $M \subseteq L$.

In view of the above discussion it is reasonable to consider the theory underlying flat filters. Such filters are easy to construct as one only needs to produce an isotone mapping γ on $\mathscr{P}$(X) such that $\gamma(X) = X$, and $\gamma(\phi) = \phi$: Having recongnized this need, it is appropriate to decide upon reasonable conditions for these mappings. The idea is to attach some spatial significance as to whether a point belongs to the output of the mapping. In other words, if $Y \subseteq X$, the decision as to whether $x \in \gamma(Y)$ should be based on all points in some small region surrounding x. This may be precisely stated by saying that the mapping γ shall be *point-based* in that for each $x \in X$ there is a subset $N(x)$ of X containing x and a mapping $\gamma_x : \mathscr{P}(N(x)) \rightarrow \{0,1\}$ such that:

$$(PB1)\ \gamma_x\ (N(x)) = 1, \text{ and } \gamma_x\ (\phi) = 0.$$

$$(PB2) \text{ For } A \in \mathscr{P}(X),\ x \in \gamma(A) \Longleftrightarrow \gamma_x\ (A \cap N(x)) = 1.$$

So that γ will be independent of the polarity of the image, it turns out to be useful to have γ *preserve complements* in that

$$(PB3) \text{ For } A \in \mathscr{P}\ (X),\ \gamma\ (X \backslash A) = X \backslash \gamma(A).$$

Remark 6. For a point-based isotone mapping γ on $\mathscr{P}$(X) to preserve complements, it is clearly necessary and sufficient that γ_x have the property that for every proper subset A of N(x), exactly one of $\gamma_x(A)$ and $\gamma_x(N(x) \backslash A)$ shall equal 1.

Now let γ be point-based with an associated family of neighborhoods $N(x)(x \in X)$. If M(x) is a second such family, it is evident that so is M (x) $\cap$ N (x). In view of this, there is no harm in assuming

$(PB4)$ If $M(x)$ $(x \in X)$ satisfies$(PB1)$ and $(PB2)$, then $N(x) \subseteq M(x)$.

Such a minimal family of subsets of X will be called the *system of neighborhoods associated with* γ. *Unless otherwise specified, when we speak of a point-based isotone mapping* γ, *it will be assumed that the family* $\{N(x):x \in X\}$ *represents this minimal system of neighborhoods.*

In order for γ to interact properly with planar rigid motions, we need to consider the effect of translations. These are mappings of the form $T_{p,q}$ where p,q are fixed integers and $T_{p,q}(i,j) = (p + i, q + j)$. Unless $p = 0 = q$ the domain and range of $T_{p,q}$ will be proper subsets of X. We agree to call the point-based isotone mapping *translation — invariant* provided it satisfies

(PB5) If N(x) is contained in the domain of the translation T, if $N(x) \neq \{x\}$, and if $y = T(x)$, then $N(y) = T(N(x))$ and $\gamma_y = \gamma_x \circ T^{-1}$.

Remark 7. The reason for the concern about whether $N(x) = \{x\}$ is that around the edge of an image there may not be enough room to form the desired neighborhoods, so that one might simply take $N(\lambda) = \{x\}$.

Definition. Let γ be a point-based, translation-invariant isotone mapping on $\mathscr{P}$(X). To say that γ is *frequency-defined* is to say that there exists an integer j such that if $N(x) \neq \{x\}, \gamma_x(A \cap N(x)) = 1$ if and only if $\#(A \cap N(x)) \geqslant j$. Here $\#Y$ denotes the cardinality of the set Y.

Theorem 8. Let γ satisfy (PB1) through (PB5). Let x be chosen so that $N(x) \neq \{x\}$ and let $k = \#N(x)$. Let j be the least cardinal number for which $\gamma_x A = 1$ for some subset A of N (x) with $j = \#A$. Necessary and sufficient conditions for γ to be frequency-defined are that k be odd and $j = (1 + k)/2$.

Proof: Let γ be frequency-defined. If $\#A \leqslant k/2$, then $\#(N(x) \setminus A) \leqslant k \setminus 2$, so there is a subset B of $N(x)/A$ with $\#B = \#A$. Since $\gamma_x(B) = 0$, this is a contradiction. If k is even, taking $B \subseteq N(x)$ with cardinality k/2 will produce a similar contradiction. Thus k is odd and $\#A \geqslant k/2$. If now $\#B = (1 + k)/2$, then $\#B > \#N(x) \setminus B$. It follows that $\gamma_x(B) = 1$ and consequently that $j = (1 + k)/2$.

Suppose conversely that k is odd and $j = (1 + k)/2$. We must show that $\gamma_x(B) = 1$ for all $B \subseteq N(x)$ for which $\#B \geqslant j$. Since $\#B \geqslant j$ implies $\#(N(x) \setminus B) < j$, it follows that $\gamma_x(N(x) \setminus B) = 0$ and consequently $\gamma_x(B) = 1$.

The mapping γ is called a *join homomorphism* is case $\gamma(S \cup T) = \gamma(S) \cup \gamma(T)$ for all $S,T \in \mathscr{P}(X)$. There is a dual notion of *meet homomorphism*, and to say that γ is a *homomorphism* is to say that it is both a join and a meet homomorphism. The next theorem shows that these conditions are extremely powerful, and will be only met in trivial circumstances.

Theorem 9. Let γ be a point-based isotone mapping on $\mathscr{P}(X)$. Then $(1a) \Longleftrightarrow (1b)$, $(2a) \Longleftrightarrow (2b)$ and $(3a) \Longleftrightarrow (3b)$.

(1a) γ is a join homomorphism.

(1b) For each $x \in X$ there is a subset A_x of N(x) such that $x \in \gamma(S) \Longleftrightarrow S \cap A_x \neq \phi$.

(2a) γ is a meet homomorphism.

(2b) For each $x \in X$ there is a subset B_x of N(x) such that $x \in \gamma(S) \Longleftrightarrow S \supseteq B_x$.

(3a) γ is a homomorphism.

(3b) For each x there is an element y_x of N(x) such that $x \in \gamma(S) \Longleftrightarrow y_x \in S$.

Proof. (1a) $\Longrightarrow$ (1b) One simply takes C_x to be the union of all subsets of N(x) that are mapped to 0 by γ_x and notes that $\gamma_x(C_x) = 0$. The set $N(x) \setminus C_x$ then serves as A_x.

(1b) $\Longrightarrow$ (1a) If $x \in \gamma(S \cup T)$, then $(S \cup T) \cap A_x \neq \phi$, so $S \cap A_x \neq \phi$ or $T \cap A_x \neq \phi$. Hence $x \in \gamma(S) \cup \gamma(T)$.

(2a) $\Longrightarrow$ (2b) One takes B_x to be the intersection of all subsets C of N(x) for which $\gamma(C) = 1$.

(2b) $\Longrightarrow$ (2a) Is obvious, as is (3b) $\Longrightarrow$ (3a).

(3a) $\Longrightarrow$ (3b) If B_x is defined as in the proof of (2a) $\Longrightarrow$ (2b), then $\gamma_x(B_x) = 1$ but $\gamma_x(C) = 0$ for all $C \subset B_x$. It is immediate that B_x is a singleton set, and (3b) follows.

Remark 10. If γ is translation-invariant, it is tempting to say that if A is in the domain of the translation T, then $\gamma[T(a)] = T[\gamma(a)]$. This is only true if both A and $T(A)$ lie in that portion of X on which $N(x) \neq \{x\}$. To illustrate this, let γ be defined by 3 by 3 neighborhoods by the $\gamma_x(A) = 1$ if $\#A \geqslant 5$ for $A \in P(N(x))$. Let $X = \{1,2,3,4,5\} \times \}1,2,3,4,5\}$, $x = (2,2)$ and $T = T_{1,1}$. Then if A is the subset indicated by 1's in Figure (a), one sees from Figs. (c) and (d) that $T[\gamma(A)] \neq \gamma[T(A)]$.

11100	00000	11100	00000
11100	01110	11100	00100
11100	01110	11000	01110
00000	01110	00000	00100
00000	00000	00000	00000
A	T(A)	$\gamma(A)$	$\gamma[T(A)]$
(a)	(b)	(c)	(d)

4 UNDERLYING STATISTICAL CONSIDERATIONS

Suppose that the true input data $d^*: X \rightarrow L$ has been corrupted by additive or multiplicative noise so as to produce an actual input $d: X \rightarrow L$. One wants to construct a flat filter F that somehow smooths the noise without blurring the picture. Thus at each level h, one wishes to estimate

$$A_h{}^* = \{x : d^*(x) \leqslant h\}$$

by means of looking at

$$A_h = \{x : d(x) \leqslant h\}.$$

Hopefully, F(d) will provide a good estimate of A_h by way of

$$\{x: (Fd)(x) \leqslant h\}.$$

To put the problem in a specific framework, suppose that A_h and $A_h{}^*$ are related by the assumption that for some fixed a priori probability $p (p > 0.5)$ one has

$$(S1) x \in A_h{}^* \Longrightarrow x \in A_h \text{ with probability } p,$$

$$(S2) x \notin A_h{}^* \Longrightarrow x \notin A_h \text{ with probability } p,$$

and that these probabilities are independent of the location of x. The goal is to define γ on $\mathscr{P}(X)$ so that (1) γ is isotone and (2) γ somehow maximizes the probability of correct classification of points. In other words, in some specific sense we wish to maximize the a posterior probability that

$$(a) \text{ If } x \in \gamma(A_h) \text{ then } x \in A_h{}^*,$$

and

$$(b) \text{ If } x \notin \gamma(A_h) \text{ then } x \notin A_h{}^*.$$

In view of the discussion of §3, it seems reasonable to restrict our attention to translation-invariant, point-based isotone mappings having a fixed system $\{N(x) : x \in X\}$ of neighborhoods of points. Because of this, we need only consider a fixed point x for which $N(x) \neq \{x\}$, and the mapping $\gamma_x : \mathscr{P}(N(x)) \rightarrow \{0,1\}$ defined by

$$\gamma_x(A) = 1 \Longleftrightarrow x \in \gamma(A)$$

for $A \subseteq N(x)$.

For a given subset $A_h{}^*$ of X, one can define a probability measure P' on $N(x)$ by the rule that for $A \subseteq N(x)$, $P'(A)$ shall be the probability of observing A, given the true input data $A_h{}^* \cap N(x)$. For example, if $N(x)$ is a 3 by 3 neighborhood cantered on x, if $A_h{}^* \cap N(x)$ and A are given by the matrices

111		101
011	and	000
000		001

where 1 denotes membership in a given set and 0 membership in its complement, we have $P'(A) = p^5(1-p)^4$. A numerical measure of the ability of γ_x to distinguish between $A_h{}^*$ and its complement is provided by the *gain* $G(\gamma_x)$. For $x \in A_h{}^*$, this is defined to be the probability that for a randomly chosen subset B of $N(x)$, $\gamma_x(B) = 1$; added to the probability that if the true input data were the complement of $A_h{}^*$, we would have $\gamma_x(B) = 0$ for B a randomly chosen subset of $N(x)$. In connection with this it is convenient to let

$$T_\gamma = \{B : B \subseteq N(x), \gamma_x(B) = 1\}$$
$$F_\gamma = \{B : B \subseteq N(x), \gamma_x(B) = 0\}.$$

It follows that

$$G(\gamma_x) = \sum \{p(A) : A \in T_\gamma\} + \sum \{p(N(x) \setminus A) : \in F_\gamma\}.$$

The reason for this is that if P'' is the probability measure associated with the complement of $A_h{}^*$, then for $B \subseteq N(x)$,

$$P''(B) = P'(N(x) \setminus B).$$

One may now search for those mappings γ such that γ_x has maximal gain for a given type of input data. The search is aided by the following result:

Theorem 11. Let P' be a probability measure defined on $N(x)$ Suppose P' has the property that

(1) for $A \subseteq N(x), P'(A) \neq P'(N(x) \setminus A)$, and

(2) if $P'(A) > P'(N(x) \setminus A)$ and if $A \subset B \subseteq N(x)$,
then

$$P'(B) > P'(r(x)/B)$$

Define $\gamma_x' : p(N(x)) \rightarrow \{0,1\}$ by the rule

$$\gamma_x'(A) = \begin{cases} 1 & P'(A) > P'(N(x) \setminus A) \\ 0 & \text{otherwise.} \end{cases}$$

Then γ_x' is isotone; furthermore, $G(\gamma_x') \geqslant G(\gamma_x'')$ for any other isotone mapping $\gamma_x'' \rightarrow \{0,1\}$.

Proof. The fact that γ_x' is isotone comes from condition (2) of the theorem. The assertion regarding $G(\gamma_x') \geqslant G(\gamma_x'')$ follows from an examination of the sets $T_{\gamma'}$, $F_{\gamma''}$, $T_{\gamma''}$ and $F_{\gamma''}$. Any changes in the sums for $G(\gamma_x')$ and $G(\gamma_x')$ must clearly come from members of

$$(T_{\gamma'} \cap F_{\gamma''}) \cup (T_{\gamma''} \cap F_{\gamma'}).$$

For $A \in T_{\gamma'} \cap F_{\gamma''}$, the term in $G(\gamma_x')$ corresponding to A is $P'(A)$, while in $G(\gamma'')$ it is $P'(N(x) \setminus A)$. By construction of γ', we have $P'(A) > P'(N(x) \setminus A)$. Similarly, for $B \in T_{\gamma''} \cap F_{\gamma'}$, the term in $G(\gamma')$ is

$P'(N(x) \setminus B)$ while the one in $G(\gamma'')$ must be $P'(B)$. Again by construction of γ', $P'(N(x) \setminus B) > P'(B)$.

Corollary 12. If $N(x) \subseteq A_h{}^*$ and $k = \#N(x)$ is odd, then the mapping $\gamma_x{}'$ described in the theorem is given by

$$\gamma'_x(A) = 1 \text{ if } k/2 < \#A.$$

Remark 13. The mapping in Corollary 12 may be implemented by means of applying a suitable median filter to d. This says that within the framework of the current model, the best way of smoothing noise in the interior is to use a median filter.

The paper will conclude by carefully examining the situation where $N(x)$ is a 3 by 3 window centered on x. The members of $A_h{}^*$ will be denoted by 1's, those of its complement by 0's, and we shall restrict our attention to a fixed x. Notice that the mapping produced by Theorem 11 is necessarily complement preserving. This is desirable because one wants the output of a filter to be independent of the polarity of the input data. In view of this, we shall concentrate on input data for which $x \in A_h{}^*$ and assume that $5 \leqslant \#(N(x) \cap A_h{}^*)$. We shall begin by looking at 5 types of input data. They will be denoted type i $(i = 0,1,2,3,4)$, according to whether or not $N(x)$ has $i = 0,1,2,3$ or 4 member of $X \setminus A_h{}^*$. Before doing the theoretical analysis, let us examine some calculated probabilities of correct classifications. This is done in Table 3. There are two items of interest here. For the type 4 case, the filters all seem to decrease the probability of correct classification. This is hardly surprising since in this instance the 8 immediate neighbors of the given pixel all have equal chances of representing members of $A_h{}^*$ as opposed to its complement. Thus no additional information can be obtained by looking at the 8 neighbors of x. The second observation is that in the type 3 case the answer seems to depend on the numerical value p of the a priori probability of correct classification.

Table 3 Probabilities of correct classification.

Prob.		Type				
Correct	Method	0	1	2	3	4
.9	A	.9991	.9954	.9787	.9128	.7071
	B	.9962	.9861	.9590	.9112	.8717
	C	.9914	.9515	.9104	.9014	.8980
.8	A	.9804	.9529	.8944	.7839	.6079
	B	.9584	.9233	.8736	.8150	.7487
	C	.9006	.8564	.8212	.8046	.7929
.7	A	.9012	.8467	.7715	.6747	.5606
	B	.8576	.8125	.7629	.7100	.6519
	C	.7757	.7450	.7222	.7057	.6906
.6	A	.7334	.6870	.6367	.5833	.5280
	B	.6963	.6661	.6354	.6043	.5722
	C	.6374	.6250	.6139	.6036	.5936

Method A. $x \in \gamma_A (A_h)$ if $5 \leqslant \# A_h \cap N(x)$.

Method B. $x \in \gamma_B (A_h)$ if $6 \leqslant \# A_h \cap N(x)$ or if $\# A_h \cap N(x) \in \{4,5\}$ and $x \in A_h{}^*$

Method C. $\# (A \cap N(x)) \in \{3,4,5,6\}$ and $x \in A_h{}^*$

Define $\gamma_x{}' : P(N(x)) \rightarrow \{0,1\}$ by the rule

$$\gamma_x{}'(A) = \begin{cases} 1 & P'(A) > P'(N(x) \setminus A) \\ 0 & \text{otherwise} \end{cases}.$$

Then $\gamma_x{}'$ is isotone; furthermore, $G(\gamma_x{}') \geqslant (\gamma_x{}'')$ for any other isotone mapping $\gamma_x{}'' \rightarrow \{0,1\}$.

One can bring Theorem 11 to bear on the problem by simply calculating probabilities of occurrence of each of the 2^9 possible subsets of $N(x)$, given the a priori assumption regarding the nature of $Al_h{}^* \cap N(x)$. This has been done but the rather tedious details will not be repeated here. Rather, we shall simply list the results in Table 4.

Table 4 Summary of "best" flat filters based on 3 by 3 neighborhoods.

DATA TYPE	Best Isotone Method
0	γ_A of Table 3
1	γ_A of Table 3
2	γ_A of Table 3
3	There is critical value p_0 of $p (p_0 \approx .8960)$ such that for $p > p_0$, γ_A is the best method and for $p < p_0$, γ_B of Table 3 is best. p_0 is determined by solving the equation $p = x(1 - p)$ where x is the root of the equation $x^6 - 9x^5 + 6^4 - 24x^3 + 6x^2 - 4x + 1 = 0$ lying between $x = 8.618$ and $x = 8.619$
4	$x \in \gamma(A_h)$ if $x \in A_h$.

Up to this point it has been assumed that the center pixel x was a member of $A_h{}^*$, and that any other members of $N(x) \cap A_h{}^*$ where randomly scattered through $N(x) \setminus \{x\}$. The situation changes dramatically if more information is known. To see this, note that $N(x)$ consists of three data types as indicated below:

```
c  b  c
b  a  b
c  b  c
```

The center pixel is type a, its North, South, East, West neighbors are type b, and the remaining four corner pixels are type c. Our basic assumption will now be that $x \in A_h{}^*$, and that we know the number of type b as well as the number of type c pixels that are in the complement of $A_h{}^*$ but not their precise location. Thus a data type may be specificed by an ordered pair of integers that denotes the number of pixels of types b and c that lie in the complement of $A_h{}^*$. For example, data type (1,2) indicates the presence of one type b pixel and two type c pixels in the complement of $A_h{}^*$. The calculations involved in applying Theorem 11 are again tedious, so they will be omitted. By way of illustration, however, the type (0,1)- case will be considered in some detail.

Recall that p denotes a priori probability of the input data being correct, and set $q = 1 - p$. The probability of observing various numbers of 1's in the corners is then seen to be

#1's	Probabilities
4	p^3q
3	$p^4 + 3p^2q^2$
2	$3p^3q + 3pq^3$
1	$3p^2q^2 + q^4$
0	pq^3

Since by assumption the type a and b pixels all lie in $A_h{}^*$, they can lumped together into a single category. There are now $6 \times 5 = 30$ data types to analyze. This involves examining 15 pairs of subsets. But this can be cut down to 6 pairs if one realizes two things:

1. If there are more 1's than 0's in both data types, the result is clear.

2. If there are two 1's in the type c data, then the decision rests entirely on the type a and b.

DATA	TYPE		
A	$N(x) \setminus A$	$P'(A)$	$P'(N(x) \setminus A)$
(5,0)	(0,4)	$\cdot\ p^6q^3$	p^3q^6
(4,0)	(1,4)	$\cdot\ p^5q^4$	p^4q^5
(3,0)	(2,4)	p^4q^5	$\cdot\ p^5q^4$
(5,1)	(0,3)	$\cdot\ p^5q^4 + 3p^7q^2$	$p^4q^5 + 3p^2q^7$
(4,1)	(1,3)	$\cdot\ p^4q^5 + 3p^6q^3$	$p^5q^4 + 3p^3q^6$
(3,1)	(2,3)	$3p^5q^4 + p^3q^6$	$\cdot\ 3p^4q^5 + p^6q^3$

NOTE: The dots indicate the higher of the two probabilities, in each row. The actual results are summarized in Table 5.

As a final item, let us see how well the "best" flat filters do with various types of data.

Table 5 — Summary of "best" flat filters based on 3 by 3 neighborhoods.

DATA TYPE	BEST ISOTONE METHOD $\gamma_x(A) = 1$ if
(0,1)	$6 \leqslant \#N(x) \cap A$ or $5 = \#N(x) \cap A$ unless $N(x) \cap A$ is type (1,4) or $4 = \#N(x) \cap A$ provided $N(x) \cap A$ is type (4,0)
(0,2)	$A \cap N(x)$ has at least 3 members that are type a or b
(1,2)	Same as (0,2)
(1,1)	$6 \leqslant \#N(x) \cap A$ or $5 = \#N(x) \cap A$ unless $N(x) \cap A$ is type (0,1,4) or (0,4,1) $4 = \#N(x) \cap A$ provided $N(x) \cap A$ is of type (1,3,0) or (1,0,3).

NOTE: The types (1,0), (2,0) and (2,1) are handled by symmetry.

Table 6 — A posterieri probability of correct classification for data of given type using "best" filter method as specified in Theorem 11.

DATA	A PRIORI PROBABILITY OF CORRECT DATA			
TYPE	.9	.8	.7	.6
(0,0)	.9991	.9804	.9012	.7334
(0,1)	.9957	.9545	.8498	.6886
(0,2)	.9914	.9421	.8369	.6826
(0,1)	.9802	.8979	.7715	.6371
(1,2)	.9526	.8499	.7311	.6134

REFERENCES

[1] T.S. Blyth and M.F. Janowitz, *Residuation theory*, Pergamon Press, Oxford, 1972.

[2] M.F. Janowitz, *An order theoretic model for cluster anlaysis*, SIAM J. Applied Math. 34 (1978), pp. 5572.

[3] M.F. Janowitz, *Monotone equivariant cluster methods*, SIAM J. Applied Math. 37 (1979), 148-165.

Spatial Domain Filters for Image Processing[1]

Winser E. Alexander

Department of Electrical and Computer Engineering
North Carolina State University
Raleigh, NC

ABSTRACT

The quest for high speed or real time computing has resulted in a concerted effort directed toward the development of faster computers as well as algorithms of lower computational complexity. The traditional measure of the utility of such algorithms has been computational complexity. However, the hardware implementation of these algorithms for real time applications has often been difficult to impossible because of data communications problems. This paper presents a computationally efficient algorithm for the implementation of spatial domain infinite impulse response digital filters. A technique is presented for the simple design of approximately circularly symmetric lowpass, highpass, bandpass, edge enhancement and contrast enhancement digital filters. Problems and proposed solutions associated with the practical application of these filters for real time and near real time image processing are also presented.

1 INTRODUCTION

Recent advances in computer systems hardware and software have made digital image processing practical for many applications such as target tracking, target recognition, earth resources analysis from satellite images, restoration of images from space probes, analysis of radiographic images, visual recognition of objects by industrial robots, digital techniques for data bandwidth reduction in televi-

[1] This work was supported by the Office of Naval Research under contract N 00014-83-K-0138 and by the Naval Surface Weapons Center under contract N 60921-83-D-A012.

sion transmissions and teleconferencing, image formation using synthetic aperture radar and infrared imaging. Many of these applications involve the collection of data in real time or require the processing of very large images. Many of the recent advances in computer circuit and systems design have been made possible because of the implementation of very large scale integration (VLSI). Very large scale integration offers a new medium for computer system architecture that may be effectively used to increase the use of parallel or pipeline structures to increase the computational rate of computer systems. A new approach is required in algorithm development in order to take advantage of this new medium. The computational requirements of real time processing are such that real time image processing may only be accomplished with the use of special purpose hardware and computationally efficient algorithms which are optimized for this application. Spatial domain digital filters offer greatest promise for filtering of images in real time. This potential is continually increasing as new advances are made in computer system architecture and very large scale integration.

1.1 Requirements for Real Time Image Processing

The computational requirements for real time filtering of images are so great that it will not be feasible in the near future to use general purpose computers for real time image processing. The minimum practical size for real time interactive image analysis is approximately 500 rows by 500 pixels per row. The refresh rate should be greater than or equal to the standard rate for commercial television which is 30 frames per second. This means that a real time image processing system must be able to process approximately 7.5 million pixels per second. This is a minimum throughput rate of approximately 130 nanoseconds per pixel. A reasonable design goal is a throughput rate of 100 nanoseconds per pixel.

In most applications, a raster scan technique is used for image collection and display. This suggest that a real time image processor must be able to accept pixel data inputs at a rate of approximately 100 nanoseconds between pixels and generate output pixels at the same rate. This effective sampling interval compares favorable with the time required to perform a single multiplication or a single data transfer from memory on a general purpose computer. It is obvious that we must use parallel or pipeline computational structures to realize real time image processing. Very large scale integration offers the potential of implementing a system with multiple processing units in a cost effective manner. Thus, the development of a real time image processing system that uses multiple processing units may be practical using VLSI as a medium.

1.2 Comparison of Candidate Algorithms

Digital filters may be classified as being of two basic types: transform domain filters and signal domain filters. Transform domain filters require that the transform of the input signal be taken. The filtering process is then performed by obtaining a product of the transformed input signal with the transform domain representation of the desired filter [1]. The output is obtained by taking the inverse transform of this product. Thus the transform domain filtering operation involves two transform operations and a multiplication operation. The Discrete Fourier Transform (DFT) is used for most transform domain filtering operations and the Fast Fourier Transform (FFT) algorithm provides a means of implementating the DFT in a computationally efficient manner.

If very large images are to be filtered, then it is practical to use either a block mode filtering procedure or the two dimensional DFT is obtained by using a one dimensional FFT algorithm first on the rows and then on the columns. In the block mode filtering procedure, a large image is separately processed in adjacent overlapped blocks. Considerable overlap is required and the background level of the blocks must be adjusted to avoid a checkerboard effect. If the image is too large to fit into fit into the computer, then it is common to transpose the image so that the DFT of the columns may be computed when the second procedure is used. In either case, the computational complexity is significantly increased for large images because of the increased data communications requirements.

In contrast, signal domain digital filters do not require a transform process. The filtering is done by taking a weighted average of input and past output values to compute the current output. One dimensional signal domain digital filters are generally called time domain filters and two dimensional signal domain digital filters are generally called spatial domain filters. This is consistent with the most frequent applications. However, the filter algorithms are appropriate for any single independent variable or double independent variable application respectively.

Spatial domain filters may be categorized as being of two basic types: nonrecursive or finite impulse response (FIR) digital filters and recursive or infinite impulse response (IIR) filters. Spatial domain IIR filter offer savings in computation time and memory requirements over transform domain filters to achieve the same filtering process [2]. This is accomplished for many filtering applications with no sacrifice in the quality of the output. Therefore, IIR filters provide advantages for certain applications where time of processing or hardware limitations are a consideration. This is of practical significance for real time or near real time signal or image processing using dedicated hardware.

Spatial domain FIR filters offer advantages over both IIR filters and transform domain filters when the number of required filter coefficients is relatively small. However, the applications for such filters with a small number of coefficients is very limited. In general, it requires a significantly greater number of filter coefficients to realize a particular impulse response for FIR digital filters than for IIR digital filters.

2 MATHEMATICAL DESCRIPTION OF THE ALGORITHM

A digitized image may be represented as a bivariate function, $f(m,n)$, where m and n cannot have negative values and are finite. If the image in question has N rows and M colums, then m has a range from 0 to $(M - 1)$ and n has a range of 0 to $(N - 1)$. The index, m, represents column m and the index, n, represents row n. The function, $f(m,n)$ then represents the value of the pixel for column m and row n.

In a similar manner, the output of the filtering process is an image and can be represented by the bivariate function, $g(m,n)$. The general bivariate difference equation for the causal, quarter plane, spatial domain IIR digital filter is then given by [3]

$$g(m,n) = \sum_{j=0}^{L} \sum_{k=0}^{L} a(j,k) f(m-j,n-k) - \sum_{\substack{j=0 \\ j+k>0}}^{L} \sum_{k=0}^{L} b(j,k) g(m-j,n-k) \tag{2.1}$$

where $a(j,k)$ and $b(j,k)$ represent appropriate filter coefficients. The parameter, L, is relatively small for practical filters (usually less than 10).

Equation (2.1) can also represent the causal FIR spatial domain filter if we let all of the values of $b(j,k)$ be equal to zero. In this case, the output is computed as a weighted sum of only the input pixel values. Thus the bivariate difference equation for the FIR spatial domain filter is given by

$$g(m,n) = \sum_{j=0}^{L} \sum_{k=0}^{L} a(j,k) f(m-j,n-k). \tag{2.2}$$

The spatial domain filter can also be represented by the two dimensional Z transform. The two dimensional Z transform for the IIR spatial domain filter is given by

$$H(z,w) = \frac{\sum_{j=0}^{L} \sum_{k=0}^{L} a(j,k) z^{-j} w^{-k}}{1 + \sum_{\substack{j=0 \\ j+k>0}}^{L} \sum_{k=0}^{L} b(j,k) z^{-j} w^{-k}} \tag{2.3}$$

The two dimensional Z transform for the causal FIR spatial domain filter is given by

$$H(z,w) = \sum_{j=0}^{L} \sum_{k=0}^{L} a(j,k) z^{-j} w^{-k}. \tag{2.4}$$

3 STABILITY ANALYSIS OF IIR SPATIAL DOMAIN FILTERS

Infinite impulse response spatial domain filters use feedback of previously computed output values to compute the current coutput. Thus, stability analysis is an important part of the design of these filters. The design problem consists of determining the filter coefficients so that the required frequency response is realized. If the resulting filter is to be useful, it must be bounded-input-bounded-output (BIBO) stable.

Let a spatial domain filter be designated by the two dimensional Z transform, $H(z,w)$, and let the input to this filter be given by a single pixel of unit magnitude at the origin ($m = 0$ and $n = 0$) and with all other input pixels having a magnitude of zero. Let us designate the output from this single pixel input as the sequence $[h(m,n)]$. The requirement that this filter is BIBO stable is equivalent to the requirement that

$$\sum_{m=0}^{\infty} \sum_{n=0}^{\infty} |h(m,n)| < \infty. \tag{3.1}$$

The IIR spatial domain filter can be represented by a state space model [4,5,6]. Practical theorems for stability analysis of quarter plane IIR spatial domain filters have been developed by Alexander and Pruess [6] using a state space model. They begin with the state space representation of the filter as given by

$$G_{m,n} = \underline{B}_1 G_{m-1,n} + \underline{B}_2 G_{m,n-1} + \underline{A} F_{m,n}$$
$$g(m,n) = \underline{D} G_{m,n}. \tag{3.2}$$

$G_{m,n}$ is a column vector such that its elements are the outputs, $g(m-j,n-k)$ in (2.1) including $g(m,n)$. $F_{m,n}$ is a column vector such that its elements are inputs, $f(m-j,n-k)$ in (2.1). $\underline{B}_1$,$\underline{B}_2$ and $\underline{A}$ are $(L+1)^2$by $(L+1)^2$ matrices and $\underline{D}$is a $(L+1)^2$ row vector such that (2.1) and (3.2) are equivalent.

Stability analysis can then be accomplished by evaluating the spectral radii of $\underline{B}_1$, $\underline{B}_2$, and the spectral radius of the matrix sum of $\underline{B}_1$and $\underline{B}_2$. If we designate the spectral radius of the general matrix $\underline{Q}$ as $\rho(\underline{Q})$,then the quarter plane IIR spatial domain filter is unstable if either $\rho(\underline{B}_1)$,$\rho(\underline{B}_2)$ or $\rho(\underline{B}_1 + \underline{B}_2)$ is greater than or equal to one.

4 SYNTHESIS OF CIRCULARLY SYMMETRIC IIR FILTERS

In this section, we consider the synthesis of frequency selective IIR spatial domain filters. We wish to design filters for which the output is invariant with respect to the orientation of the objects within the image. This means that the designed filters should have a circularly symmetric frequency response [7].

4.1 Low Pass Filter Design

The magnitude squared characteristic for the one dimensional Butterworth prototype analog filter is given by

$$|H(s)|^2 = \frac{1}{1 + \epsilon^2(-1)^n (s/\omega_c)^{2n}}. \tag{4.1}$$

This function can be converted to a two dimensional radial function by letting ω_c represent a radial cutoff frequency and by letting the Laplace transform variable, s, represent a radial Laplace transform variable where

$$s^2 = s_1^2 + s_2^2. \tag{4.2}$$

The parameter, s_1 is the Laplace transform variable for the horizontal direction and s_2 is the Laplace transform variable for the vertical direction. We then have

$$|H(s_1,s_2)|^2 = \frac{1}{1 + \epsilon^2(-1)^n [(s_1^2 + s_2^2)/\omega_c^2]^n}. \tag{4.3}$$

We can use the bilinear transformation to map $|H(s_1,s_2)|$ into the discrete space domain as follows:

$$\omega_c \rightarrow \tan (\omega_c T/2) \tag{4.4}$$

$$s_1 \rightarrow (z - 1)/(z + 1) \tag{4.5}$$

$$s_2 \rightarrow (w - 1)/(w + 1)\cdot \tag{4.6}$$

Then we have

$$H(z,w) = \frac{1}{1 + \epsilon^2(-1)^n [P(z,w)/\tan (\omega_c T/2)]^n} \tag{4.7}$$

where

$$P(z,w) = [(z-1)/(z+1)]^2 + [(w-1)/(w+1)]^2\cdot \tag{4.8}$$

We can also represent $H(z,w)$ as a ratio of two polynomials as follows:

$$H(z,w) = \frac{N(z,w)}{D(z,w)}\cdot \tag{4.9}$$

In general, both $N(z,w)$ and $D(z,w)$ have zeros outside the region defined by $|z| \leqslant$ and $|w| \leqslant 1$. The surface given by $|z|$ equal one a $|w|$ equal one is commonly referred to as the unit bidisc. $N(z,w)$ may have zeros outside the unit bidisc and $H(z,w)$ may still be a stable filter. However, any filter which has a denominator polynomial with zeros outside the unit bidisc is BIBO unstable. The procedure for design of a one dimensional filter would be to factor $|H(z)|^2$ into a transfer function with poles located only outside the unit circle where $|z|$ is greater than one and a transfer function which has poles only inside the unit circle [8]. The stable transfer function with poles inside the unit circle would then be used as the filter transfer function. This procedure does not work in two dimensions because in general, a bivariate polynomial does not necessarily have a finite number of roots.

We can approximate $D(z,w)$ as a product separable function by factoring $D(z,w)$ along the plane where $z = w$ [9]. The poles thus acquired are then used to form a polynomial in each variable. Using this approach, we have

$$D(z,z) = D_1(z)D_1(z^{-1})\cdot \tag{4.10}$$

We then form the approximation to $H(z,w)$ as

$$H_L(z,w) = \frac{N(z,w)}{D_1(z)D_1(w)}\cdot \tag{4.11}$$

Note that with this approximation, the magnitude response is correct for the plane where $z = w$ and is an approximation to the circularly symmetric filter along other planes. Thus we can write

$$|H_L(z,z)| = |H(z,z)|\cdot \tag{4.12}$$

4.2 Design of Other Frequency Selective Filters

Once we have an acceptable low pass filter, we can use that filter to design other frequency selective fiters. Consider the stable low pass IIR filter given by

$$H_L(z,w) = \frac{N_L(z,w)}{D_L(z,w)}\cdot \tag{4.13}$$

We can obtain a boost filter using the relationship:

$$|H(z,w)| = a + b|H_L(z,w)|^2 . \tag{4.14}$$

This can be written in the alternative form:

$$H(z,w) = a + \frac{bN_L(z,w)N_L(z^{-1},w^{-1})}{D_L(z,w)D_L(z^{-1},w^{-1})}. \tag{4.15}$$

Note that $H(z,w)$ is an unstable filter. However, we can implement $H(z,w)$ in cascade form with one filter recursing in the foward direction and one filter recursing in the reverse direction [10]. Thus, we let

$$H_1(z,w) = \frac{a|D_L(z,w)|^2 + b|N_L(z,w)|^2}{D_L(z,w)} \tag{4.16}$$

and

$$H_2(z,w) = \frac{1}{D_L(z,w)}. \tag{4.17}$$

We can implement $H(z,w)$ by shifting $H_1(z,w)$ to obtain a causal output and then implementing $H_2(z,w)$ with a reverse recursion in both row and column.

If we desire a high pass filter, then we let a equal to one and b equal to minus one. If we desire a high frequency boost filter with gain of A in the boost region, then we can let a be equal to A and b be equal to one minus A. If we desire a low frequency boost filter with gain of A in the boost region, then we can let a be equal to one and let b be equal to A minus one. Thus, we can obtain circularly symmetric high pass, high frequency boost and low frequency boost filters from a circularly symmetric low pass filter.

5 IMPLEMENTATION

As previously mentioned, the implementation of image processing algorithms in real time will require the use of parallel and/or pipeline computational structures. It is also important to minimize the number of data transfers required to implement the algorithm. If we are to take advantage of the new capabilities provided by VLSI, we must minimize the number of times that data transfers occur between data registers, memory, etc. on chip and other data registers, memory, etc. off chip. Very large scale integration offers the potential of including some intermediate memory on chip along with the arithmetic logic unit which is used for multiplication and addition. The primary operations required for image filtering applications are multiplication, addition and data transfer. It is important to minimize each of these in order to realize real time image filtering.

5.1 State Space Representation

The general form of the state space representation for the IIR spatial domain filter is given in (3.2). We want to use a second order filter stage (L = 2) to minimize roundoff errors and coefficient truncation errors. This second order stage is simple enough to make its implementation as a single chip processor practical. Higher order filters could then be implemented by cascading these second order filters. Let the two dimensional Z transform of the input sequence, $[f(m,n)]$, be designated as $F(z,w)$ and the two dimensional Z transform of the output sequence, $[g(m,n)]$ be designated as $G(z,w)$. Then we can write the transfer function of the second order filter stage, $H(z,w)$, as the ratio of the output and input transforms as follows:

$$H(z,w) = \frac{G(z,w)}{F(z,w)} = \frac{\sum_{j=0}^{2} \sum_{k=0}^{2} a(j,k) z^{-j} w^{-k}}{1 + \sum_{\substack{j=0 \\ j+k>0}}^{2} \sum_{k=0}^{2} b(j,k) z^{-j} w^{-k}} . \tag{5.1}$$

A difference equation representation of the output, $g(m,n)$, as a function of the input, $f(m,n)$, and past values of the input and output is given by

$$\begin{aligned} g(m,n) = a(0,0)f(m,n) &+ \sum_{j=1}^{2}[a(j,0)f(m-j,0) - b(j,0)g(m-j,0)] \\ &- \sum_{j=0}^{2}[a(j,1)f(m-j,n-1) - b(j,1)g(m-j,n-1)] \\ &- \sum_{j=0}^{2}[a(j,2)f(m-j,n-2) - b(j,2)g(m-j,n-2)] \end{aligned} \tag{5.2}$$

The flow diagram symbols for representation of two dimensional difference equations are given in Fig. 5.1. The flow diagram corresponding to (5.2) is given in Fig. 5.2. We define state variables to be the input nodes for the delays. Note that this assignment is arbitrary and is chosen to minimize the data communications requirements. The state equation for this second order filter can then be written as

$$\begin{aligned} G_{m,n} &= \underline{B}_1 G_{m-1,n} + \underline{B}_2 G_{m,n-1} + \underline{A} f(m,n) \\ g(m,n) &= \underline{D} G_{m,n} + \underline{E} f(m,n) \end{aligned} \tag{5.3}$$

Note that (5.3) differs slightly from (3.1) in that (3.1) uses an input vector, $F_{m,n}$ and (5.3) uses only the single input, $f(m,n)$. This single input structure is more efficient with regard to data transfers and the associated data storage.

x(m,n)	——> z^{-1}	;	y(m,n) = x(m−1,n)
x(m,n)	——> w^{-1}	:	y(m,n) = x(m,n−1)
x(m,n)	——> a	;	y(m,n) = ax(m,n)

Fig. 5.1 Flow diagram symbols for two dimensional difference equation

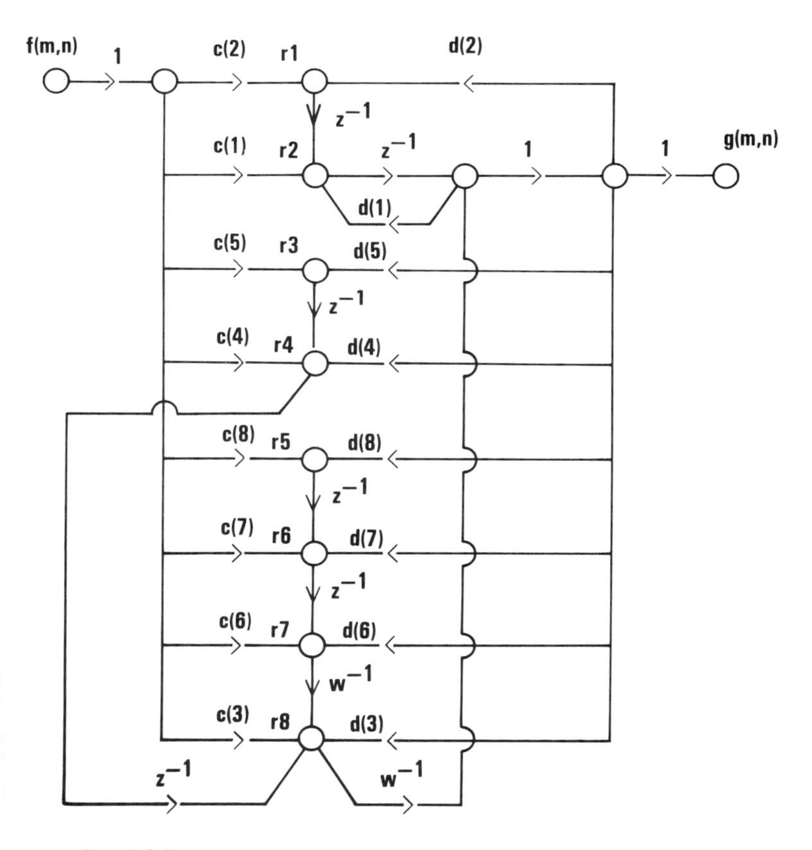

Fig. 5.2 Flow diagram of second order IIR Spatial Domain Filter

The structure given in (5.2) only requires two state variables along the vertical direction (delays associated with w). This is important because each state variable along the vertical direction requires the storage of a row of intermediate data. The state space representation has the advantage that the input values are used only once by the algorithm and do not have to be stored for later use. In addition, the number of storage registers required to implement the filter is minimized. The state variables are updated during the computation of each output and the output is computed as a linear combination of the current values of the state variables and the current input [4,5,6].

5.2 Transient Suppression for IIR Spatial Domain Filters

An image can be considered to be a finite duration data set with truncation occuring at the edges of the image. Thus, an image can be considered to be the product of a two dimensional rectangular window function and an infinite duration, two dimensional function. The study of the effects of this data set truncation is related to the study of the convergence of Fourier series. The most familiar concept from this theory as related to one dimensional data is the "Gibbs phenomenon".

The artifacts associated with data truncation are limited to the edge of the image for two dimensional FIR filters. However, the resulting edge transients can propogate throughout the output image with two dimensional IIR filters. These transient effects can completely obscure the output image when filters with a long duration impulse response such as high pass or high frequency boost filters are used.

We represent an image having N rows and M pixels per row as a two dimensional data set, $[f(m,n)]$, where $f(m,n)$ is the pixel in column m of row n. In order to compute appropriate initial conditions, we can extend the image as follows:

$$\begin{aligned} f(m,n) &= f(m,0);\ n < 0 \\ f(m,n) &= f(0,n);\ < 0 \\ f(m,n) &= f(M-1,n); m > M-1 \\ f(m,n) &= f(m,N-1); > N-1 \end{aligned} \tag{5.4}$$

We assume that the infinite impulse filter with two dimensional Z transform, $H(z,w)$, is linear and shift invariant. With this assumption and with the above extensions of the image, the outputs are in a steady state prior to the first input values at the edges of the image. In addition, there are no changes in the inputs at the edges of the image. The result is that the first values of the outputs at the edges of the image are equal to the steady state values prior to the first inputs and no transients occur.

We can compute the outputs at the edges of the image (initial conditions) with the use of the two dimensional equivalent of the final value theorem. We designate the output sequence, $[g(m,n)]$. Then we have

$$g(0,0) = \lim_{\substack{w \to 1 \\ z \to 1}} (1-z)(1-w)H(z,w) \sum_{m=0}^{\infty} \sum_{n=0}^{\infty} f(0,0) z^{-m} w^{-n}. \tag{5.5}$$

The result is given by

$$g(0,0) = H(1,1) f(0,0) \cdot \tag{5.6}$$

We can compute the output for the first line as follows:

$$\sum_{m=0}^{M} g(m,1) z^{-m} = \lim_{w \to 1} (1-w)H(z,w) \sum_{m=1}^{M} \sum_{n=0}^{\infty} f(m,0) z^{-m} w^{-n} \tag{5.7}$$

If we treat the first row as a separate data set and designate the one dimensional Z transform of this row as

$$F_1(z) = \sum_{m=1}^{M} f(m,0) z^{-m} \cdot \tag{5.8}$$

Then (5.7) results in the following relationship:

$$G_1(z) = H(z,1) F_1(z) \tag{5.9}$$

where G_1 (z) is the one dimensional Z transform of the desired output for the first row. Thus, we can obtain the output sequence, $[g(m,0)]$, by filtering the input sequence, $[f(m,0)]$, with the one dimensional filter, H(z,1). Similarly, we can compute the initial conditions for the first column which is the output sequence, $[g(0,n)]$, by filtering the input sequence, $[f(0,n)]$, with the one dimensional filter, $H(1,w)$.

5.3 Linear Phase Filtering With IIR Spatial Domain Filters

Infinite impulse response spatial domain filters inherently have nonlinear phase characteristics. This is true because of the use of feedback of past output values. However, linear phase is often desirable for image processing applications. Linear phase can be obtained with IIR spatial domain filters by filtering the image twice [10]. The image is filtered starting from the first row, first pixel and ending with last row, last pixel. Then the image is filtered backward starting with the last row, last pixel and ending with the first row, first pixel. The result is a filter transfer characteristic which is the magnitude squared of the original characteristic. Thus, the filter with four poles and four zeros effectively has eight poles and eight zeros and linear phase when this procedure is used.

6 APPLICATIONS

Spatial domain filters are well suited for use as frequency selective filters. Applications include image enhancement for edge extraction or other pattern recognition applications, dynamic range compression, bandwidth optimization, interpolation, prefiltering for computer assisted image registration, classification and evaluation, etc. Some of these applications are discussed in this section and examples of the use of IIR spatial domain filters are presented.

6.1 Frequency Selective Filters and Edge Enhancement

The simple design procedure described herein for the design of frequency selective filters can be used in an interactive mode to facilitate edge enhancement and other subjective image enhancement applications. A low pass or a high pass filter may be specified by the cutoff frequency and the number of filter stages desired. A high frequency enhancement or low frequency enhancement filter may be specified by a break frequency (the cutoff frequency for the low pass filter used in the design precedure) and the magnitude of the boost. Thus, the user does not have to have a knowledge of filter theory or be concerned with signal to noise considerations, etc. in order to design the desired filter. This is a very valid approach for subjective enhancement applications because decisions about the type of filter desired are usually based upon experience. Thus, it is important for the user to be supplied with several options which can be implemented with a minimum of effort and without special training. IIR spatial domain filters are well suited for this application.

Examples of the use of IIR spatial domain frequency selective filters are given in Figs. 6.1 to 6.3. Figure 6.1 gives an original SEASAT-A synthetic aperture radar (SAR) image of the Chesapeake Bay. This image was provided by the Office of Naval Research as a representative SAR image for testing image processing algorithms. Figure 6.2 is the output of a single second order stage lowpass filter with the half power cutoff frequency equal to six tenths of

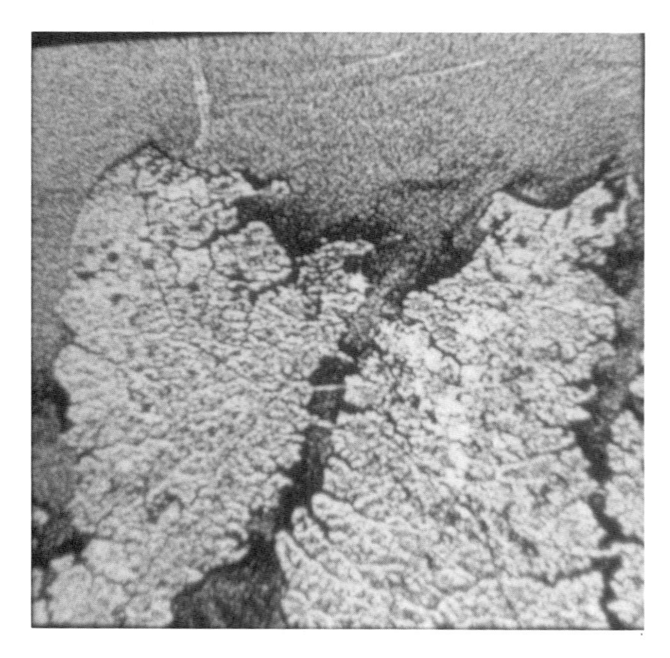

Fig. 6.1 – 512 by 512 SEASAT-A image of the Chesapeake Bay

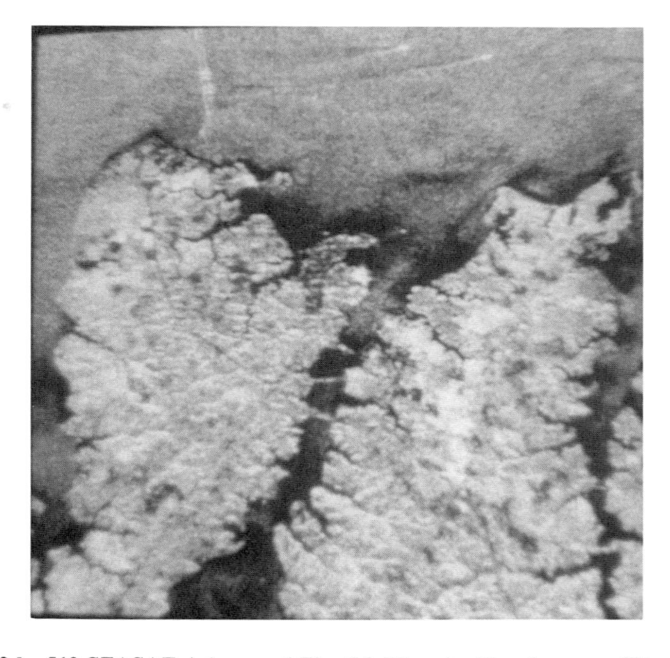

Fig. 6.2 – 512 by 512 SEASAT-A image of Fig. 6.1 filtered with a low pass filter with a normalized cutoff frequency of 0.6 π

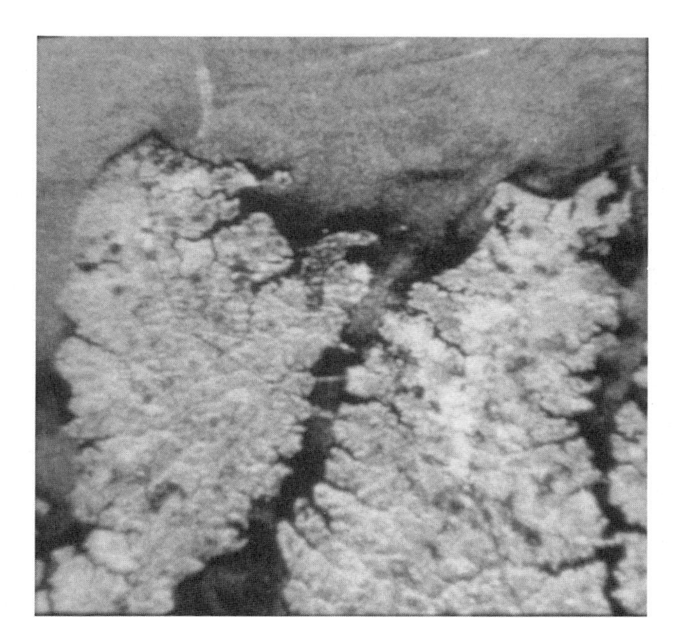

Fig. 6.3 – 512 by 512 SEASAT-A image of Fig. 6.2 filtered with a high frequency boost filter with a boost magnitude of 4.0 and a normalized break frequency of 0.25 π

the Nyquist frequency. Such a filter may be effectively used as a speckle smoothing filter. Figure 6.3 is the result of filtering the output image in Fig. 6.2 with a high frequency boost filter with boost magnitude of 4.0. The boost filter was obtained from a lowpass filter with a cutoff frequency of one fourth of the Nyquist frequency. Such a filter may be used as a prefilter for edge tracking algorithms. These examples are not presented as optimal processing of the subject image. They are only provided as a demonstration of the capability of IIR spatial domain filters.

6.2 Dynamic Range Compression

Electro-optical sensors respond to reflected or emitted radiation. A typical electro-optical imaging system uses a single detector or an array of detectors in a scanning mode to form the image. If the signal of interest is reflected radiation such as is the case for visible imaging systems, the detected signal is made up of two components: the illumination component and the reflection component. Infrared sensors typically detect radiation emitted by objects. The available dynamic range of an electro-optical imaging system is usually several orders of magnitude. This is particularly true for synthetic aperture radar imaging systems. On the other hand, display systems are usually limited to at most two orders of magnitude and human observers can only detect approximately 50 different intensity levels [1]. Therefore, it is not possible to directly display all of the information contained in many images.

The illumination component of optical images or the overall background radiation for infrared images generally has low spatial frequency content but may have a wide dynamic range. This is the case where shadows exist in the optical imagery or where hot or cold spots occur in infrared imagery. The reflected component or the emitted component of the signal is usually of priority interest and generally has higher spatial frequency content. This signal is formed by the difference in emissivity or reflectivity of each item in the image.

The detected signal is therefore a product of the illumination or background radiation and the reflectivity or emissivity at each point in the image. Homomorphic filtering using spatial domain digital filters provides an effective means of dynamic range compression by providing the capability to supress the lower frequency component of the signal and enhance the higher frequency component of the signal [11]. This procedure is accomplished by taking the logarithm of the input signal, filtering with a high frequency boost filter and exponentiating the resulting output.

ACKNOWLEDGMENT

I wish to thank Jung Kim, a graduate research assistant at North Carolina State University, for developing the computer software to implement the filtering algorithms presented in this paper.

REFERENCES

1. William K. Pratt, *Digital Image Processing,* John Wiley and Sons, Inc., Somerset, N.J., pp. 279-303, 1978.

2. Earnest L. Hall, "A Comparison of Computations for Spatial Filtering," *Proceedings of IEEE,* Vol. 60, no. 7, pp, 887-891, 1972.

3. Russel M. Mersereau and Dan Dudgeon, "Two Dimensional Digital Filtering," *Proceedings of IEEE,* Vol. 63, pp. 610-623, 1975.

4. Robert P. Roesser, "A Discrete State-Space Model for Linear Image Processing," *IEEE Transactions on Automatic Control,* Vol. AC-20, pp. 1-10, 1975.

5. E. Fornasini and G. Marchesini, "State Space Realization Theory for Two Dimensional Filters," *IEEE Transactions on Automatic Control,* Vol. AC-21, pp. 484-492, 1976.

6. Winser E. Alexander and Steven A. Pruess, "Stability Analysis of Two Dimensional Digital Recursive Filters," *IEEE Transactions on Circuits and Systems,* Vol. CAS-27, pp. 11-15, 1980.

7. A. Papoulis, *Systems and Transforms with Applications in Optics,* McGraw-Hill Book Co., New York, N.Y., p. 140, 1968.

8. Bernard Gold and Charles Rader, *Digital Processing of Signals,* McGraw-Hill Book Co., New York, N.Y., pp. 48-97, 1969.

9. Winser E. Alexander and Earnest E. Sherrod, "Two Dimensional Digital Filters for Subjective Image Processing," *Proceedings of the 13th Asilomar Conference on Circuits, Systems and Computers,* November, 1979.

10. Jose' M. Costa and Anastasios N. Venetsanopoulos, "A Group of Linear Spectral Transformations for Two Dimensional Digital Filters," *IEEE Transactions on Acoustics, Speech, and Signal Processing,* Vol. ASSP-24, pp. 424-425, 1976.

11. Thomas G. Stockham, Jr., "Image Processing in the Context of a Visual Model," *Proceedings of the IEEE,* Vol. 60, pp. 828-842, 1972.

Edge Detection by Partitioning[1]

Jong-Sen Lee

Naval Research Laboratory
Washington, DC

ABSTRACT

The objective of an edge operator is to detect the presence and location of gray level changes in an image. Various edge detection algorithms have been developed in recent years. In this paper, a new edge detector is proposed based on the idea of separating pixels in a local window into two sets of similar intensities. Then, the difference in intensity averages of these two sets is the edge magnitude. The technique of separating pixels in a local window into two sets is motivated by the sigma filter [1] which is an image noise smoothing technique based on averaging only pixels within two standard deviations from the intensity of the center pixel. The characteristics of this edge operator are its flexibility in setting magnitude of edges to be detected and also its consistency in detecting edge strength independent of edge orientations. Furthermore, it can be easily generalized for surface detection in three dimensional images. This algorithm is also computationally efficient since only simple fixed point operations are involved.

For illustration, several 256 × 256 images are processed with its algorithm and comparisons are made with the well-known Sobel operator.

1 INTRODUCTION

Edge detection is an important process employed in scene segmentation. Various edge detection algorithms have been developed in recent years. Most of these algorithms are basically local operators in the sense that edge detection is

[1] Research supported by the Office of Naval Research under work request N00014-83-WR-0027.

restricted to a small (3 × 3 or 5 × 5) sliding window. The local operators are generally more sensitive to noise, but they are computationally efficient and especially suitable for parallel implementation. Algorithms operating on a large area such as Hueckel operator [2] and Laplacian operator [3] are less sensitive to noise, but they do so at the expense of high computational burden. Most existing algorithms are modelled to detect a step edge in some optimal way. The detection of ideal (in the absence of noise) step edges would by a simple process. In real images, however, noise, texture and surface imperfections make it necessary to design algorithms capable of ignoring such minor intensity variations. Robert's [4] gradient operator using a 2 × 2 window and the Sobel [5] operator using a 3 × 3 window approximate the local edges by gradients. Another approach is to detect edges by matching templates of ideal edges in many orientations. The Prewitt [6] operator has eight 3 × 3 templates, while Nevatia and Babu [11] using a larger 5 × 5 window create 6 directional masks. Suciu and Reeves [7] studied the performance of several edge detection algorithms using ideal edges, ramp edges and noisy edges and concluded that Prewitt and Sobel operators were in general superior to the other local edge operators tested.

It seems that a natural approach to the edge detection problem would divide the pixels in the window into two groups of distinctive intensity levels, and their difference would be used to define the edge magnitude. Delp and Mitchell [8], in their statistical algorithms, divide pixels in the window into two intensity levels by selecting a suitable threshold. The threshold and the two intensity levels are chosen in such a way as to preserve the first, second and third order sample moments. Yakimosky [9], approaches the problem from the viewpoint of a statistical hypothesis test. In this paper, a simpler algorithm, separating pixels into two sets with similar intensities, will be presented.

The motivation for this algorithm stems from the concept of the sigma filter [1] which is an efficient noise smoothing procedure. The sigma filter replaces the center pixel in a window by the average of only those pixels whose intensity level is within the two noise standard deviations from the center pixel. The effect is that the procedure partitions pixels in the window into three sets: one set contains pixels within the two sigma range, and the other two sets contain pixels above and below the two sigma range. After reducing the three sets into two sets, the edge magnitude is obtained as the difference in the averages of pixel intensities of these two sets. In our test its performance and computational efficiency are comparable with the Sobel operator. In this paper we will concentrate on the algorithm defined on a (3 × 3) sliding window. For a larger window, it would be necessary to check whether the set boundary passes through the center pixel and to test the connectivity of pixels in each set.

For clarity, a brief review of the sigma filter will be given in the next section. Section III will be devoted to the development of the edge detection algorithm for a 3 × 3 window. Its characteristics will be discussed and comparison will be made with the Sobel operator.

2 THE SIGMA FILTER

The sigma filter is based on the Gaussian probability distribution for which the two-sigma probability is defined as the probability that a random variable falls within two standard deviations from its mean. The two-sigma probability for a one-dimensional Gaussian distribution is .955, which is taken to mean that 95.5% of a set of random samples lie within the range of two standard deviations. If we assume that the a priori mean is the gray level of the pixel to be estimated, we can establish a two sigma range from that gray level and include in the average only those pixels within the two sigma range. The two sigma range is generally large enough to include 95.5% of pixels from the same distribution in the window, yet in most cases, it is small enough to exclude pixels representative high-contrast edges and linear features. For images with unknown noise characteristics, the intensity range Δ, can be determined either from a rough estimation of noise standard deviation σ (Δ=2σ), or from the desirability of retaining the gray level difference between the feature of interest and its background. The sigma filter can be applied repeatedly with reduced σ after each iteration. Two or three iterations are generally sufficient to reduce the noise level significantly. A detailed discussion of the sigma filter is given in Reference [1].

3 A SIMPLE EDGE DETECTION ALGORITHM

The sigma filter divides pixels in the window into three sets: a set containing pixels within an intensity range, and the other two containing pixels above and below that intensity range. More specifically, let x_{ij} be the intensity or gray level of pixel (i, j), then the three sets are

$$A = \{x_{kl}/x_{kl} > (x_{ij} + \Delta)\},$$

$$B = \{x_{kl}/(x_{ij} + \Delta) \geqslant x_{kl} \geqslant (x_{ij} - \Delta)\},$$

$$\text{and } C = \{x_{kl}/x_{kl} < (x_{ij} - \Delta)\},$$

where x_{kl}'s are pixels within the window centered at pixel (i,j). Also let N_A, N_B and N_C be the number of pixels in the set A, set B and set C respectively, and Σ_A, Σ_B and Σ_C be the sum of intensities of all pixels in set A, set B and set C respectively. Set A and set C could be empty sets, while set B contains at least the center pixel x_{ij}. The problem is to devise a way to reduce these three sets into two sets. We shall continue to view the difference in the averages in each set as the edge magnitude.

To facilitate understanding of the development of the edge detection algorithm, we will use one dimensional examples some of the problems encountered. All one-dimensional examples in Figure 1 have the window spanning seven pixels. For a major edge as shown in Figure 1(A) with intensity increment Δ as shown, only three pixels are within the intensity range, or $N_B = 3$. The set A which contains pixels above $(x_{ij} + \Delta)$ has only one pixel; the third pixel to the right of x_{ij}. The set C contains three pixels of a lower average gray level. For a case like this, set A will be merged with set B and the edge magnitude is taken to be the difference in the averages of gray levels of these two sets. In other words, set A or set C will merged with set B, whichever has fewer pixels. For finer edges such as the one shown in Figure 1(B), the value of Δ apparently is too large to detect such an edge. To remedy this, a sequence of Δ of decreasing magnitude, say $\Delta_1, \Delta_2, \ldots, \Delta_k$ will be established beforehand

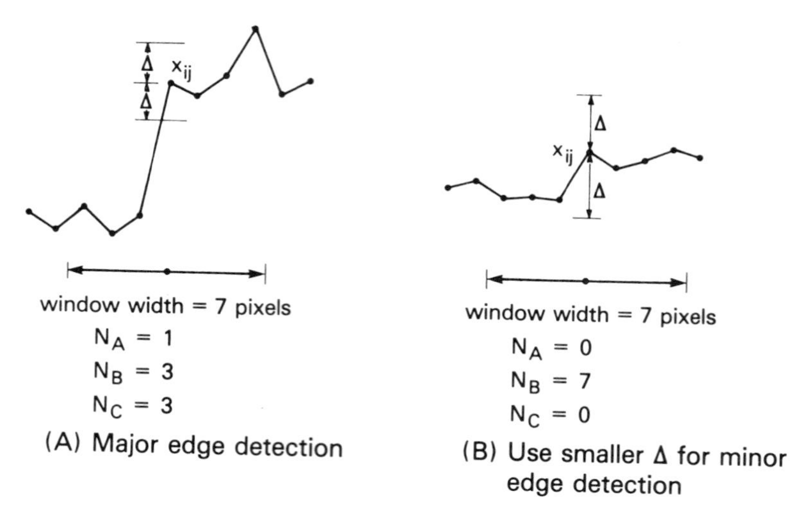

Figure 1 – The edge detection algorithm is illustrated with one dimensional examples. For a high-contracted edge and the intensity range Δ as shown in (A). The edge magnitude is the difference in averages of the set (A B) and the set C. For a low contrasted edge and the Δ value as shown in (B), all pixels fall into the intensity range. To detect this edge, a smaller Δ has to be utilized.

in which Δ_k is the finest gray level differential one wishes to detect. If the set contains all pixels in the window (in this case $N_B = 7$) we shall replace Δ_1 by Δ_2 and continue the process until either A or C is nonzero, or the reduction of Δ reaches Δ_k. In this setup we automatically eliminate noisy edges which we wish to ignore in any case. In the absence of definite knowledge of the Δ sequence, use $\Delta_{i+1} = \Delta_i/2$. Generally speaking, in our simulation with a 3×3 window, the sequence of three or four Δ's was found sufficient for most cases.

Sometimes a ramp edge and an isolated pixel in the window do present problems. For example, a ramp edge and an isolated pixel as shown in Figure 2(A) and 2(B) with the size of Δ as shown, both result in $N_B = 1$. In most cases, the isolated pixel is to be ignored, that is, we set the edge magnitude equal to zero. However, for a ramp edge shown in Figure 2(A) we would obtain erroneous results, if we set the edge magnitude to zero. Consequently, when $N_B = 1$, a test has to be made to see if either $N_A$orN_C is zero. If not, the edge magnitude of the ramp edge is the difference in the average of sets A and C.

Although the algorithm developed so far could be applied with various window sizes, care should be exercised when the window dimension is of dimension 5×5 or larger. One of the problems of large window sizes arises from the fact that the boundary of the set B does not necessarily pass through the center pixel. Consequently, the edge will be broaded or smeared. This problem occurs in many other edge detection algorithms whenever large windows are used. The main purpose of using a large window is to reduce the noise effect. To suppress this problem in our algorithm, the connectivities of pixels in the final two sets must be tested to weed out disconnected pixels. In

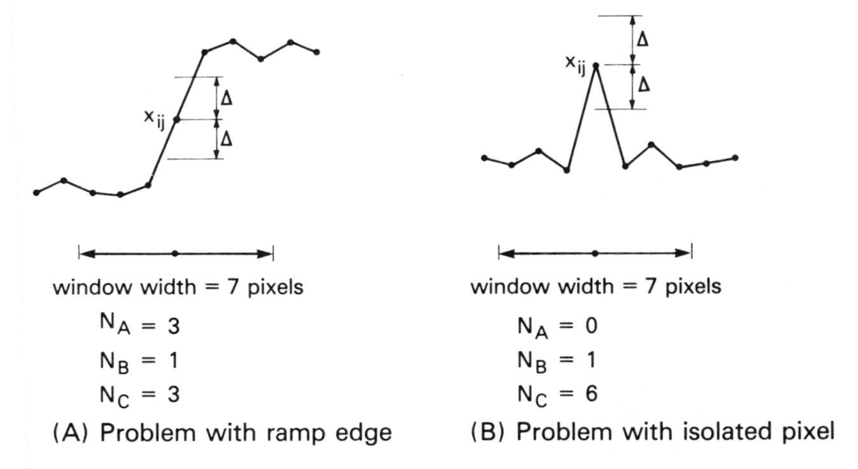

Figure 2 Problems encountered with ramp edges and isolated pixels are illustrated with one dimensional examples. With the Δ value as shown in both (A) and (B), only the pixel (i,j) itself falls in the intensity range. The edge magnitude in (A) is computed from the intensity averages of set A and set B, while in the case of (B) it is set to zero.

this case the edge magnitude will be computed only if the set boundary passes by the center pixel.

For a 3×3 window, the large window problem does not exist. However, for the algorithm to be effective, it is desirable to have an approximately equal number of pixels in the final two sets. The algorithm will be slightly modified for iterative Δ reduction to allow one or two pixels among the nine pixels lying outside the intensity range. In other words if $N_B = 8$ or 9, the Δ should be reduced to the next level.

The edge detection procedure with a 3×3 window is described as follows: Choose a Δ sequence $\Delta_1, \Delta_2, \ldots, \Delta_k$ (or use $\Delta_{i+1} = \Delta_i/2$),

(1) Starting with $\Delta = \Delta_1$ in a 3×3 sliding window

(2) Compute $(x_{ij} - \Delta, x_{ij} + \Delta)$

(3) Accumulate N_A, N_B, N_C and Σ_A, Σ_B, Σ_C over all pixels in these 3×3 windows

(4) If $N_B \geqslant 8$ (i.e., Δ too large), go to step (7). Otherwise, continue to the next step.

(5) If $N_B = 1$ (i.e., isolated pixel or ramp edge), go to step (8). Otherwise, continue to the next step.

(6) If $N_A > N_C$, the edge magnitude S is

$$S = (\Sigma_B + \Sigma_C)/(N_B + N_C) - \Sigma_A/N_A.$$

Figure 3 The edge detector is applied to the girl image (A), (i.e., upper left). The result is shown in (B) (i.e., upper right). For comparisons, the same image operated on by the Prewitt operator and Sobel operator are shown in (C) (i.e., lower left) and (D) (i.e., lower right) respectively. The better contrast in (C) is due to the use of higher normalization constant.

Otherwise, interchange subscript A and C. Compute edge magnitude for the next pixel (go back to step (1) with next window position).

(7) Replace Δ_i by the smaller Δ_{i+1} in the sequence or reduce Δ by half. If the end of Δ sequence is reached (i.e., flat area), set the edge magnitude S to zero. Otherwise, go to step (2).

(8) If $N_A = 0$ or $N_B = 0$ (i.e., isolated pixel), set the edge magnitude S to zero. Otherwise (i.e., ramp edge), the edge magnitude S is

$$S = \Sigma_A/N_A - \Sigma_B/N_B.$$

The above algorithm is to be applied to every pixel in a 3 × 3 sliding window.

To substantiate the theoretical development of this algorithm, a 256×256 image shown in Figure 3(A), is used for illustration. Figure 3(B) is the result of applying the algorithm developed in this paper, with $\Delta_1 = 30$, $\Delta_2 = 15$ and $\Delta_3 = 7$. The same image operated on by the Prewitt and Sobel operators are shown in Figure 3(C) and 3(D) respectively. The Prewitt operator results in better contrast due to the use of a higher normalization constant. These three operators seen to perform equally well judged from simple visual inspection. As far as computational efficiency is concerned, the algorithm of this paper is slightly faster than the Sobel operator in our simulations. The Prewitt operator is the slowest because of the eight directional maskings used.

4 REMARKS AND DISCUSSION

(1) The selection of the sequence of Δ is not a very critical task. It is illustrated in Figure 4(A) and Figure 4(B). The starting value of Δ_1 is 50 and 30 respectively with the maximum number of reductions equal to 5 and a reduction factor of 2 (i.e., $\Delta_{i+1} = \Delta_i/2$). The results as indicated in Figure 4(A) and 4(B) show almost identical results. The effect of reducing the maximum number of Δ reductions is shown in Figure 4(C) and 4(D) with the maximum number of reductions equal to 3 and 1, respectively. Figure 4(B) and 4(C) show very little difference. However, if the maximum number of reductions equals 1, as shown in Figure 4(D), only major edges are visible. This fact indicates that by proper selection of the Δ sequence, this algorithm can be made to disregard minor edges.

(2) The algorithm of this paper does not provide an edge direction, whereas the latter is readily available from the Sobel operator and many other edge operators. If edge directions are required, we propose that the centroids of the final two sets should be computed. The direction of the line connecting the two centroid is then the edge direction. To conserve computer time, the edge direction should be computed only for those pixels with high edge magnitude.

(3) To overcome the noise problem in an image, either edge operators with large windows are applied or the image is preprocessed with a noise filter, such as the edge-retaining sigma filter [1]. Then the 3×3 edge detection algorithm is applied to the smoothed image resulting in a considerable reduction of

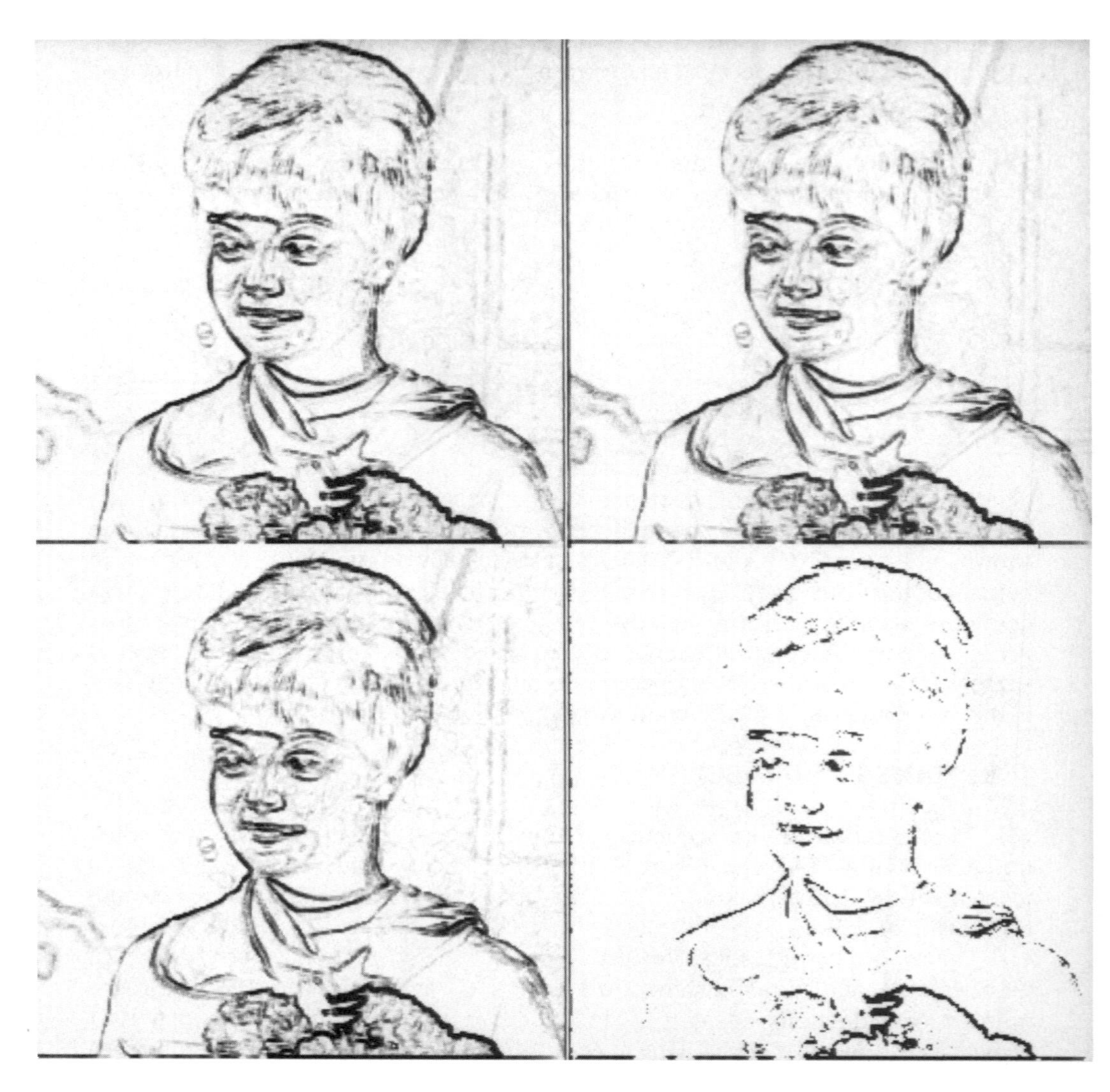

Figure 4 The selection of Δ sequence does not effect the result of this edge detection algorithm as illustrated with (A) (i.e., upper left) and (B) (i.e., upper right), where (A) and (B) have the $\Delta = 50$ and $\Delta = 30$ respectively. Fine or noisy edges can be eliminated by reducing the maximum number of Δ reduction as shown in (C) (i.e., lower left) and (D) (i.e., lower right), where the maximum number of Δ is reduced from five (B), to three (C), then to one (D).

noisy edges. A SEASAT Synthetic Aperture Radar (SAR) image of size 256×256 is used for illustration. It is well-known that SAR images are affected by the speckle noise [10]. Figure 5(A) shows an original SEASAT SAR image of Chesapeake Bay area. Numerous noisy edges are shown in Figure 5(B) which is the result of applying the 3×3 edge detection algorithm. Figure 5(C) is Figure 5(A) smoothed by the sigma filter. Figure 5(D) shows the result of

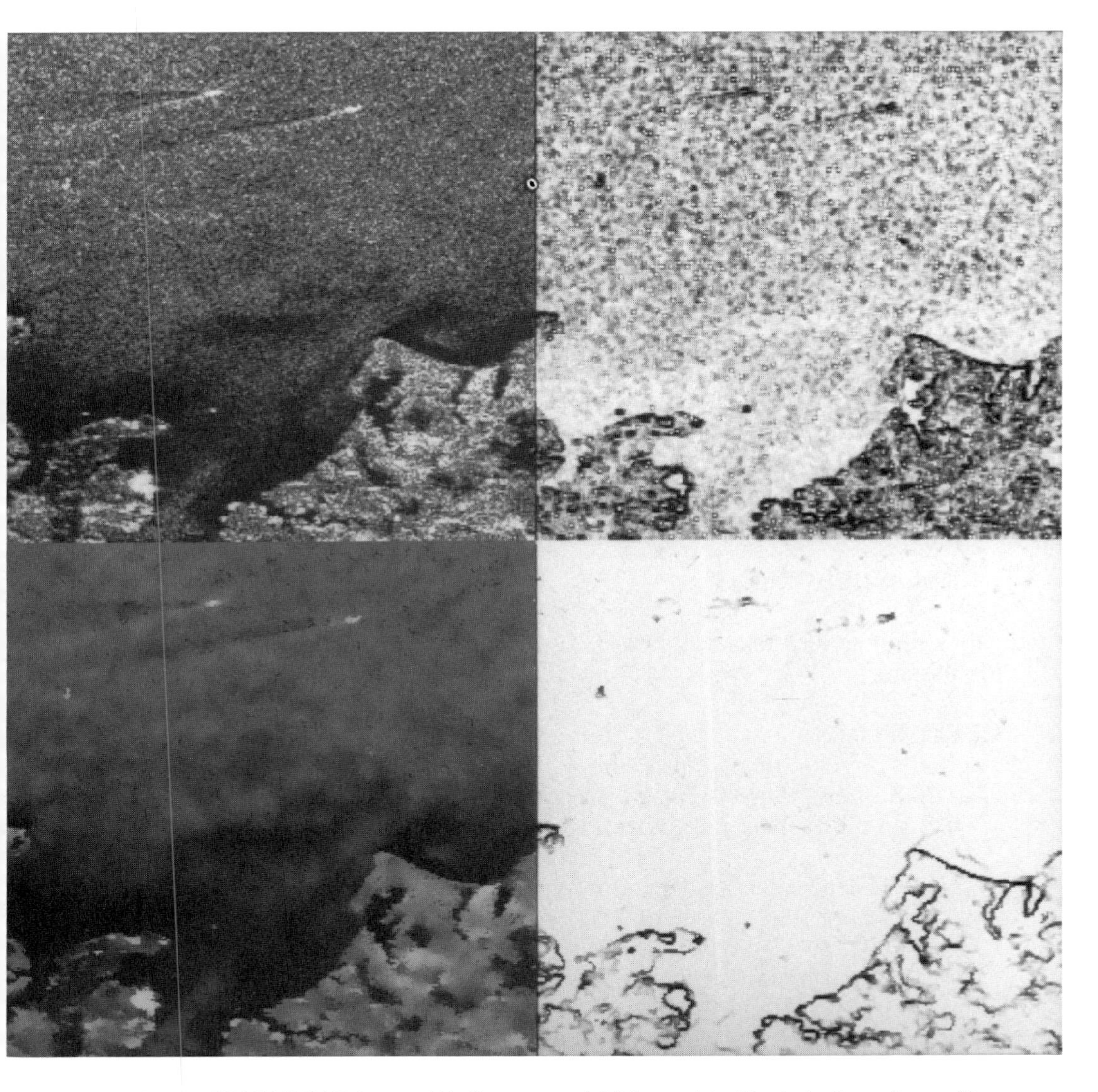

Figure 5 — A SEASAT SAR image (A) (i.e., upper left) is used to illustrate the noise problem encountered in edge detection. The 3×3 edge detector is applied to it and numerous noisy edges appears in (B) (i.e., upper right). The same image (A) is processed by the 7×7 sigma filter twice to smooth noise while retaining sharp edges as shown in (C) (i.e., lower left), and the edge detection result is shown in (D) (i.e., lower right).

applying the 3×3 edge detection to Figure 5(C). Clearly, Figure 5(D) is much cleaner edge map of features such as, ship wakes and coastlines.

(4) As indicated in the comparison study [7] of edge operators, the edge magnitudes produced by many edge operators vary with the orientation of edges. For example, the edge magnitude produced by the Sobel operator on a 45° edge could by 50% higher than that of a vertical or horizontal edge. This inconsistency in edge magnitude could present a problem in some applications. However, the edge operator developed in this paper is consistent in edge orientation even for curved edges and corners. Apparently, this is because the pixels in the window are partitioned into sets, rather than separated by differentiation or by template matching.

(5) In many image applications, such as medicine, the images are three dimensional. The edge detection algorithm of this paper can be easily generalized for surface detections. Using a 3×3×3 sliding cube, we can partition the twentyseven pixels into sets in a similar procedure as we did in the two-dimensional case.

5 CONCLUSION

An edge detection algorithm based on partitioning pixels in the window into two sets of similar intensities has been developed in this paper. The algorithm which utilizes a multi-stage decision process to separate pixels into two sets has a kind of artifical intelligence flavor. The algorithm developed in this paper for 3×3 window is by no means in its final form. Further refinement and extension to larger window sizes is both possible and desirable. The main purpose of this paper is to reveal a natural and simpler concept for edge detection.

ACKNOWLEDGMENT

The author wishes to thank Drs. I. Jurkevich and A. F. Petty for many helpful discussions.

REFERENCES

1. J. S. Lee, "Digital Image Smoothing and the Sigma Filter", Computer Vision, Graphics and Image Processing, Nov., 1983.

2. M. H. Hueckel, "An Operator Which Locates Edges in Digitized Pictures", Journal of the ACM, Vol. 18, 1971, pp. 113-115.

3. D. Marr, "Vision", Chapter 2, W. H. Freeman and Company, 1982.

4. L. G. Roberts, "Machine Perception of Three-Dimensional Solids", in Optical and Electro-Optical Information Processing, J. T. Tippet et al. (eds.), MIT Press, Cambridge, Mass., 1965, pp. 159-197.

5. W. K. Pratt, "Digital Image Processing", Wiley, New York, 1978.

6. J. M. S. Prewitt, "Object Enhancement and Extraction", in Picture Processing and Psychopictorics, B. S. Lipkin and A. Rosenfeld (eds.), Academic Press, New York 1970.

7. R. E. Suciu and A. P. Reeves, "A Comparison of Differential and Moment Based Edge Operators", Proceedings of the 1982 IEEE Pattern Recognition and Image Processing, pp. 97-102.

8. E. J. Delp and O. R. Mitchell, "Image Compression Using Block Truncation Coding", IEEE Trans. Comm. Vol. Com, 27, pp. 1335-1341.

9. Y. Yakimovsky, "Boundary and Object Detection in Real World Images", Journal of ACM, Vol. 23,1976, pp. 599-618.

10. J. S. Lee, "A Simple Speckle Smoothing Algorithm for Synthetic Aperture Radar Images", IEEE Trans. on System, MAN, and Cybernetics, Vol. SMC-13, No. 1, Jan/eb, 1983, pp. 85-89.

11. R. Nevatia and K. R. Babu, "Linear Feature Extraction and Description", Computer Graphics and Image Processing, Vol. 13, 1980, pp. 257-269.

A Syntactic Approach for SAR Image Analysis[1]

M. A. Eshera, H. S. Don,
K. Matsumoto,[2] and K.S. Fu

School of Electrical Engineering
Purdue University
West Lafayette, IN

ABSTRACT

In this paper, we investigate the application of some syntactic techniques for the segmentation of Synthetic Aperture Radar (SAR) images and for the extraction of their interesting objects. In general, the processing of aerial images is a complex task due to not only the noisy nature of these images, but also the vast diversion of the shape, the relative size, and the informative features of the important objects in them. The proposed approach performs filtering on the SAR images to improve their quality. Then it utilizes both the edge-based and the region-based neighborhood information to perform preliminary segmentation on the image. The global information in the image is used to assist in the final segmentation and to improve the information contents of the extracted features using some (deterministic or stochastic) syntactic techniques. Further analysis and interpretation of the images could be carried out by heigher level syntactic techniques for the recognition of noisy and distorted objects in the image. Experimental results are presented and discussed.

1 INTRODUCTION

The processing of aerial images, in general, and the Synthetic Aperture Radar (SAR) images, in particular, is a challenging task. This is not only due to their low signal-to-noise ratio, but also due to the vast diversion of the shape, the relative size, the nature, and the informative features of the interesting objects

[1] This work was supported by ONR Contract N00014-79-C-0574.
[2] On leave from the National Aerospace Laboratory, Science & Technology Agency, Tokyo, Japan.

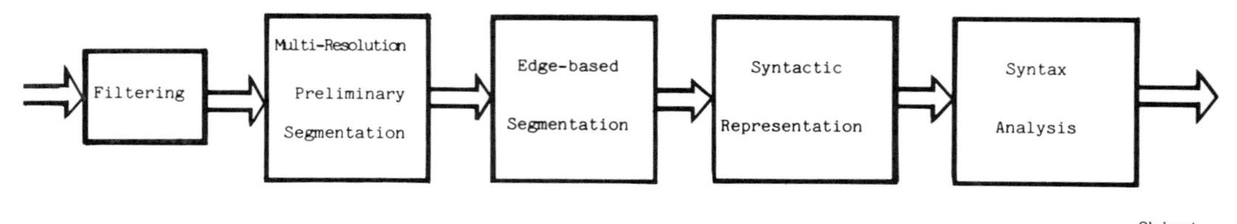

Figure 1 Syntactic Approach for SAR Image Analysis

in these image. Therefore, a relatively sophisticated system is needed for the analysis of this type of images. Such a system must utilize as much of the image information contents as possible, and must combine the powerful features of several image analysis techniques.

In this paper, we present an image analysis system for the analysis of SAR data. A block diagram of this system is shown in Fig. 1. In the early stages of the analysis, the system utilizes some efficient preprocessing techniques in order to improve the quality of the SAR images, and to extract the tentative primitives of the different objects in the images. Then in the later stages of analysis, the system apply the syntactic approach to image understanding in performing higher level interpretation and analysis of the images. In the remaining part of this section, we present a brief overview of every block in the system, while in the following sections we discuss the approaches and the algorithms used in the different parts of the system, as well as our experimental results.

The input to the system is an SAR raw image, usually in the form of 256 × 256 or 512 × 512 array of pixels, each pixel is represented by 8 bits gray scale value at that particular location in the image. Due to the coherent speckle noisy nature of this type of image, the system performs filtering, at the first stage, to upgrade the quality of the images and improve their signal-to-noise ratio. The basic idea is to remove the noise without distorting (or blurring) the image. The second stage in the system consists of multiresolution region-based segmentation. In the third stage, the system utilizes the results of the region-based segmentation in performing edge-based preliminary segmentation. The remaining stages of the system consist of applying the syntactic approach for performing further segmentation and detecting some objects in the image.

In Section 2, we discuss the preprocessing techniques used in the system. We also present a new local edge operator which has been shown to give very good results when applied to the SAR images. The utilization of multiresolution technique in the segmentation process is shown to be very useful in order to be able to handle the diversity in size of the interesting objects in the SAR images. Also in that section, we discuss how we utilize the multiresolution technique to assist in the segmentation and object detection in the images.

A higher level of image analysis demands the use of structural information contents of the images. The syntactic approach to image analysis utilizes the structural information, in the form of a set of primitives and relations among these primitives, in carrying out further analysis of the images. The primitives are representation of simple, but informative, features of the image. They take the form of either small regions or contour segments in the image. The relations between the primitives represent the physical relations between these features. In Section 3, we briefly review the basic concepts of syntactic techniques used in our system, while in Section 4, we discuss the utilization of those concepts in performing detailed segmentation and object extraction in our system. In Section 5, we present some concluding remarks and suggested work for the future.

2 PREPROCESSING AND PRELIMINARY SEGMENTATION

2.1 Preprocessing

The noisy nature of SAR images is self-evident from looking at them by human eyes, as shown in Fig. 2. It also has been studied by several researchers [3], [5], [6]. The coherent nature of the noise in this type of image handicaps most of the standard image processing techniques, in the sense that they fail to give satisfactory results. A main type of noise in the SAR images takes the form of speckles which behave as multiplicative noise. In our system the major part of the analysis is carried out by the syntactic techniques, in the later stages of the

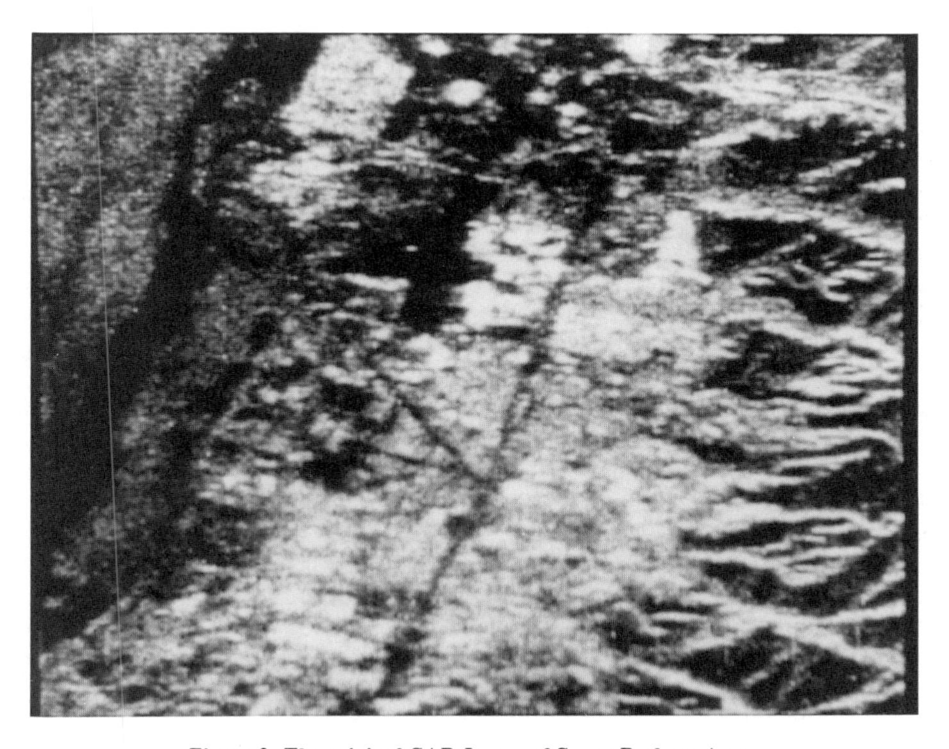

Figure 2 The original SAR Image of Santa Barbara Area

system. Nevertheless, the performance of some preprocessing operations to improve the image quality should be an appealing idea. The system utilizes very elementary preprocessing in the form of median filtering to remove the speckles. The median filter proves to perform better than most of the other standard filtering techniques in the preprocessing stage.

2.2 Preliminary Segmentation

In general, the process of segmentation is the partitioning of an image into some regions of interest. Traditionally, there have been two conventional approaches for segmentation, namely the edge-based approach and the region-based approach. The edge-based segmentation approach aims at finding edges around the different regions in the image. It relies on the fact that edges are of the most important features which are used by human eye for image interpretation. It usually utilizes some local edge operators to extract edges from the image. Due to noise and distortion that exist in almost any real image, this approach requires a relatively sophisticated technique to interpret the local edges and connect them in a proper way. The noisier the image is, the more ambiguous the results of the local edge operator will be, and consequently the more complex the interpretation process must be. In cases like aerial images, this process is too complex for this technique to be applied alone. Figure 3 shows the results of applying the Sobel edge operator on the Santa Barbara image.

Since the objective of segmentation is to find the meaningful regions in the image, the region-based segmentation approach aims at finding the interesting regions directly, rather than finding the edges around them. This approach utilizes the image properties in mapping individual image pixels into regions or mapping small regions in the image into larger regions. Some local techniques of this approach classify the pixels based on the properties of these particular pixels or the properties of their close neighbors, while more global techniques base the classification decision on the properties of larger regions of pixels. More sophisticated region segmentation techniques use the idea of split-and-merge of the image subregion to perform the segmentation. We will discuss this concept in more detail later in this section. For some very simple images, the region based segmentation techniques may yield satisfactory segmentation results. However, for any relatively complex image, e.g., a simple outdoor scene, even the most sophisticated region segmentation techniques still would not give satisfactory results.

In cases like SAR images, where the image quality is very poor, both region-based and edge-based segmentation approaches must be utilized to cooperate together in yielding more satisfactory results. In Section 2.2.1, we propose a new local edge operator which is used for the detection of local edges in the images. In Section 2.2.2, we present the utilization of the multiresolution technique in a split-and-merge region-based segmentation. Our system combines both region-based and edge-based concepts in performing the segmentation. Usually it performs region-based preliminary segmentation as discussed in Section 2.2.2, and then utilizes the results of a local edge operator to perform the syntax analysis and produce the final segmentation.

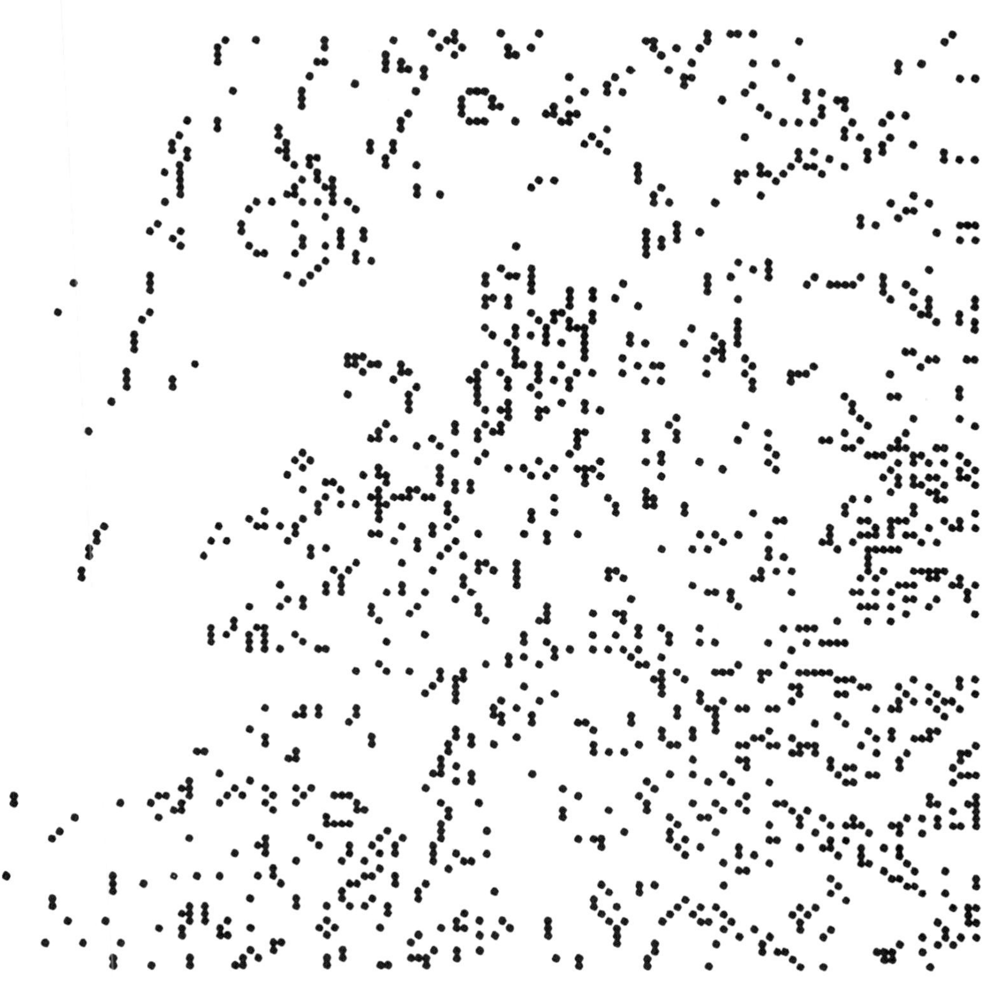

Figure 3 Results of Sobel Edge-Operator for Santa Barbara SAR Image, thresholded at 85 percentile

2.2.1 Variance Edge Operator

There are several major reasons that restrict the use of many conventaional edge operators in SAR images. The first of which, is the coherent noise in these images which acts as multiplicative rather than additive noise. The second reason is that most conventional edge operators assume certain edge orientation in the local neighborhood. Moreover, most conventional edge operators assume that if there is noise in the image, it must be Gaussian white noise. In contrary to this assumption, the noise in SAR image can rather be modeled by Gamma distribution. Also, most conventional edge operators are very sensitive to noise, because they are basically derivative operators. Since the signal-to-noise ratio of SAR images is 1, therefore the performance of any of these edge operators on SAR images is simply unsatisfactory [3].

A new edge operator is proposed which is specially designed to suit SAR images, since it assumes Gamma distribution model of noise. It also does not assume any orientation of edges in the local neighborhood. The Gamma probability density function takes the form of

$$f(x;\alpha,\beta) = \frac{x^{\alpha-1}\beta^{-\alpha}e^{-x/\beta}}{\Gamma(\alpha)}$$

where

- α is a system parameter equal to the number of independent samples averaged
- β is the first moment of the Gamma distribution
- $\Gamma(\alpha)$ is the Gamma function.

The operator uses arbitrary moving window Ω of N pixels. The edge value γ at the center pixel in the window is defined by:

$$\gamma = \frac{\overline{x^2}}{2\bar{x}^2}$$

where

$$\overline{x^2} = \frac{1}{N}\sum_{\Omega} x^2$$

$$\bar{x} = \frac{1}{N}\sum_{\Omega} x \ .$$

Every local neighborhood area, i.e., window Ω, is assumed to belong to one of two classes: those areas which contain single category, only, i.e., contain no edges, and those which contain two categories i.e., contain edges. If Ω contains a single category, then

$$E[\gamma_1] = 1 + \frac{1}{\alpha} \ .$$

If Ω contains two categories i and j, then

$$E[\gamma_2] = \left(1 + \frac{1}{\alpha}\right)\frac{(\beta_i^2 + \beta_j^2)}{(\beta_i + \beta_j)^2} \ .$$

For the SAR image of the Santa Barbara area, we can assume that all pixels are independent, and the system parameter $\alpha = 64$. If there is no edge in a window, then $E[\alpha] = 1.0156$. The result of applying this edge operator on the same image is shown in Fig. 4.

2.2.2 MultiResolution Split-and-Merge Region Segmentation

The multiresolution technique has been used in several image representation and processing tasks. The idea is to process the image at several levels of resolution, or at certain levels only in the case that we know the approximate size

Figure 4 Results of Variance Edge-operator for Santa Barbara SAR Image, thresholded at 85 percentile

of the objects. The resulting information is then used to guide the rest of the processing. This concept is very appealing in the processing of aerial images, since one of the major characteristics of these images is the vast diversity of the objects that may exist in them. Some different objects may have the same shape but different sizes or different relative dimensions. A clear example of these is the difference between highways and airport runways, where the size and the ratio of width-to-length become major discriminative features. Our system utilizes the multiresolution technique in performing preliminary region-based segmentation using split-and-merge approach.

We assume that only the approximate size of objects in terms of the area represented by each pixel is known. Starting from the image with a suitable rough resolution, we use a simple thresholding, e.g., thresholding over the gray scale or the variance of the subimages, to obtain a rough segmentation of the image. The results of this rough segmentation is then used for further segmentation at finer levels of resolution. The formal algorithm for this technique is given below; the utilization of this approach to the detection of objects from the Santa Barbara SAR image is discussed in Section 4.

ALGORITHM : MultiResolution Region-Based Segmentation

Purpose:

To perform split-and-merge multiresolution region-based preliminary segmentation

Input:

A gray scale image

Output:

Segmented image containing the candidate objects

Method:

I. Obtain a suitable rough resolution image based on the approximate size of the interesting objects.
II. Obtain a segmented image I_1 using some simple criterion, e.g., simple thresholding of the gray scale values.
III. Obtain a finer resolution image.
IV. Obtain a segmented image I_2.
V. Mask I_2 by I_1 in I_3.
VI. From I_2, pick up the pixels which are adjacent to object pixels in I_3 and move them into I_3.
VII. Repeat Step VI, until no more pixels can be added to I_3.

3 SYNTACTIC APPROACH

3.1 Syntactic Representation of Images

One of the main concepts of the syntactic approach to image analysis and pattern recognition is the idea of decomposing an image into subimages, or a pattern into subpatterns, which are simpler to analyze than the orginal image. By recursively applying this decomposition concept, complex images can be broken down into very simple primitives. Images, or objects in the images, are represented in terms of a set of primitives and the relationships between primitives.

There has been several techniques for image representation in the syntactic approach, which can be classified according to the form of relationship between the primitives. The first major class of techniques represents images by strings, which are formed by left-to-right ordered concatenation of primitives. The primitives, in this case, are assumed to possess two attachment points. In the second major class, images are represented by means of trees which represent a hierarchical form. The third, and the most general, major class of syntactic image representation techniques utilizes graphs, which are collections of nodes representing the image primitives (or subimages) and branches representing the relations between the different subimages. In the remaining part of this paper, we limit our work to the string case, since its analysis is computationally the most efficient.

An example of representing an object by a string of primitives is shown in Fig. 5. The primitives are chosen, in this example, as small line segments of the boundary of the object, and coded by Freeman's chain code which is shown in the same figure. A set of similar objects is represented by a set of strings that is called a language. Phrase-structure grammars are convenient and compact mathematical model for the generation and representation of string languages. They are defined as 4-tuple of the form:

$$G = (V_T, V_N, P, S)$$

where

V_T: is a set of terminal symbols (or primitives)
V_N: is a set of nonterminal symbols, with $V = V_T \cup V_N$, $V_T \cap V_N = \Phi$.
P: is a set of production rules of the form: $\alpha \rightarrow \beta$, where α and β are strings over V, and α contains at least one symbol of V_N.
$S \epsilon V_N$: is a designated nonterminal symbol called the Starting Symbol.

The language generated by a grammar G, denoted as $L(G)$, is the set of strings formed from terminal symbols starting from the starting symbol, S, via the repeating applications of production rules from P, i.e.,

$$L(G) = \{x \mid x \epsilon V_T^* \text{ such that } S \overset{*}{\longrightarrow} x\}$$

where

V_T^*: is the set of all possible strings of symbols in V_T.

Phrase-structure grammars and languages are classified into four types, namely finite-state, context-free, context-sensitive, and unrestricted, based on the type of their production rules. When the left-hand side of each production in P contains only a single nonterminal, the grammar is called context-free. The significance and the details of this classification can be found in several references on syntactic pattern recognition as well as on the theory of formal languages [4].

Some even more sophisticated and powerful versions of phrase-structure grammars are called stochastic grammars, in which probability measures are assigned to their production rules. The production rules of stochastic grammars are of the form $\alpha_i \overset{p_{ij}}{\longrightarrow} \beta_j$, where p_{ij} is the probability measure assigned to this production rule. The stochastic string languages generated by this type of grammars are used to represent patterns with uncertainty and randomness, which usually exist in practical applications. Every string in the language of such a grammar has an associated probability measure, which is the product of

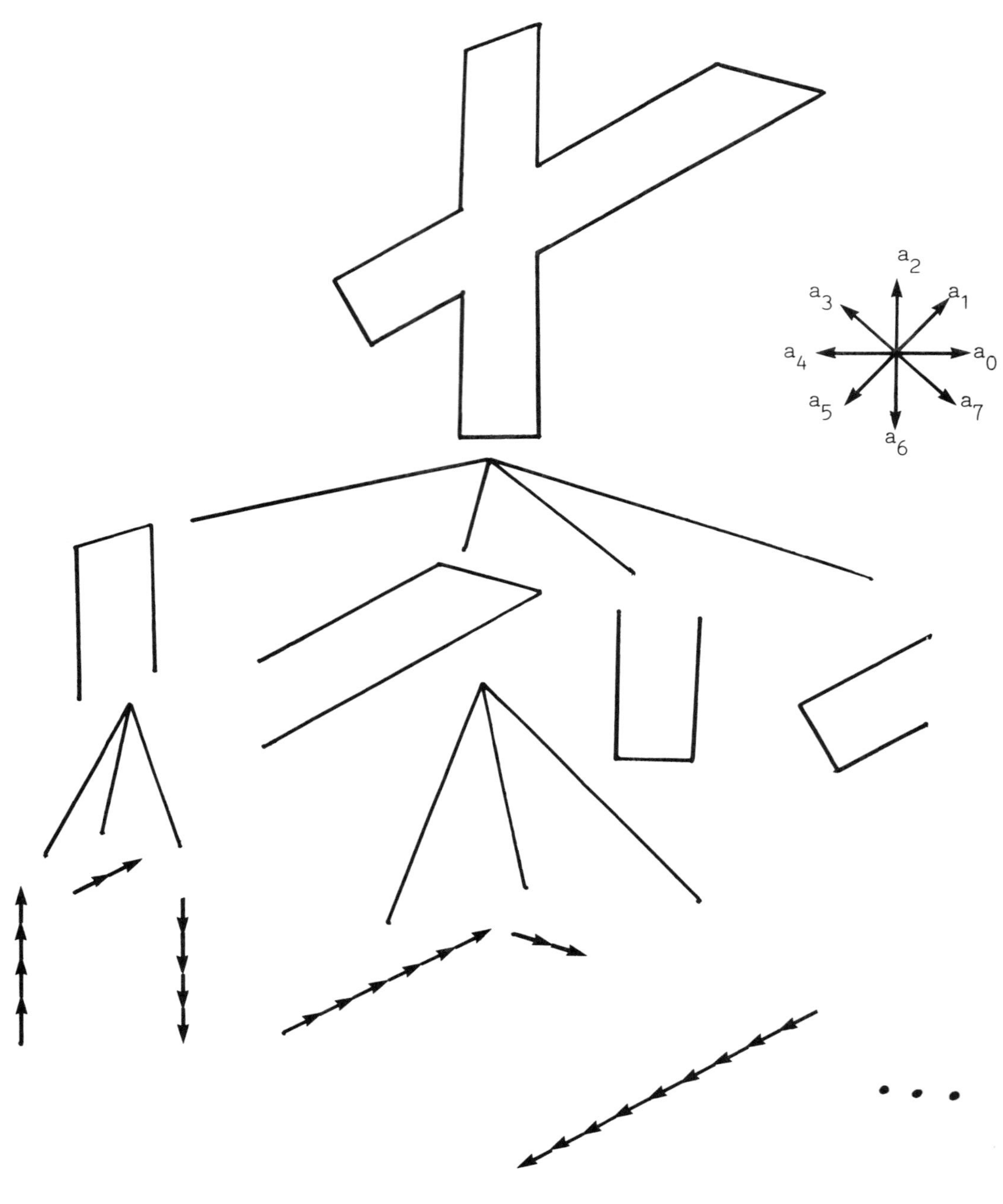

Figure 5 Object decomposition into primitives

the probabilities assigned to all the production rules which are used to generate this string. An application example of this grammar is discussed in details in Section 4, and is utilized for the extraction of the sea-coast boundary from the Santa Barbara SAR image.

3.2 Syntax Analysis For Object Detection

In the syntactic approach to image analysis, after we represent a class of objects by a string language and construct a context-free grammar which generates this language, the next step is to use a recognizer for the recognition of objects of this class. The recognizer designed for a particular grammar will recognize only the objects represented by strings in the language of this grammar. For an unknown object described by a string x, the basic problem of recognition is essentially answering the question of whether x belongs to the language generated by a particular grammar. If yes, then the object described by x is in the class generated by this grammar. If no, then this object does not belong to this class of objects.

Another more interesting question about the object concerns the details of the decomposition of this object into simpler components and primitives. Both questions, of the string membership in the language and of the detailed decomposition of the object, are handled by some efficient parsing techniques [1], [4]. Parsing algorithms for finite-state and context-free languages, has been implemented on VLSI chips and perform the parsing in time linearly proportional to the length of the string [2].

A grammar is constructed for every class of objects that are expected to exist in the image. Such a grammar generates a language covering all the possible versions of a certain object with different size, orientation, disposition etc. Moreover, for the recognition of noisy and distorted patterns, error-correcting parsers have been developed [4]. In Section 2, we have shown how some candidate objects are extracted from an image. In the next section, we discuss our results of performing the syntax analysis on the candidate objects in order to extract the exact objects from the SAR image of Santa Barbara.

4 EXPERIMENTAL RESULTS

4.1 Sea-Coast Boundary Extraction

In this experiment our interest is to find the sea-coast boundary in the Santa Barbara SAR image. As it can be seen in this image, even a human can hardly tell where the real sea-coast boundary is. The method contains two steps. In the first step, global information about the sea-coast boundary is extracted by roughly locating where the sea-coast boundary is. In the second step, local information is then used to trace the sea-coast boundary using the guidance of the global information obtained in the first step. These two steps are described in more details in the following.

In the first step, we need to roughly locate where the sea-coast boundary is. Using the terminology of syntactic approach, we are trying to infer a grammar for the description of the structure of the given sea-coast boundary. Examining the image carefully, we can find that sea and land have different

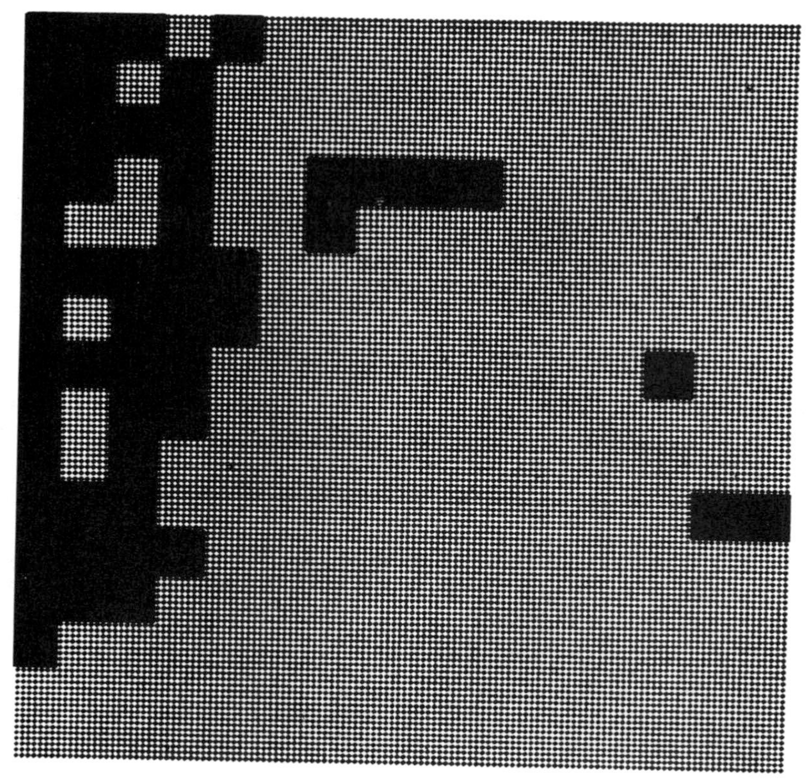

Figure 6 Thresholded Image that roughly separates Sea and Land

characteristics. The most significant difference is their texture busyness. The land looks much more busy than the sea. Thus the image is first divided into 256 16 × 16 subimages and then the variance for each subimage is computed. By a simple thresholding, we obtain the binary image as shown in Fig. 6. In this figure, regions of sea and land are roughly separated except that there are some noise blocks in the image. Assuming that both sea and land are connected regions, we can remove these isolated or dangling blocks. The result is shown in Fig. 7. The shape of the sea-coast boundary is roughly sketched out in Fig. 7 and will be used as the global information.

In Fig. 7, the sea-coast boundary is roughly sketched by a string of line segments. Each line segment may have different lengths. Starting from the top, we assign a nonterminal with attributes of length and direction to each line segment. This is shown in Fig. 8. The stochastic finite-state grammar which describe the sea-coast boundary is as follows:

$$S \xrightarrow{p=1} A_1$$

$$A_1 \xrightarrow{P_{ij}} a_j A_1, \quad j = 0.1,...,7$$

$$A_1 \xrightarrow{P_{1x}} A_2, \quad \text{if segment end is reached}$$

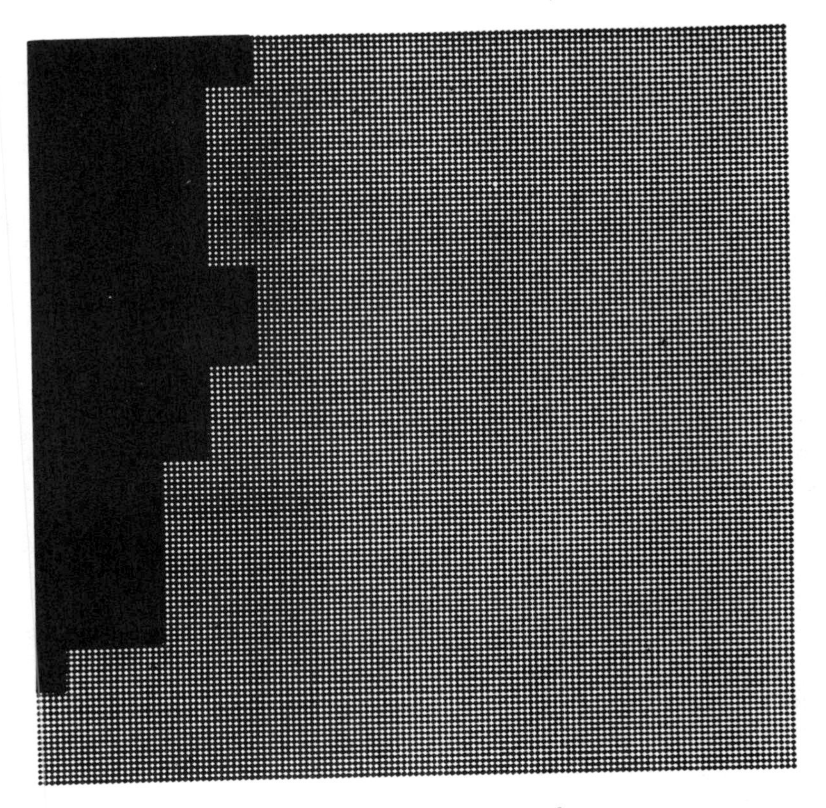

Figure 7 Rough Sea-coast boundary

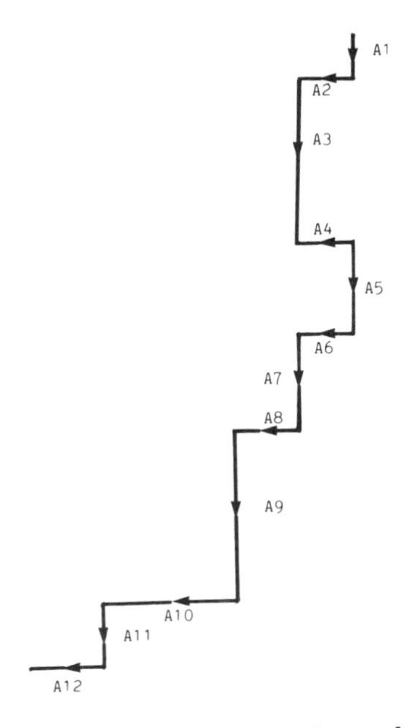

Figure 8 Assignment of a nonterminal to each line segment

$$A_2 \xrightarrow{P_{2j}} a_j A_2, \qquad j = 0.1,...,7$$

$$A_2 \xrightarrow{p_{1x}} A_3, \qquad \text{if segment end is reached}$$

.

.

.

$$A_{12} \xrightarrow{P_{12j}} a_j A_{12}, \qquad j = 0.1,...,7$$

$$A_{12} \xrightarrow{P_{12x}} b, \qquad \text{if boundary end is reached}$$

where each nonterminal $A_i, i = 1,...12$ has attributes as defined in Fig. 8 and the primitives $a_j, j = 0,1,...,7$ are defined in Fig. 9. Primitive b is a null primitive. The probability p_{ij} is assigned according to each nonterminal A_i. For example in Fig. 8, A_3 has the direction downward. So we assign the highest probability to a_6 and the second highest probability to a_5 and a_7 and so on. In this experiment, we first assign $p_{36} = 0.9$, $p_{35} = p_{37} = 0.7$, $p_{30} = p_{34} = 0.5$, $p_{31} = p_{33} = 0.3$, $p_{32} = 0$ and then normalize them so that we have $\sum_{j=0}^{7} p_{3j} = 1$. The production rules with probabilities p_{ij}, $j = 0,... ,7$ are used when the segment end has not been reached, i.e., these are conditional probabilities. When the segment end is reached, we apply the second production rule of each nonterminal with $p_{ix} = 1$. A line is drawn which bisects the next corner. If the boundary follower reaches this line, the second production rule is applied. This is illustrated in Fig. 10.

The above inferred grammar is a global description about sea-coast boundary. To trace the true sea-coast boundary, we also need local information, i.e., the local boundary strength obtained by applying a local operator to each pixel. In this experiment Sobel operator is used since this edge operator also gives the edge direction information which we need. Consider the situation as shown in Fig. 11 and suppose that we have reached pixel $x(i,j)$. The possible next boundary points are shown as dots in Fig. 11. The grammar indicates that pixel $x(i + 1,j)$ has the highest probability to be the next boundary point. The

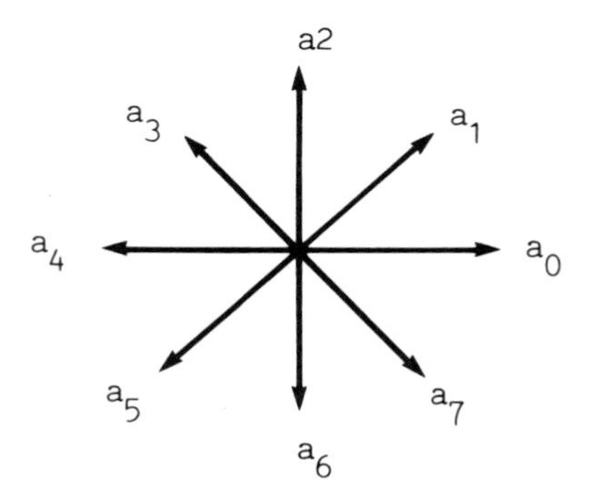

Figure 9 The 8 primitives for sea-coast boundary

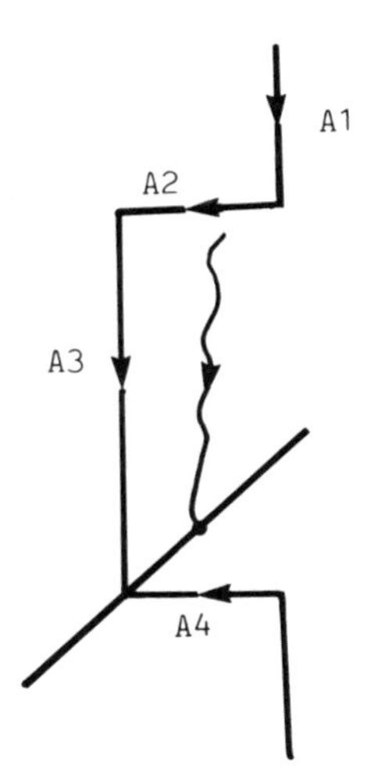

Figure 10 A line which bisects the next corner is used to indicate the end of a line segment

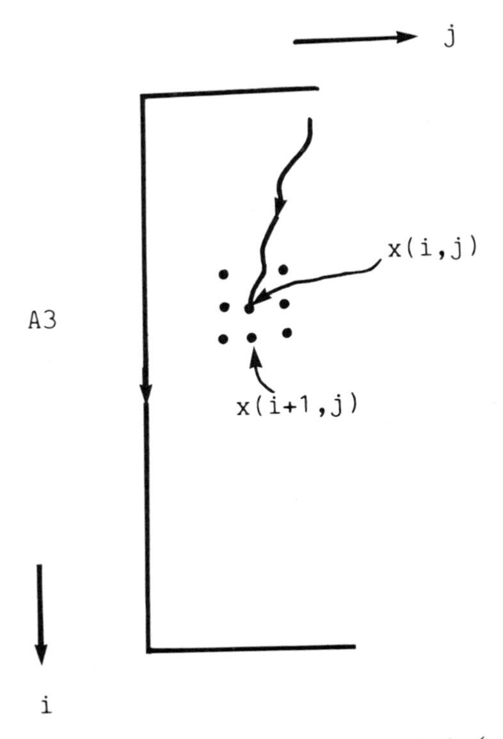

Figure 11 The possible next pixels for pixel $x(i,j)$

other points have lower probabilities. However, the local information does not necessarily match the global information. We will use pixel $x(i+1,j)$ to illustrate this. As shown in Figure 9, for each direction, we consider that the gray-levels at the left-hand side of each direction is higher (darker) than those at the right-hand side. For A_3, if the local information at $x(i+1,j)$ is a_6, then we say that $x(i+1,j)$ best matches the global information since the global information is also a_6. Obviously a_5 and a_7 match the global information less than a_6. We can assign probability to pixel $x(i+1,j)$ to reflect how this pixel is likely to be the next boundary point by (1) increasing its probability if the local information matches the global information or (2) decreasing its probability if the local information does not match the global information. After computing the weighted probability, we then multiply them with the local edge strength obtained by applying Sobel operator. The three pixels with the highest values are selected for further processing. The method becomes practical only if the search space can be reduced to a reasonable level. In this experiment, since the global information is available, we need only keep several paths with highest probability. Each line segment is processed sequentially. At the beginning, we select a point with the highest local boundary strength within a range of 15 pixels (horizontally) from the top point of line segment A_1. The ending point of each line segment will be used as the starting point for the succeeding line segment. The final result is shown in Fig. 12. As we can see, the result matches

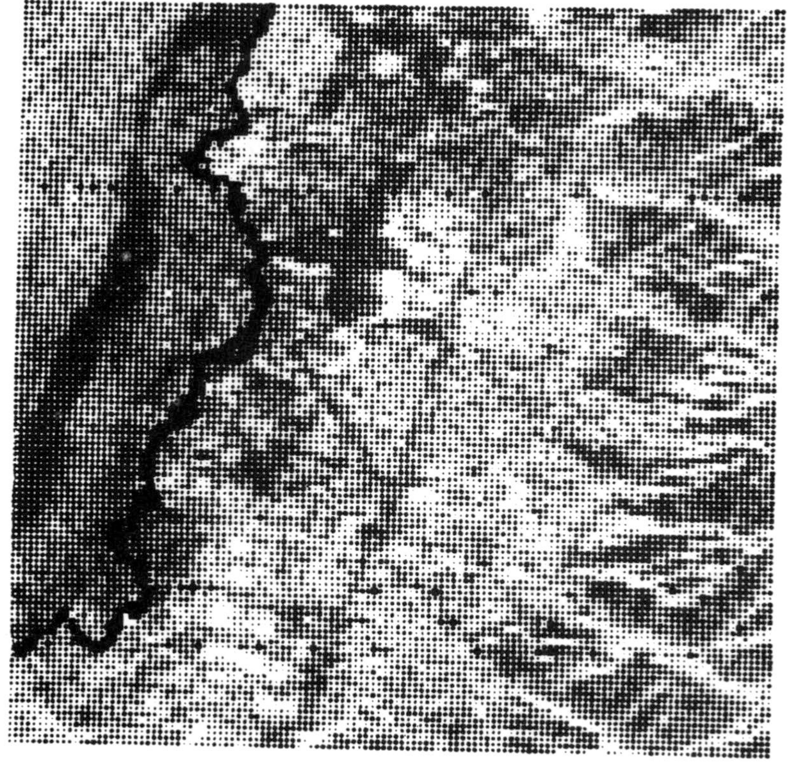

Figure 12 The extracted sea-coast boundary

very good at the top and the bottom portions of the sea-coast boundary, but get lost at some places in the middle part. Since global information has been used to guide the boundary following process, the deviation from the true boundary will not be too large even if the program gets lost at some places.

4.2 Object (Airport) Detection

In this section, we present the results obtained from using our system for detecting some objects of interest. Specifically this experiment is for the detection of an airport in the image. In general, airports are characterized, from their region-based features, as relatively large and uniform flat regions which exhibit high reflectance. From their structural, mainly edge-based features, they are characterized by containing some runways, usually two or more, which take the form of relatively long but not too narrow flat areas.

The system starts by filtering out the speckles from the image, as we discussed earlier in Section 2.1. Then by knowing the approximate size of the interesting objects in terms of the area represented by the image pixels, the system performs the multiresolution region-based segmentation as discussed in Section 2.2.2. The algorithm for this preliminary segmentation is also given in Section 2.2.2. The results of applying that algorithm to the Santa Barbara SAR image is shown in Figs. 13 through 15. The resulting image contains the candidate objects based on the region preliminary segmentation. Then the system

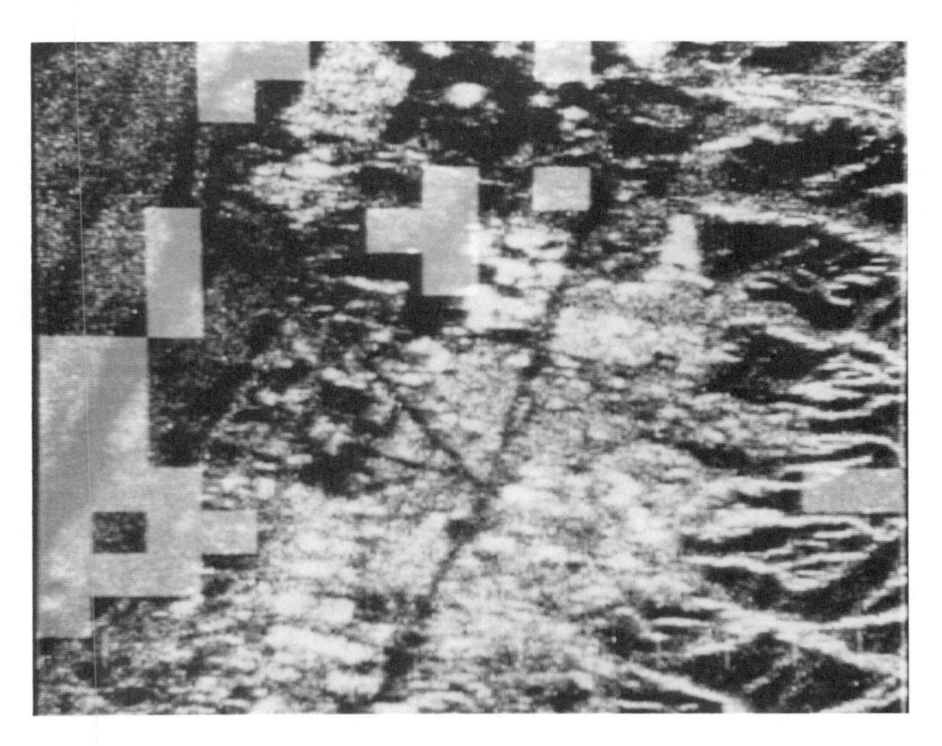

Figure 13 Rough resolution Image for the Santa Barbara Image

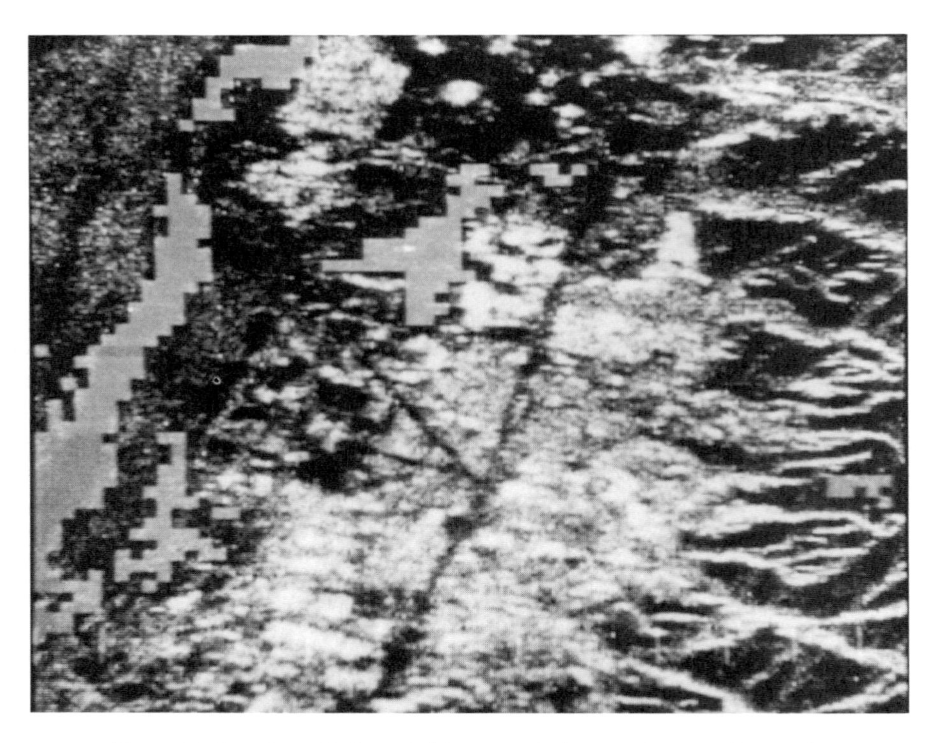

Figure 14 Results of Split-and-Merge for the Image shown in Figure 13

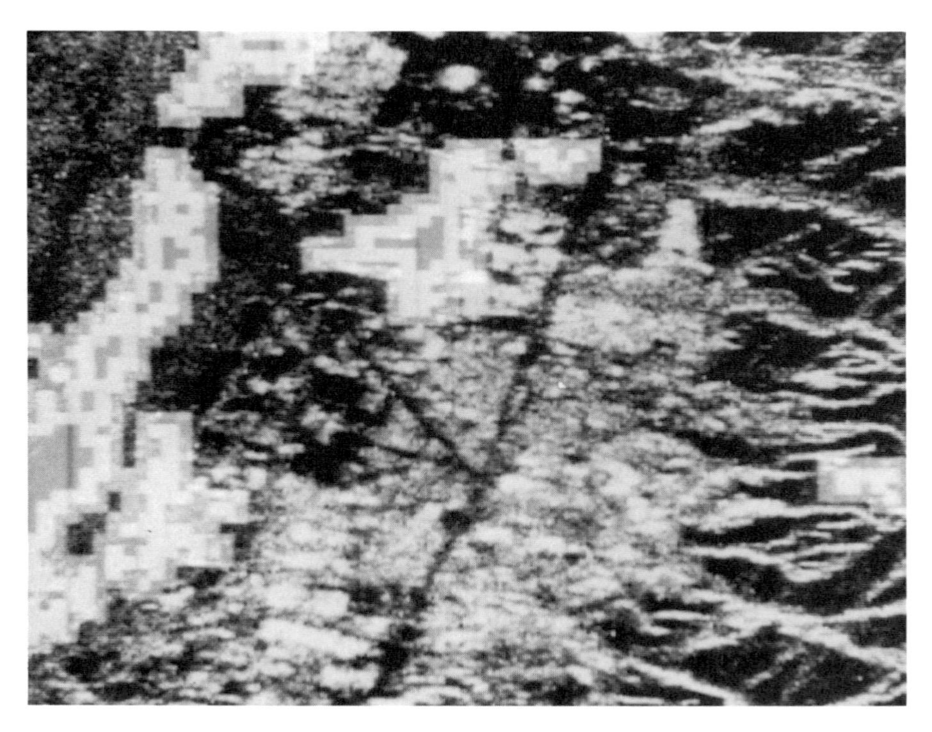

Figure 15 Masking around candidate objects for the Santa Barbara Image

concentrates on some mask areas around the candidate objects, and restrict the edge extraction within these areas. In Fig. 16, we show the result of applying the variance edge operator, as proposed in Section 2.2.1, to extract the actual edge in these regions. For this particular experiment, it can be seen from Fig. 16 that the actual edges of the objects in these regions are very noisy. Therefore, for now we may consider the edges of the segmented regions, as shown in Figs. 17 and 18, as a close approximation of the objects boundaries, and use them for the syntax analysis on the candidate objects, as we discussed in Section 3.2. This problem will be the subject of further investigation. We will apply the technique presented in Section 4.1 for sea-coast boundary extraction to extract the detailed boundary of the airport and utilize error-correcting parsing to perform the syntax analysis of the object.

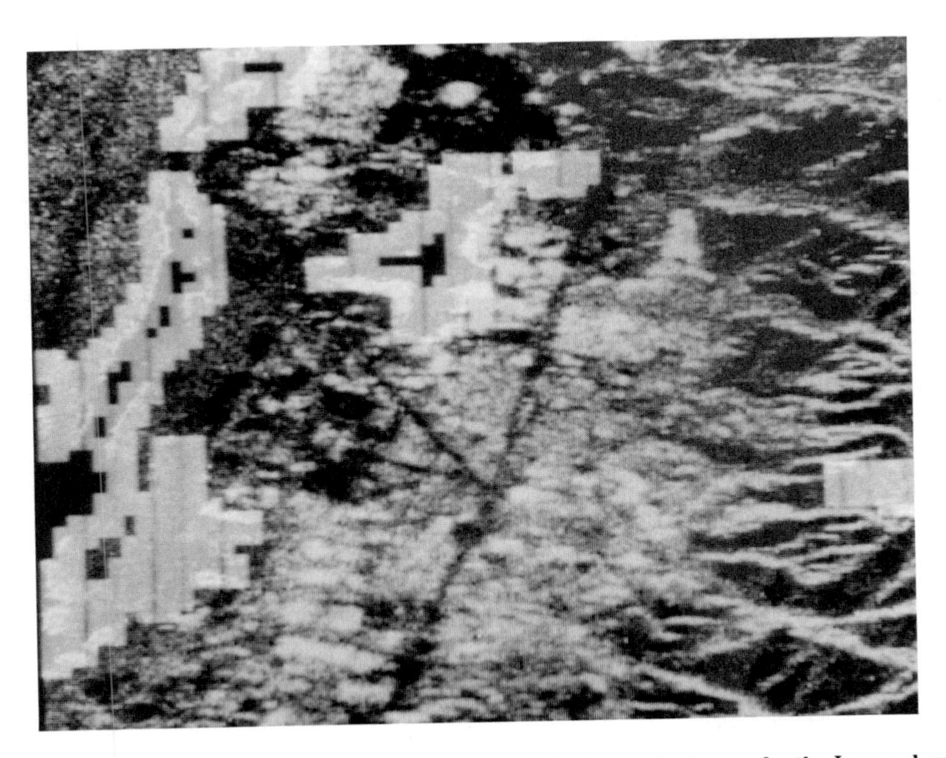

Figure 16 Results of the Variance Edge-Operator within the masked areas for the Image shown in Figure 15

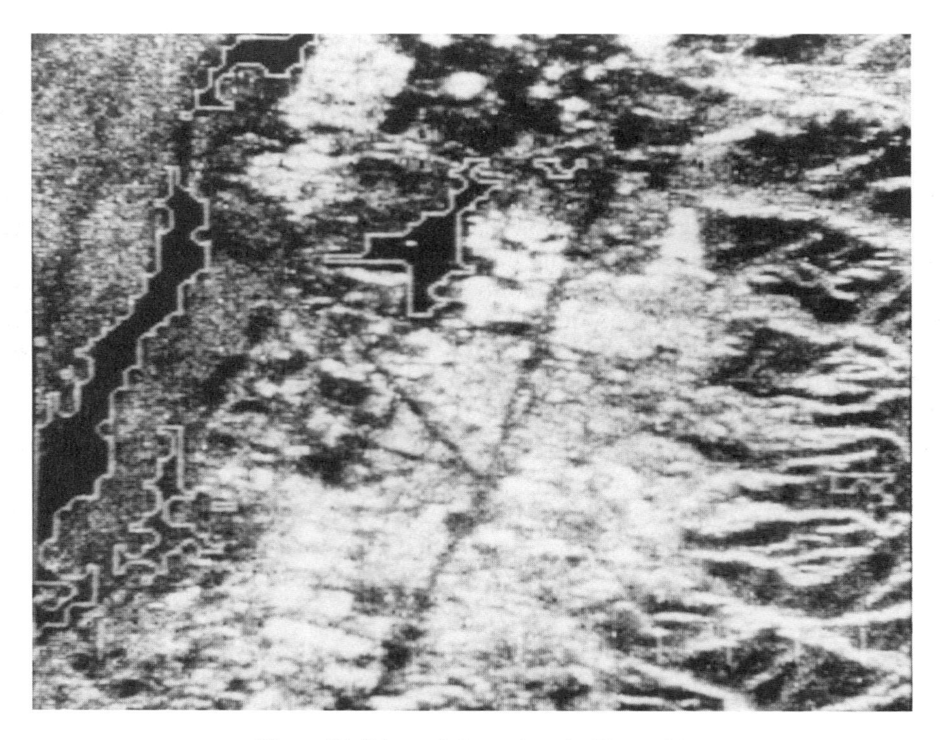

Figure 17 Edges of the regions in Figure 14

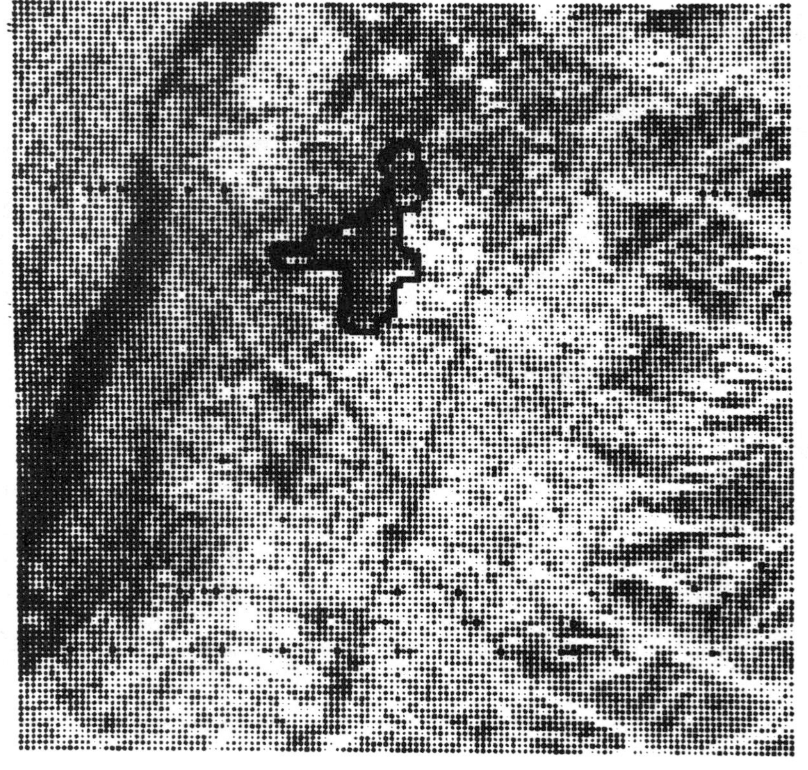

Figure 18 The region boundary as an approximation to the Airport boundary

5 CONCLUDING REMARKS

It has been shown in the literature that most of the conventional image processing techniques fall short when facing the task of SAR image analysis and do not provide satisfactory results. There are several reasons for this handicap which we have briefly discussed in this paper. Therefore, the stage is being left open for any more powerful technique to challenge this task. The syntactic approach is a powerful approach for image analysis. It enjoys several desirable capabilities, both in the representation and in the processing aspects of image analysis. It exhibits a great flexibility in both issues, which appears to be very much needed to cope with the vast diversity in shape, relative size, nature of the informative features of different objects, as well as the noisy nature of SAR images.

In this paper, we have investigated some of the capabilities, which the syntactic approach has to offer in challenging the SAR image analysis task. We focused our attention on two major tasks, the first is to perform refined segmentation of an image by extracting the precise boundaries of regions in the image. The second task is to detect objects of interest, which possess some common (usually structural) properties. The implementation of both techniques was performed and some experimental results are presented and discussed, in Section 4. The Seasat SAR image of Santa Barbara was used in both experiments. In the first experiment, we aim at extracting the sea-coast boundary, while in the second experiment we applied the object detection approach for locating the airport in the image.

The preliminary results obtained, so far, from both experiments are presented and discussed in Section 4. The approach shows very good and potentially promising results. In the sea-coast boundary extraction experiment, a stochastic finite-state grammar is used, both as a global descriptor of the sea-coast boundary and as a guide for the extraction of the precise boundary.

In the object detection experiment, the multiresolution split-and-merge region preliminary segmentation is used to locate candidate objects. The boundaries of the extracted candidate objects are used as an approximation of their exact boundaries. The extraction of more precise boundaries will be carried out in future work, using an approach similar to the one used in the sea-coast boundary extraction experiment. That is, a global syntactic descriptor will be utilized to overcome the distortion of the locally extracted boundaries, with the help of error-correcting syntax analysis.

In summary, we have shown in this paper that the syntactic approach to image analysis possesses a great potential and offers several of its superior features to meet the challenge of SAR image analysis. Preliminary results have shown that the proposed system can handle the diversity of size and shape of objects through the high descriptive power of its representation techniques, and the flexibility of its grammars. It copes with the noisy nature of these images using stochastic grammars and error-correcting syntax analysis, where many other image processing techniques fall short due to the coherent non-Gaussian nonadditive nature of the noise.

REFERENCES

1. A.V. Aho and J.D. Ullman, *The Theory of Parsing, Translation, and Compiling,* Prentice-Hall, Englewood Cliffs, N.J., 1972.

2. Y.T. Chiang and K.S. Fu, "VLSI Architectures for Syntactic Pattern Recognition," *Proceedings of CVPR Conference,* June 21-23, Arlington, VA , 1983.

3. V.S. Frost, J.A. Stiles, K.S. Shanmugan, and J.C. Holtzman, "A Model for Radar Images and Its Application to Adaptive Digital Filtering for Multiplicative Noise," *IEEE Transaction on Pattern Analysis and Machine Intelligence,* Vol. PAMI-4, No. 2, pp. 157-166, March 1982.

4. K.S. Fu, *Syntactic Pattern Recognition and Applications,* Prentice-Hall, Englewood Cliffs, N.J. 1982.

5. J.S. Lee, "Speckle Analysis and Smoothing of Synthetic Aperture Radar Images," *Computer Graphics and Image Processing,* Vol. 17, pp. 24-32, 1981.

6. J.S. Lee, "A Simple Speckle Smoothing Algorithm for SAR Images," *IEEE Transaction on Systems, Man, and Cybernetics,* Vol. SMC-13, pp. 85-89, January/February 1983.

Parametric Techniques for SAR Image Compression[1]

Benjamin Friedlander

Systems Control Technology, Inc.
Palo Alto, CA

ABSTRACT

Satellite based Synthetic Aperture Radar (SAR) provides a high-resolution all-weather imaging capability for remote sensing and surveillance in the ocean environment. Some of the issues related to efficient representation of SAR images are discussed. Several compression techniques based on autoregressive modeling of the video signals are presented. Autoregressive models with both constant and time-varying parameters are considered. Some examples are presented to illustrate the performance of these techniques.

1 INTRODUCTION

Satellite based Synthetic Aperture Radar (SAR) provides a high-resolution all-weather imaging capability for a variety of remote sensing and surveillance applications in the ocean environment [1],[2]. Ocean imaging by spaceborne SAR is a relatively recent development, and considerable work remains to be done before the potential of these systems is fully realized. A continually operating SAR system generates very large amounts of data (in the order of 100 million bits-per-second for SEASAT [3]). The processing, transmission and storage/retrieval of these data represents a challenging technological problem. While SAR sensor systems (i.e., the radar systems) are well-developed by

[1] This work was supported by the Office of Naval Research under Contract No. N00014-81-C-0300.

now, techniques for processing and handling the data collected by them are still under active investigation.

In this paper we consider some issues related to SAR image compression. Efficient representation of SAR data is necessary for reducing transmission bandwidth in the satellite/ground-station link and in the communication network used to distribute data, and for reducing the cost of data storage/retrieval systems. Many techniques are available for compression of general images. These techniques include: pulse-code-modulation (PCM), statistical coding, predictive coding and various transform coding techniques, see e.g. [4]-[6]. However, these techniques do not utilize the special features of SAR imagery. The techniques presented here were designed specifically for the SAR problem.

The degree to which an image can be compressed depends on the redundancy inherent in the original image. SAR images of land are usually very complex and do not lend themselves well to compression. SAR images of the ocean are of a different nature. Images of the ocean surface are highly regular and contain relatively few features such as waves, sea swells and ship wakes. The regularity and relative simplicity of these images makes it possible, at least in principle, to achieve high compression ratios.

The structure of the paper is as follows. A brief review of the processing required to generate a SAR image is presented in Section 2. It is shown that SAR image formation is closely related to spectral analysis of nonstationary processes. A parametric compression technique based on autoregressive (AR) modeling is described in Section 3. An extension of that technique to AR models with time-varying coefficients is given in Section 4. Some examples illustrating the performance of the proposed compression schemes are presented in Section 5.

2 SAR IMAGE FORMATION

A number of books and review papers describing the principles of SAR are currently available [7]-[15]. In this section we briefly review the operation of a SAR system.

A synthetic aperture radar is a coherent radar system located on a moving platform. The radar transmits a narrow pulse of electromagnetic energy, which is reflected from an area on the ground and is returned to the receiver. Consider for simplicity the case of a SAR system observing a single scatterer, as depicted in Figure 1. The time delay between the transmission of the radar signal and its return will be proportional to the distance R(t) between the radar and the scatterer. For a radar moving at a constant velocity V, the radar-scatterer range will be approximately a quadratic function of time. Let R_0 denote the distance from the scatterer to the line of flight, and t_0 be the time at which the platform is at the closest point to the scatterer. Then,

$$R(t) = [R_0^2 + V^2 (t - t_0)^2]^{1/2} \approx R_0 + V^2(t - t_0)^2/2R_0. \tag{1}$$

We make here the assumption that $V|t - t_0| << R_0$. The delay or phase of the return signal will also be a quadratic function of time:

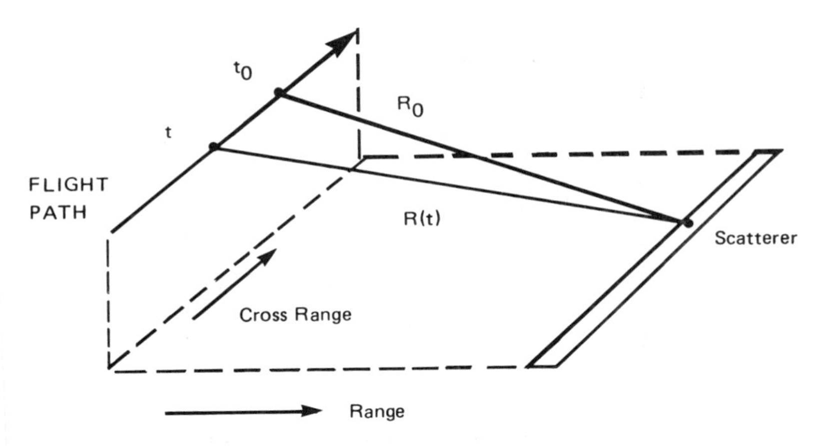

Figure 1 SAR geometry.

$$\Phi(t) = 4\pi\ R(t)/\lambda = \Phi_0 + 2\pi\ V^2(t - t_0)^2/\lambda\ R_0 \tag{2}$$

where λ is the wavelength of the radar transmission, and $\Phi(t)$ is the phase of the return signal relative to the transmitted signal. The frequency of the return signal is the first derivative of its phase and will, therefore, be a linear function of time:

$$\frac{d\Phi(t)}{dt} = 4\pi\ V^2\ (t - t_0)/\lambda R_0. \tag{3}$$

One of the first steps in conventional SAR image formation is the application of a quadratic phase correction to the return signal. (In effect, subtracting $2\pi V^2 t^2/\lambda$ from $\Phi(t)$). This demodulation step converts the linearly time-varying frequency of the radar return into a fixed frequency

$$f_0 = 4\pi\ V^2\ t_0/\lambda R_0. \tag{4}$$

The magnitude of this frequency is proportional to the location of the scatterer in the azimuth or cross-range direction. The energy of the signal at frequency f_0 is proportional to the radar reflectivity of the scatterer at the corresponding azimuth location. Thus, by performing a spectral analysis of the phase-corrected signal it is possible to determine the distribution of scatterers in the cross-range detection.

Returns from different ranges are separated on the basis of the time of the return pulse. The location of a given scatterer is therefore completely determined by the frequency and timing of the return signal, as depicted in Figure 2. The complete SAR image is formed by computing the power spectrum of the signal at the output of each range cell. This image may be considered to be the time-varying spectrum of the nonstationary process consisting of the sampled return signal, organized in a particular order. Note that the actual order in which the radar signal arrives corresponds to the rows of the array in Figure 2A, while the processing takes place along its columns. This reordering of the original time series is called in the SAR literature "corner turning".

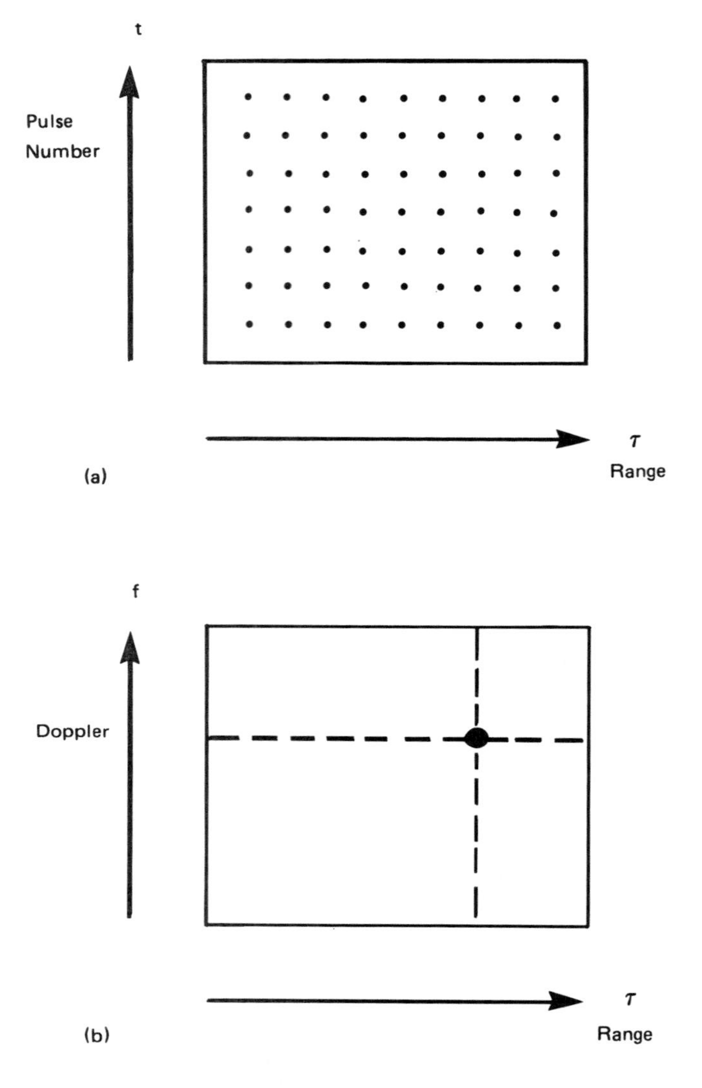

Figure 2 The radar return from a single scatterer. (a) The phase-corrected video array. (b) The column-by-column spectrum of the array data.

The interpretation of SAR image formation as spectral analysis of the phase-corrected and corner-turned radar returns motivates the study of the compression techniques considered in this paper. The basic idea is to find an efficient spectral representation for the video signal. A large amount of data is then reduced to a relatively small number of parameters, from which the signal spectrum (i.e., the SAR image) can be reconstructed. A unique feature of this approach is that it can be applied directly to the radar signals, rather than to the SAR image. Standard compression techniques require the formation of the

SAR image prior to image compression. Since SAR image formation requires a large amount of computation, the possibility of operating directly on the radar signal has practical advantages.

Autoregressive and autoregressive moving-average (ARMA) techniques have been widely used for modeling stationary time-series. These techniques can be adapted to the SAR image compression problem, as will be shown in the following two sections. We have chosen these models because of their relative simplicity and the availability of algorithms for fitting such models to data. It is quite possible that alternative models can be developed to provide more efficient representations of SAR data. Very few results seem to be available on the analysis and modeling of nonstationary processes of this type.

3 AUTOREGRESSIVE MODELS WITH CONSTANT PARAMETERS

A zero-mean stationary time-series y_t is said to be an AR process of order N, if it obeys the following stochastic difference equation:

$$y_t = - \sum_{i=1}^{N} A_i y_{t-i} + e_t \tag{5}$$

where e_t is a zero-mean white noise process with variance σ^2 [16],[17]. The power spectral density function $S(\omega)$ of this process is given by

$$S(\omega) = \sigma^2 / |A(e^{j\omega})|^2, \tag{6a}$$

where

$$A(e^{j\omega}) = 1 + A_1 e^{-j\omega} + \cdots + A_N e^{-jN\omega}. \tag{6b}$$

Autoregressive models can be used to efficiently represent the spectrum of time series. Consider the case of an autoregressive time series y_t observed over a finite interval $\{t = 0, \ldots T\}$. A number of techniques are available for estimating a set of autoregressive parameters from the observed data, see e.g., [16]-[20]. The estimated AR parameters $\{\hat{A}_1, \ldots, \hat{A}_N, \hat{\sigma}^2\}$ can be used to approximately reconstruct the spectral density function of the process. Thus, instead of transmitting or storing the original data, we can transmit or store the AR parameters. When the number of parameters $(N + 1)$ is small compared to the number of data points (T), considerable data reduction or compression is achieved.

Autoregressive modeling has been used extensively in speech compression applications. The linear-predictive-coding (LPC) method of speech analysis uses AR parameters and pitch information to synthesize speech signals [21]. The SAR compression technique considered is similar in spirit to the LPC technique. The main differences are: (i) the SAR data is more complex and less structed than speed data, (ii) phase information plays an important role in speech synthesis, while only magnitude (power spectrum) information is needed to generate the SAR image.

The SAR image compression technique consists of the following steps: (i) A set of AR parameters is estimated from the video signal at the output of each range cell. These parameters are quantized prior to transmission or storage, (ii). Each set of quantized is used to reconstruct one line of the SAR image, via equation (6). The complete image is thus reconstructed line-by-line, where each line is represented by a different set of AR parameters. This method is inefficient since it makes no use of the possible correlation between adjacent lines of the image. More efficient compression techniques will be discussed in the next section.

Estimating the AR Parameters

Let $y_{t,k}$ denote the complex video signal at the output of the k-th range cell. The AR representation of this signal is estimated by solving the following set of overdetermined linear equations:

$$\underbrace{\begin{bmatrix} y_{N,k} \\ y_{N-1,k} \\ \cdot \\ \cdot \\ \cdot \\ y_{T,k} \end{bmatrix}}_{\underline{y}_k} = - \underbrace{\begin{bmatrix} y_{N-1,k} \cdots\cdots y_{0,k} \\ y_{N-2,k} \cdots\cdots y_{1,k} \\ \cdot \qquad\qquad \cdot \\ \cdot \qquad\qquad \cdot \\ \cdot \qquad\qquad \cdot \\ y_{T-1,k} \cdots\cdots y_{T-N,k} \end{bmatrix}}_{Y_K} \begin{bmatrix} A_{1,k} \\ \cdot \\ \cdot \\ \cdot \\ \cdot \\ A_{N,K} \end{bmatrix} . \tag{7}$$

The AR parameters are chosen to minimize the sum of squared prediction errors

$$\hat{\sigma}^2 = \frac{1}{T-N} \sum_{t=N}^{T} \epsilon_{t,k}^2$$

where

$$\epsilon_{t,k} = y_{t,k} + \sum_{i=1}^{N} \hat{A}_{i,k}\, y_{t-i,k}, t = N, \ldots, T.$$

A large number of robust, computationally efficient least-squares techniques are available for solving equation (7). We refer to [22]-[24] for details. Here we note only that the least-squares solution can be written explicitly as

$$\begin{bmatrix} \hat{A}_{1,k} \\ \cdot \\ \cdot \\ \cdot \\ \hat{A}_{N,k} \end{bmatrix} = -\,(Y_k^T\, Y_k)^{-1} Y_k^{\prime T} \underline{y}_k . \tag{8}$$

The more robust least-squares techniques obtain the estimates without explicitly computing the sample covariance matrix $(Y_k{}' Y_k)$ or its inverse.

The presence of additive measurement noise in the data $y_{t,k}$ introduces a bias in the AR parameter estimates and distorts the resulting image. For finitely correlated noise this problem can be alleviated by replacing equation (7) with

$$(Z_k' \, Y_k) \begin{bmatrix} A_{1,k} \\ . \\ . \\ . \\ A_{N,k} \end{bmatrix} = - Z_k^{T} \underline{y}_k \tag{9}$$

where Z_k is a shifted version of the data matrix Y_k,

$$Z_k = \begin{bmatrix} 0 & \cdots & 0 \\ . & \cdots & . \\ . & \cdots & . \\ 0 & \cdots & 0 \\ y_{N-1,k} & \cdots & y_{0,k} \\ . & \cdots & . \\ . & \cdots & . \\ y_{T-M-1,k} & \cdots & y_{T-M-N,k} \end{bmatrix} \tag{10}$$

Equations (8) and (9) are called the Yule-Walker and Modified Yule-Walker equations, respectively. See [25] for a more detailed discussion of these equations and their use in parametric spectral estimation.

Note that the AR parameters are computed directly from the video signal. Thus, there is no need to form a SAR image at the transmitting end, prior to compression. The quantized AR parameters contain the information needed to reconstruct a compressed SAR image at the receiving end. Reduction of the processing required at the transmitter is especially useful for spaceborne applications, where on-board processing is very expensive compared to processing on the ground.

The compression ratio achievable by this AR technique can be evaluated as follows. Let Q_y be the number of bits required to represent each sample of the complex video signal $y_{t,k}$, and let Q_A be the number of bits used for storing the complex AR parameters $A_{i,k}$. Each column of the original data array requires $T \cdot Q_y$ bits, while its AR representation required $N \cdot Q_A$ bits. The compression ratio β is, therefore,

$$\beta = \frac{T}{N} \cdot \frac{Q_y}{Q_A} \tag{11}$$

The choice of the AR model order (N) and the number of quantization levels (Q_A) will effect both the compression ratio and the image quality. Higher compression ratios will generally lead to poorer image quality. The tradeoff between those two factors is difficult to analyze and needs to be evaluated by experimentation. Some examples will be presented in Section 5.

The shape of an autoregressive spectrum is very sensitive to the values of the AR parameters. Thus, a relatively large number of quantization levels Q_A will generally be needed to obtain good spectral fit. A better parametrization of

AR processes is given by the so-called reflection or partial-correlation coefficients [21],[24]. There is a one-to-one correspondence between AR parameters and reflection coefficients. However, the latter can be quantized more coarsely, for the same quality of spectral fit. In other words, higher compression ratios can be achieved for a given image quality by using this alternative parametrization. Reflection coefficients and the related lattice structures are widely used in speech compression systems because of this property. See [21],[24] for a more detailed discussion.

The AR technique presented above can be implemented in many different ways. Both batch and recursive implementations are possible. These estimation algorithms lend themselves to a high degree of parallelism, since different lines can be processed independently and simultaneously. Efficient implementations are available requiring in the order of KNT multiplications and additions.

4 TIME-VARYING AUTOREGRESSIVE MODELS

Efficient image compression techniques utilize the correlation between adjacent lines as well as within each line of the image. The technique introduced in the previous section can not achieve a high degree of compression since it completely ignores line-to-line correlation. However, it is possible to modify this approach in one of several ways to increase the achievable compression ratio.

Correlation between adjacent lines in the image translates into correlation between the AR parameters representing these lines. In other words, the sequence of AR parameter vectors is correlated. It is possible to apply various vector coding techniques to reduce the redundancy in this sequence. Such techniques have been applied successfully in speech compression applications [26].

Another approach, which has some interesting possibilities, is to represent the entire image by a time-varying autoregressive model. Let y_t be a nonstationary process, obeying the following stochastic difference equation:

$$y_t = - \sum_{i=1}^{N} A_i(t)\, y_{t-i} + e_t \tag{12}$$

where e_t is a white noise process with variance $\sigma^2(t)$. We can formally associate with this process a time-varying spectral density function $S_t(\omega)$,

$$S_t(\omega) = \frac{\sigma^2(t)}{|A\,(e^{j\omega};t)|^2} \tag{13a}$$

where

$$A(e^{j\omega};t) = 1 + A_1(t)\, e^{-j\omega} + \ldots + A_N(t) e^{-jN\omega} \tag{13b}$$

It can be shown that for a certain class of nonstationary processes this definition is a reasonable extension of the power spectrum defined for stationary processes. For further discussion and some examples see [27].

Recalling our interpretation of the SAR image as a time-varying spectrum, it follows that a set of time-varying AR parameters can be used to represent the complete image. This suggests a global compression technique based on fitting

The variance $\sigma^2(k)$ is estimated by averaging the squared prediction errors related to the k-th line of the image. As before, any robust least-squares technique (e.g., [22][23]) can be used to solve Eq. (16). The least-squares solution can be explicitly written as

$$\begin{bmatrix} \hat{\theta}_1^0 \\ \cdot \\ \cdot \\ \hat{\theta}_1 \end{bmatrix} = -\ (X'X)^{-1}\ X^T \underline{x}\ \cdot \tag{17}$$

This is the equivalent of the Yule-Walker equations used in the stationary case. The equivalent of the Modified Yule-Walker equations can be similarly derived; see [27] for details.

The SAR image is reconstructed from the quantized parameters $\{\hat{\theta}_1, \ldots, \hat{\theta}_N, \sigma^2\}$. The k-th line of the image is given by

$$\hat{S}\ (\omega{:}k) = \frac{\hat{\sigma}^2(k)}{|\hat{A}\ (e^{j\omega};k)|} \tag{18a}$$

where

$$\hat{A}\ (e^{j\omega}{:}k) = 1 + \hat{A}_1(k)\ e^{-j\omega} + \ldots + \hat{A}_N(k)\ e^{-jN\omega} \tag{18b}$$

$$\hat{A}_1(k) = \sum_{l=0}^{L-1} \hat{a}_{1,l} k^l . \tag{18c}$$

Assuming that the time-varying AR parameters are quantized to Q_θ bits, the total image will be represented by NLQ_θ bits. The original video array contains TKQ_y bits. Thus the compression ratio is given by

$$\beta = \frac{T}{N}\ \frac{K}{L}\ \frac{Q_y}{Q_\theta} \tag{19}$$

This compression ratio is larger than that of the constant parameter AR technique (cf. (11)) by a factor of K/L. Thus, the time-varying AR technique has the potential for achieving significantly higher compression ratios than the constant parameter technique.

The main difficulty with the time-varying AR technique described above is that it involves the solution a high dimensional estimation problem. Note that the parameter vector in Eq. (16) is of dimension NL, which may be very large. High dimensional problems lead to numerical and computational difficulties. Recall that the constant parameter AR techniques involves the solution of K independent N-dimensional problems.

The problem of dimensionality may be alleviated in several ways. An obvious method is to divide the video array into a number of narrow strips and fit a time-varying AR model to each strip. This reduces the required polynomial order L, and thus the dimension of the parameter vector.

Another approach is to first fit a constant parameter AR model to each line, as described in Section 3, and then fit a polynomial function to each parameter sequence. In other words, first estimate $\{\hat{A}_{1,k}, \ldots, \hat{A}_{N,K},$

a time-varying AR model to the entire video array. The image is then reconstructed from the quantized AR parameters via equation (13).

To make the time-varying AR model useful for compression purposes, it is necessary to find a compact parametrization for the time variation of the coefficients $A_i(t)$. We may assume, for example, that the AR parameters are a polynomial function of time:

$$A_i(t) = \sum_{l=0}^{L-1} a_{i,l} t^2. \tag{14}$$

Equation (12) can be written in this case as

$$y_t = - \sum_{i=1}^{N} e_i^T x_{t-1} + e_t \tag{15a}$$

where

$$\theta_1 = [a_{i,0}, \ldots, a_{iL-1}]^T \tag{15b}$$

$$x_t = y_t \, [1, t, \ldots, t^{L-1}]^T \tag{15c}$$

Note that Eq. (15) has the form of a standard linear regression problem. Thus, any of the techniques used for estimating the coefficients of a constant parameter AR model can be applied here as well.

Based on the previous discussion we can now summarize the equations for the SAR image compression problem. Given the video array $\{y_{t,k}, 0 < t < T, 1 < k < K\}$, the time-varying AR parameters are given by the least-squares solution of the following overdetermined set of linear equations:

$$\begin{bmatrix} y_{N,1} \\ \vdots \\ y_{T,1} \\ \hline \vdots \\ \hline y_{N,K} \\ \vdots \\ y_{T,K} \end{bmatrix} = - \begin{bmatrix} x_{N-1,1}^T & \cdots & x_{0.1}^T \\ \vdots & & \vdots \\ x_{T-1,1}^T & \cdots & x_{T-N,1}^T \\ \hline \vdots & & \vdots \\ \hline x_{N-1,K}^T & & x_{0.K}^T \\ \vdots & & \vdots \\ x_{T-1,K}^T & \cdots & x_{T-N,K}^T \end{bmatrix} \begin{bmatrix} \theta_1 \\ \vdots \\ \theta_N \end{bmatrix} \tag{16a}$$

where

$$x_{i,k} = y_{i,k}\ [1,k,\ldots,k^{L-1}]^T \qquad (16b)$$

$$\theta_i = [a_{i,0},\ldots,a_{i,L-1}]^T \qquad (16c)$$

$k = 1,\ldots,K\}$. Then solve the following set of linear equations for $i = 1,\ldots,N$

$$\begin{bmatrix} \hat{A}_{i,1} \\ \hat{A}_{i,2} \\ \cdot \\ \hat{A}_{i,K} \end{bmatrix} = \begin{bmatrix} 1 & 1 & \ldots & 1 \\ 1 & 2 & \ldots & 2^{L-1} \\ \cdot & \cdot & & \cdot \\ \cdot & \cdot & & \cdot \\ \cdot & \cdot & & \cdot \\ 1 & K & \ldots & K^{L-1} \end{bmatrix} \begin{bmatrix} a_{i,0} \\ a_{i,1} \\ \cdot \\ \cdot \\ \cdot \\ a_{i,L-1} \end{bmatrix}. \qquad (20)$$

The image will be reconstructed from the quantized parameters $\{\hat{a}_{i,0},\ldots \hat{a}_{i,L-1}\}$ as indicated in Eq. (18).

This two-step procedure involves the solution of K N-dimensional problems and N L-dimensional problems. It makes use of the line-to-line correlation in the image without performing global optimization as in the time-varying AR technique described earlier. A detailed performance comparison of these compression methods is not yet available.

Finally, we note that the AR estimation procedure described above is not statistically efficient. The resulting estimates may be quite inaccurate if the data is very noisy. Improved procedures based on maximum likelihood estimation can be developed. However, such estimates will be considerably more complex than the one presented here.

5 SOME EXAMPLES

To illustrate the performance of autoregressive compression techniques we present a few examples. Figures 3-5 were generated using the constant parameter AR technique described in Section 3. Each of these images consists of 128 × 128 pixels. Figure 3 is a synthetically generated image, while Figures 4-5 are SEASAT SAR images provided by the Navy Research Laboratory. Note the improvement in the compressed image quality as the AR model is increased. These figures indicate that the constant parameter AR technique is capable of providing compressed images with good quality for relatively simple scenes. See also [6] for an interesting example.

The time-varying AR technique described in section 4 has been tested so far only on some very simple images containing linear features. Figure 6 and 7 are two such examples. In these figures, triangles depict the original image and dots depict the compressed image. These figures present a somewhat sparsely sampled version of a 128 × 128 pixel image. See [27] for further examples and a more detailed discussion. Note that the compressed 128 × 128 image is represented by NL=12 and NL=18 parameters, in Figures 4 and 5 respectively. More complicated images will require, of course, a much larger number of parameters.

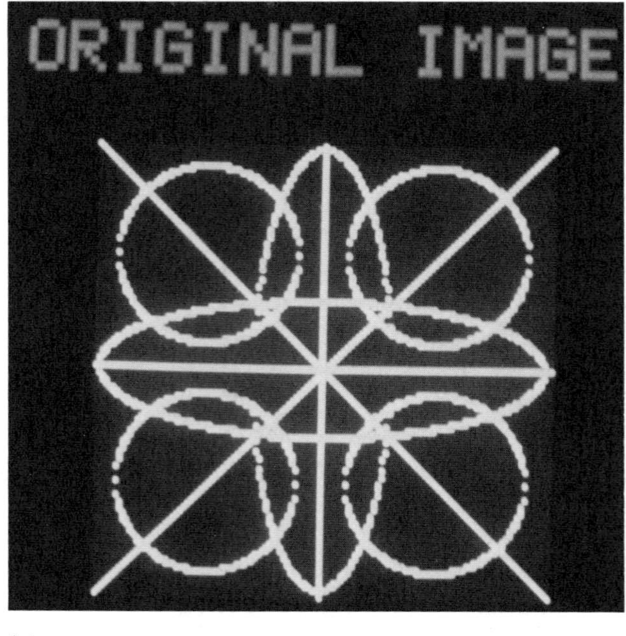

(a)

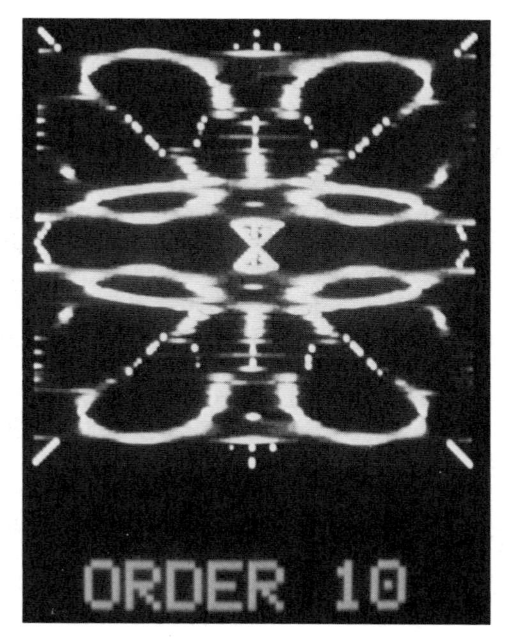

(b)

Figure 3 Constant parameter autoregressive compression of a synthetic image. (a) Original image. (b) Compressed image, N = 10. (c) Compressed image, N = 20. (d) Compressed image, N = 30.

(c)

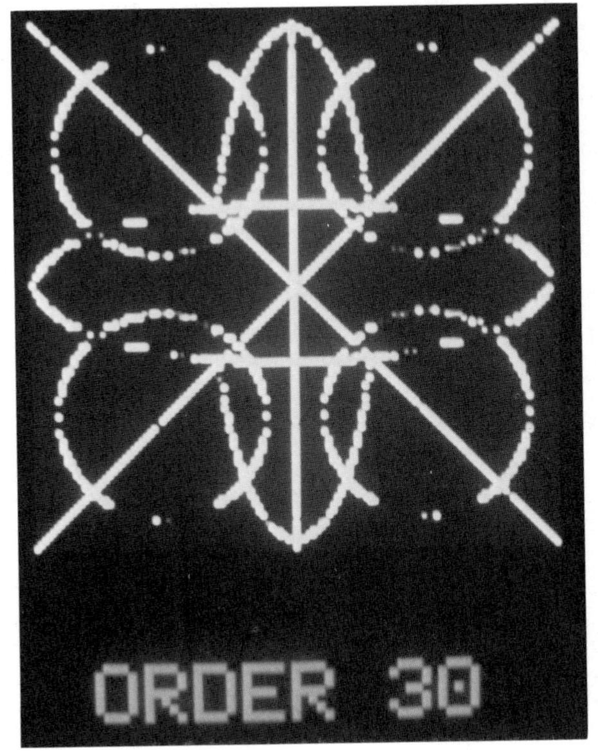

(d)

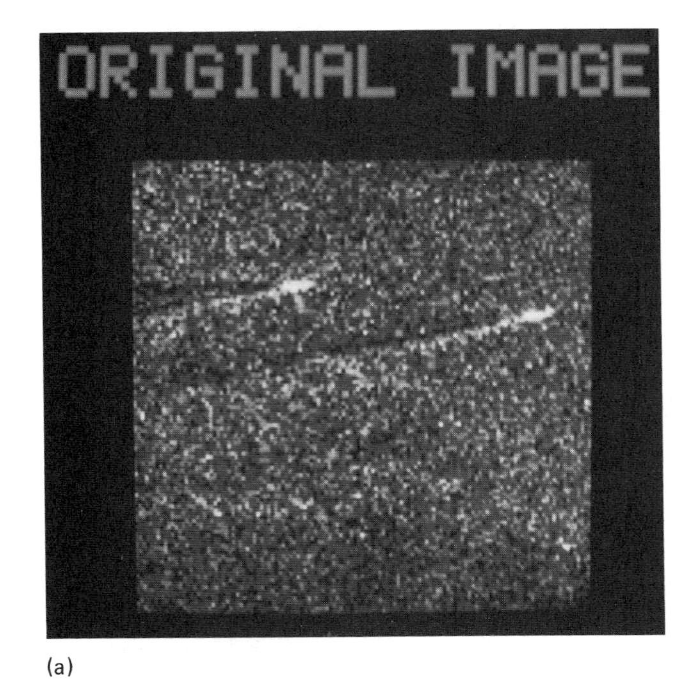

(a)

(b)

Figure 4 Constant parameter autoregressive compression of SEASAT SAR image of a ship wake, Chesapeake Bay. (a) Original image. (b) Compressed image, N = 10. (c) Compressed image, N = 20. (d) Compressed image, N = 30.

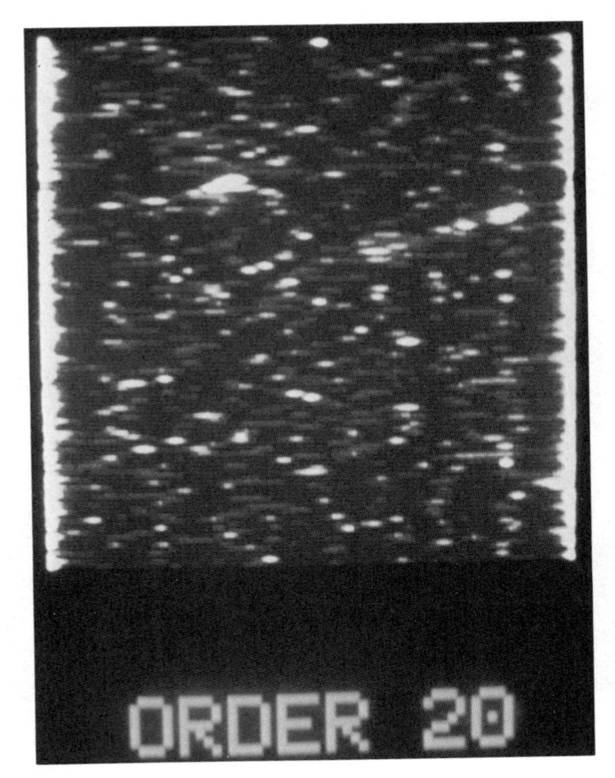

(c)

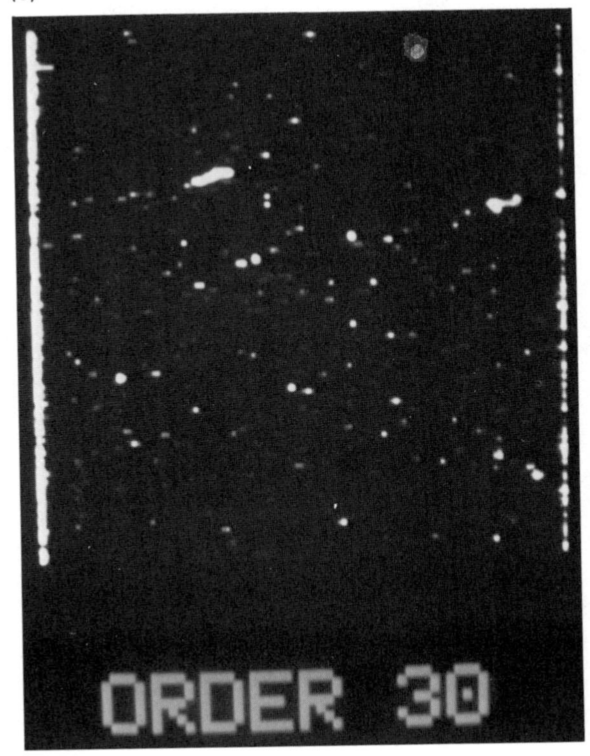

(d)

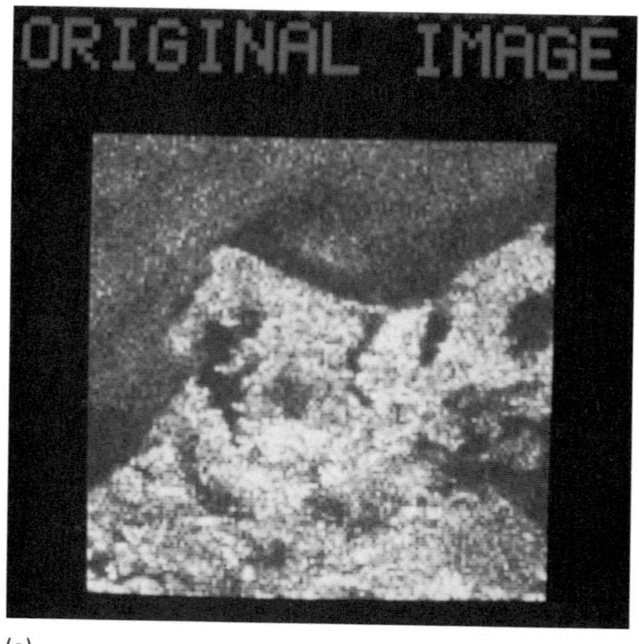

(a)

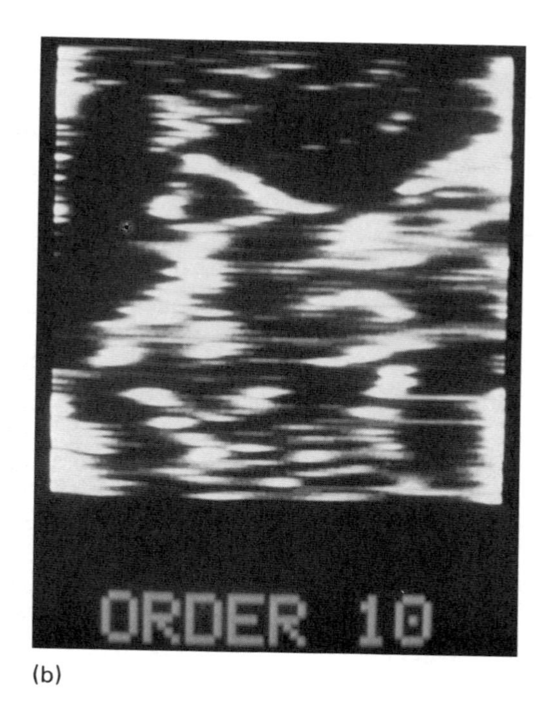

(b)

Figure 5 Constant parameter AR compression of SEASAT SAR image of a shoreline in Chesapeake Bay. (a) Original image. (b) Compressed image, N = 10. (c) Compressed image, N = 20. (d) Compressed image, N = 30.

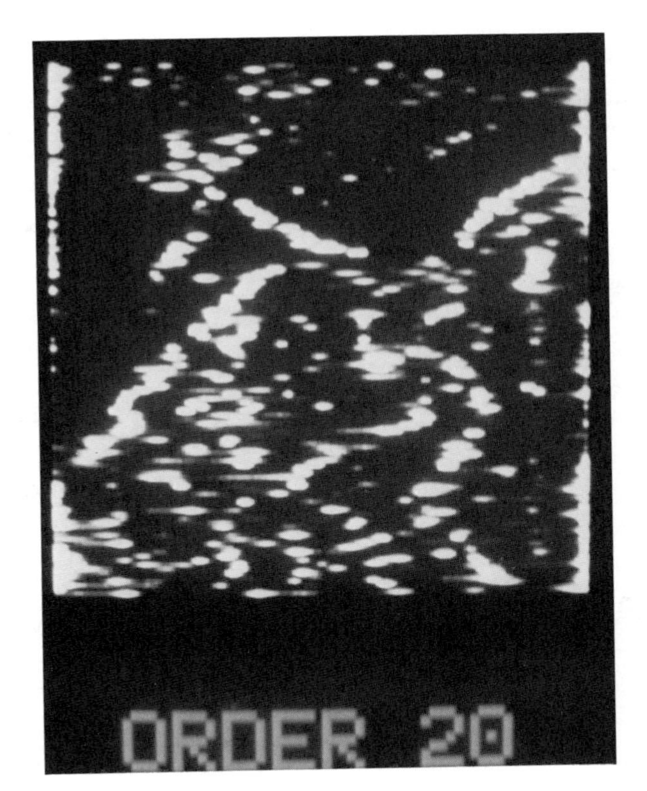

(c)

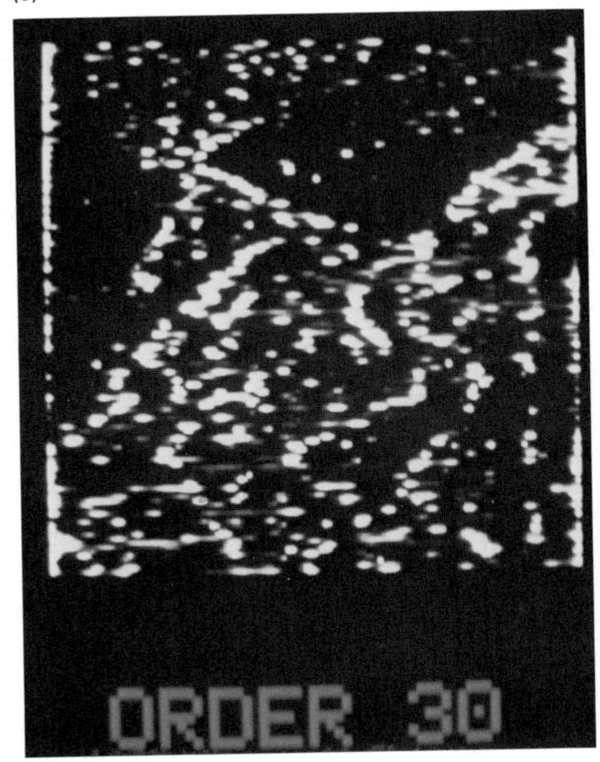

(d)

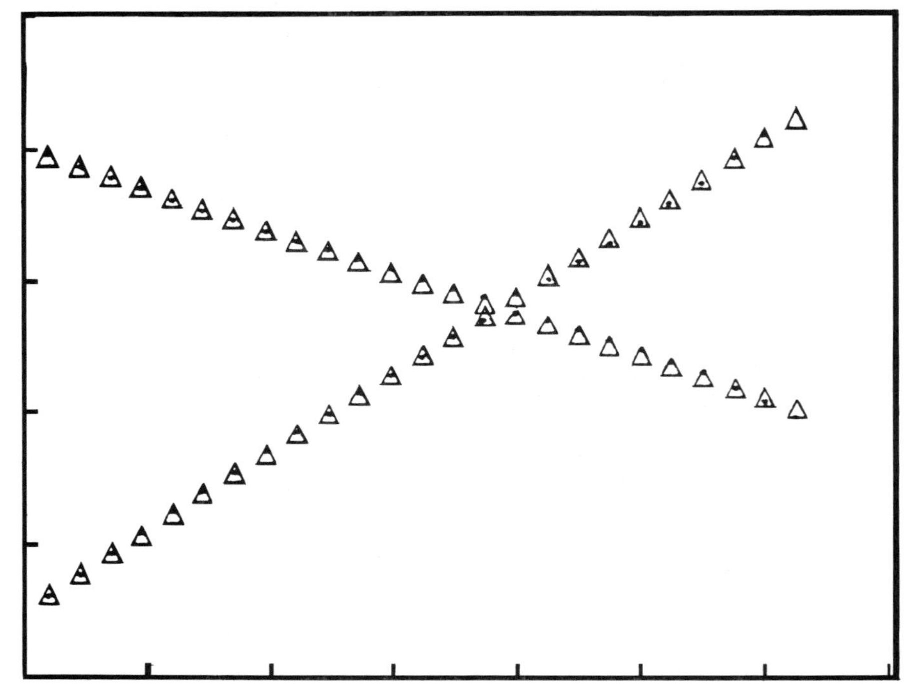

Figure 6 Time-varying AR modeling of a synthetic image, N = 4, L = 3.

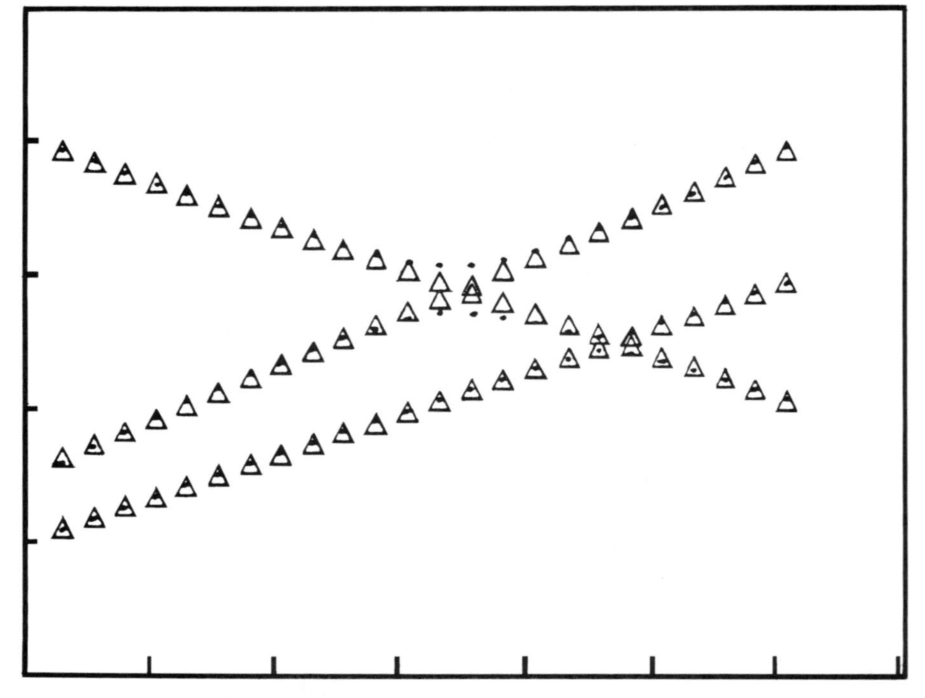

Figure 7 Time-varying AR modeling of a synthetic image, N = 6, L = 3.

6 CONCLUSIONS

Several parametric techniques for SAR image compression were presented in this paper. These techniques are based on the interpretation of a SAR image as the time-varying spectrum of a nonstationary process. The proposed approach has the useful property of performing compression without prior image formation. It also raises some interesting questions about spectral representation of nonstationary processes, such as: what is the class of processes that can be exactly represented by a finite dimensional time-varying AR model? Is it possible to perform spectral analysis on the video signal prior to corner-turning? (The re-ordering of the video signal requires a large and costly memory). Are there more efficient representations of a time-varying spectrum?

Finally we note that the proposed techniques are not yet fully developed, and their performance has not been adequately evaluated. Further work is required to define performance measures for SAR images and to develop cost-effective implementations of compression algorithms.

ACKNOWLEDGMENTS

The author gratefully acknowledges the contributions of K. Lashkari and K.C. Sharman who provided the results presented in Section 5.

REFERENCES

1. Elachi, C., "Spaceborne Imaging Radar: Geologic and Oceanographic Applications," *Science*, Vol. 209, pp 1073-1082, 5 September 1980.

2. Beal, R.C., P.S. De Leonibus and I. Katz, eds., *Spaceborne Synthetic Aperature Radar for Oceanography*, the Johns Hopkins University Press, 1981.

3. Jordan, R.L., "The SEASAT: A Synthetic Aperture Radar System," *IEE J. Oceanic Eng.,* Vol. OE-5, no. 2, pp 154-164, April 1980.

4. Jain, A.K., "Image Data Compression: A Review," *Proc. IEEE*, vol. 69, no. 3, pp 349-389, March 1981.

5. Pratt, W.K., *Image Transmission Techniques,* New York: Academic Press, 1979.

6. Jain, A.K., "Advances in Mathematical Models for Image Processing," *Proc. IEEE*, Vol. 69, No. 5, pp 502-528, May 1981.

7. Harger, R.O. *Synthetic Aperture Radar Systems: Theory and Design*, Academic Press, New York, 1970.

8. Kovaly, J.J., *Synthetic Aperture Radar*, Dedham, MA: Artech House, 1976.

9. Brown, W.M., and L.J. Porcello, "An Introduction to Synthetic Aperture Radar," *IEEE Spectrum*, pp 52-62, September 1969.

10. Cutrona, L.J., "*Synthetic Aperture Radar*," in Radar Handbook, M.I. Skolnik, Ed., New York: McGraw-Hill, 1970, Ch 23, 23.1-23.25.

11. Rihaczek, A.W., *Principles of High-Resolution Radar*, New York: McGraw-Hill, 1969.

12. Brookner, E., *Radar Technology*, Artech House, 1977.

13. Tomiyasu, K., "Tutorial Review of Synthetic-Aperture Radar (SAR) with Applications to Image of the Ocean Surface," *Proc. IEEE*, Vol. 66, No. 5, pp 563-583, May 1978.

14. Hovanessian, S.A., *Introduction to Synthetic Array and Imaging Radars,* Artech House Inc., 1980.

15. Kirk, J.C., Jr., "A Discussion of Digital Processing in Synthetic Aperture Radar," *IEEE Trans. Aero. Sys.,* Vol. AES-11, No. 3, pp 325-337, May 1975.

16. Box, G.E.P. and G.M. Jenkins, *Time Series Analysis, Forecasting and Control*, Holden-Day 1970.

17. Anderson, T.W., *The Statistical Analysis of Time Series,* John Wiley & Sons, 1971.

18. Childers, D.G., *Modern Spectral Analysis*, IEEE Press, 1978.

19. Kay, S.M. and S.L. Marple, Jr., "Spectrum Analysis — A Modern Perspective," *Proc. IEEE* Vol. 69, No. 11, pp 1380-1419, November 1981.

20. Special issue on Spectral Estimation, *Proc. IEEE*, Vol. 70, No. 9, September 1982.

21. Markel, J.D. and A.H. Gray, *Linear Prediction of Speech*, Springer-Verlag, 1976.

22. Lawson, C.L. and R.J. Hanson, *Solving Least Squares Problems*, Prentice Hall, Englewood Cliffs, NJ, 1974.

23. Bierman, G.J., *Factorization Methods for Discrete Sequential Estimation,* Academic Press, New York, 1977.

24. Friedlander, B., "Lattice Filters for Adaptive Processing," *Proc. IEEE*, Vol. 70, No. 8, pp 829-867, August 1982.

25. Friedlander, B., "Instrumental Variable Methods for ARMA Spectral Estimation," *IEEE Trans. Acoustics Speech and Signal Processing*, vol. ASSP-31, no. 2, pp 404-415, April 1983.

26. Buzo, A., et al., "Speech Coding Based Upon Vector Quantization," *IEEE Trans. Acoustics Speech and Signal Processing*, Vol. ASSP-28, No. 5, pp. 562-574, October 1980.

27. Sharman, K.C., and B. Friedlander, "Time-varying Autoregressive Modeling of Nonstationary Signals," Technical Report TM-5448-01, Systems Control Technology, Inc., March 1983.

Data Compression of a First Order Intermittently Excited AR Process[1]

Jerry D. Gibson

Department of Electrical Engineering
Texas A&M University
College Station, TX

ABSTRACT

A Laplacian, intermittently excited, first order autoregressive (IEAR(1)) process for image source modeling in a data compression task is described. Some properties of the Laplacian IEAR(1) process are investigated, and a new differential encoding system structure that is well-matched to IEAR processes is defined and studied. The new data compression system is shown to offer better performance than classical differential pulse code modulation at the same transmitted data rate. When the predictor coefficient is unknown, two simple techniques which take advantage of the IEAR(1) "zero defect" property can be used to estimate the predictor coefficient very quickly.

1 INTRODUCTION

The statement made by Oliver over thirty years ago [1], "In present day communication systems we ignore the past and pretend each sample is a complete surprise," is still true for most communications systems today. Perhaps the most pervasive example of this situation is the use of pulse code modulation (PCM) to obtain a digital representation of signals such as speech and images. An important research area for about the past twenty years has been the design and analysis of data compression systems. The goal of data compression systems is to utilize source information, such as spectra or probability distribu-

[1]This work was supported by the Office of Naval Research Statistics and Probability Program under Contract No. N00014-81-K-0299.

tions, to obtain an efficient digital representation of the source (speech, images, etc.) with, at most, an acceptable loss in quality. Because of their reliance on spectral and statistical information, it is not surprising that the design of high performance data compression systems is highly source model dependent.

Images and speech are often modeled by standard autoregressive (AR) processes or autoregressive-moving average (ARMA) processes. However, under appropriate conditions, these processes are approximately Gaussian [2], and since speech and images are not usually Gaussian-distributed, these classical models are suspect. Exponential autoregressive (EAR), exponential moving average (EMA), and exponential autoregressive-moving average (EARMA) processes have been constructed and studied recently by several authors [3-9]. This work extends the standard AR, MA, and ARMA models to certain non-Gaussian processes, the variants of which may prove useful for image modeling.

Perhaps more important than the fact that the processes can be made non-Gaussian is the particular structure of the models in [3-9]. Specifically, from [7], a first order, exponential AR process is given by

$$X_k = \begin{cases} pX_{k-1} & \text{w.p. } p, \\ pX_{k-1} + E_k & \text{w.p. } 1-p, \end{cases} \tag{1}$$

where $0 \leqslant p < 1$ and E_k is a sequence of i.i.d. exponentially distributed random variables with parameter λ. Irrespective of the distributions of $\{E_k\}$ and $\{X_k\}$, it is felt that AR processes as in Eq. (1), which have no driving term with some probability (p here), may be very useful for image (and speech) modeling. The advantages of a model with a structure similar to Eq. (1) are: (i) The autocorrelation function, and hence, the spectral properties, of Eq. (1) are the same as the usual Gaussian AR process with coefficient p, (ii) for appropriate choices of p (perhaps time-varying), the no-excitation state can accurately model solid areas and slowly changing areas better than a randomly driven model, and (iii) like the standard AR model, the driving term distribution can be specified. For descriptive purposes and for lack of a better name at present, we call AR processes with the general structure of Eq. (1) intermittently-excited autoregressive processes (IEAR). Of course, one can consider higher order IEAR processes as well as intermittently excited ARMA (IEARMA) processes and intermittently excited MA (IEMA) processes.

This paper is concerned with the data compression of a first order, intermittently excited autoregressive process, with the eventual goal of image compression. The data compression of such sources has never been investigated, and the application of these random processes to image modeling and compression is also new.

2 IEAR PROCESSES

A random process of the form in Eq. (1) can be derived following Gaver and Lewis [7]. We begin with the classical first order autoregressive process

$$X_k = \alpha X_{k-1} + \epsilon_k \tag{2}$$

with $|\alpha| < 1$ and where X_k is independent of ϵ_j for $j > k$. Therefore, the characteristic function of the random variable X_k is

$$\begin{aligned}\phi_{X_k}(j\omega) &= E\{\exp[-j\omega X_k]\}\\ &= E\{\exp[-j\omega(\alpha X_{k-1} + \epsilon_k)]\}\\ &= \phi_{X_{k-1}}(j\alpha\omega)\phi_{\epsilon_k}(j\omega).\end{aligned} \tag{3}$$

Solving for $\phi_{\epsilon_k}(j\omega)$ and assuming that X_k is marginally stationary, we have

$$\phi_\epsilon(j\omega) = \frac{\phi_X(j\alpha\omega)}{\phi_X(j\omega)}. \tag{4}$$

We are now in the position to specify $\phi_X(j\omega)$, and hence the distribution of X_k in Eq. (2), subject to existence conditions on $\phi_\epsilon(j\omega)$ in Eq. (3).

Letting

$$\phi_X(j\omega) = \frac{1}{1 + \left(\dfrac{\omega}{\lambda}\right)^2}, \tag{5}$$

which is the characteristic function of a double-sided exponential or Laplacian random variable with parameter λ, we find that

$$\phi_\epsilon(j\omega) = \alpha^2 + (1 - \alpha^2)\left[\frac{1}{1 + \left(\dfrac{\omega}{\lambda}\right)^2}\right]. \tag{6}$$

From Eq. (6) we conclude that the sequence ϵ_k takes on the value 0 with probability α^2 and is a Laplacian-distributed random variable with probability $1 - \alpha^2$. Hence, we can write the difference equation generating the sequence $\{X_k\}$ as

$$\begin{aligned}X_k &= \alpha X_{k-1} + \epsilon_k\\ &= \begin{cases}\alpha X_{k-1}, & \text{w.p. } \alpha^2\\ \alpha X_{k-1} + w_k, & \text{w.p. } 1 - \alpha^2\end{cases}\end{aligned} \tag{7}$$

where $\{w_k\}$ is an independent and identically distributed (i.i.d.) sequence of Laplacian random variables.

It is of interest to examine a slightly more general model with the same basic structure as Eq. (7). For example, consider the random process

$$X_k = \begin{cases}\alpha X_{k-1}, & \text{w.p. } p\\ \alpha X_{k-1} + w_k, & \text{w.p. } 1 - p,\end{cases} \tag{8}$$

where we desire that the sequence $\{X_k\}$ be Laplacian with parameter λ. From Eq. (4) we thus have that

$$\phi_\epsilon(j\omega) = \frac{1 + \left(\dfrac{\alpha\omega}{\lambda}\right)^2}{1 + \left(\dfrac{\omega}{\lambda}\right)^2}$$

$$= p + (1 - p)\left[\frac{1 + \dfrac{(\alpha^2 - p)}{(1 - p)}\left(\dfrac{\omega}{\lambda}\right)^2}{1 + \left(\dfrac{\omega}{\lambda}\right)^2}\right] \tag{9}$$

It is evident from Eq. (9) that the sequence $\{w_k\}$ in Eq. (8) is not generally Laplacian, when $p \neq \alpha^2$, if $\{X_k\}$ is Laplacian. The significance of this result is simply that we are attempting to construct a random process with certain properties and that we do not have complete freedom in choosing the various characteristics and parameters. Therefore, we must be judicious in our selections in order to model the salient properties of the source.

The result in Eq. (7) shows that we can construct an AR process which has a Laplacian distribution, and the same technique used to obtain Eq. (7) can be used to generate other non-Gaussian AR processes. The question is, which distribution do we want? For image modeling, the actual distribution of pixel amplitudes is extremely image dependent and can vary from uniform, Gaussian, and Rayleigh to different bimodal distributions. It is likely, however, that by suitably restricting the image class under consideration, such as for example, to high altitude photographs of the earth, to low altitude photographs of farmland, or to head-and-shoulders images, reasonable assumptions on the pixel amplitudes may be possible. For data compression purposes, another point of interest is the result in Eq. (9) that the sequence $\{w_k\}$ is not Laplacian when $\{X_k\}$ is Laplacian if $p \neq \alpha^2$. Previous experiments on the differential encoding of images indicate that images with widely varying distributions of pixel amplitudes often have a prediction error that is Laplacian. Hence, if $p \neq \alpha^2$, there is some question as to whether to specify the distributions of $\{X_k\}$ or to specify the distribution of $\{w_k\}$. This same dichotomy holds, in general, whenever p in Eq. (8) cannot be chosen as needed to fix the distribution of $\{X_k\}$ and $\{w_k\}$.

As a final point in this section, it is noted that for the random process in Eq. (7)

$$E\{X_k X_{k+j}\} = \alpha^j E\{X_k^2\}. \tag{10}$$

The intermittently excited process in Eq. (7) thus retains the usual spectral properties of the classical AR process.

3 PROPOSED DATA COMPRESSION SYSTEM

Let us assume that we wish to obtain an efficient digital representation of the source sequence

$$s_k = \begin{cases} \alpha s_{k-1}, & \text{w.p. } p \\ \alpha s_{k-1} + w_k, & \text{w.p. } 1 - p \end{cases} \tag{11}$$

which we designate as an IEAR(1) source. A data compression system particularly well-suited to this task is shown in Fig. 1. For an input sequence of the form in Eq. (11), the operation of the system would ideally proceed as follows. If at time k, $s_k = \alpha s_{k-1}$, then for $\hat{s}_{k-1} = s_{k-1}$, we can achieve $e_k = 0$. The coder communicates this fact to the receiver via a suitably chosen (short) binary sequence. At the receiver, since $\hat{s}_{k-1} = s_{k-1}$ and $\hat{e}_k = 0$, we have $\hat{s}_k = \alpha s_{k-1} = s_k$, which is ideal encoding (zero distortion). Alternatively, if at time k, $s_k = \alpha s_{k-1} + w_k$ and $\hat{s}_{k-1} = s_{k-1}$, then $e_k = w_k$. If the coder can achieve $\hat{e}_k = w_k$, then we will again have $\hat{s}_k = \alpha s_{k-1} + w_k = s_k$, which is ideal encoding. Of course, $\hat{e}_k = w_k$ is not achievable in general with a finite bit rate; however, negligible error in the representation may be possible with a finite number of bits.

To see how the system in Fig. 1 can achieve data compression and near ideal encoding of IEAR sources, consider the following. Let

n = number of bits used to represent e_k when $e_k \neq 0$, called the "transmission" condition,
m = number of bits used to represent the $e_k = 0$ condition, called the "no transmission" condition,
$P[T]$ = probability of "transmission,"
$P[NT]$ = probability of "no transmission,"
f_s = sampling rate (samples/sec), and
R = total average transmitted data rate (bits/sec).

Then

$$\begin{aligned} R &= \{mP[NT] + nP[T]\}f_s \\ &= \{m(1 - P[T]) + nP[T]\}f_s \\ &= \{m + (n - m)P[T]\}f_s. \end{aligned} \tag{12}$$

Since $n > m$, to minimize R we need to minimize $P[T]$. If we let $P[T] = 0.2$, then for $R = 2f_s$ (2 bits/sample), we have

m	0	0.25	0.5	1.0
n	10	9	8	6

Thus, at two bits/sample, we can achieve any m, n pair in the table when $P[T] = 0.2$. A rate of 2 bits/sample is sufficiently low to be of major interest. The two questions remaining are: (i) is $P[T] = 0.2$ reasonable for physical sources, and (ii) will any admissible m, n pair yield near ideal encoding?

First, we consider possible values for $P[T]$. For this purpose we construct the random process model

$$s_k = \begin{cases} \alpha s_{k-1}, & \text{w.p. } \alpha^2 \\ \alpha s_{k-1} + w_k, & \text{w.p. } 1 - \alpha^2 \end{cases} \tag{13}$$

where w_k is assumed to be Laplacian. If we match Eq. (13) to the first order autocorrelation of typical images we would find $\alpha = 0.9$. Therefore, we note that $\alpha^2 = 0.81$ and $1 - \alpha^2 = 0.19$, but further $P[T] = P[e_k \neq 0] = 0.19$. The choice of $P[T] = 0.2$ thus seems reasonable.

To examine the question of whether an admissible m, n pair can achieve near ideal encoding, let us consider the pair $m = 0.5$ bit/sample and $n = 8$ bits/sample. The $m = 0.5$ bit/sample average rate can likely be attained using variable-length coding or run-length coding techniques. The question is then whether $n = 8$ bits/sample is sufficient to accurately represent the sequence $\{w_k\}$. It can be shown that memoryless 8 bit quantization of a random variable can achieve a minimum mean squared error of about 5×10^{-6}, which is a signal-to-noise ratio of more than 50 dB. Since SNR values above 35 dB are considered excellent, we feel that near-ideal encoding is possible with the system in Fig. 1 for the IEAR process in Eq. (13) with $\alpha = 0.9$.

The foregoing discussion indicates that the system in Fig. 1 will be quite effective for data compression of an IEAR source as in Eq. (13). We now wish to examine the performance of this system in additional detail. A standard performance indicator for differential encoding systems is the signal-to-noise ratio (SNR) defined by

$$SNR \triangleq \frac{E\{s_k^2\}}{E\{[s_k - \hat{s}_k]^2\}} \tag{14}$$

where the reconstruction error, $s_k - \hat{s}_k$, is called quantization noise and is denoted by q_k.

To compute the SNR in Eq. (14), we need to develop equations which describe the behavior of the system in Fig. 1. First, we note that for the source in Eq. (8), e_k will be zero with probability p^{k-1} and nonzero with probability $1 - p^{k-1}$. Hence, as k gets large, we will generally find that $e_k \neq 0$ even though e_k will likely be very small in most instances. As a result, we write our equations for Fig. 1 to include a quantization noise term. It should be noted that since the Fig. 1 system is not a finite state device, there are many ways that $e_k \neq 0$ can occur. In fact, at time k, there are 2^{k-1} states and only one state is guaranteed to be zero.

From Fig. 1, the prediction error is given by

$$e_k = s_k - \hat{s}_{k|k-1} \tag{15}$$

where $\hat{s}_{k|k-1} = \alpha \hat{s}_k$. Defining

$$\hat{e}_k = e_k + q_k \tag{16}$$

then

$$\begin{aligned} \hat{s}_k &= \hat{s}_{k|k-1} + \hat{e}_k \\ &= \hat{s}_{k|k-1} + e_k + q_k \\ &= s_k + q_k, \end{aligned} \tag{17}$$

consistent with the earlier discussion of the SNR in Eq. (14). Using Eq. (17), the predicted value is

$$\hat{s}_{k|k-1} = \alpha \hat{s}_k = \alpha s_k + \alpha q_k. \tag{18}$$

Using Eq. (8), we calculate

$$\begin{aligned} E\{s_k^2\} &= E\{s_k^2 | s_k = \alpha s_{k-1}\} p + \\ &\qquad E\{s_k^2 | s_k = \alpha s_{k-1} + w_k\}(1 - p) \\ &= \alpha^2 E\{s_{k-1}^2\} + (1 - p) E\{w_k^2\}. \end{aligned} \tag{19}$$

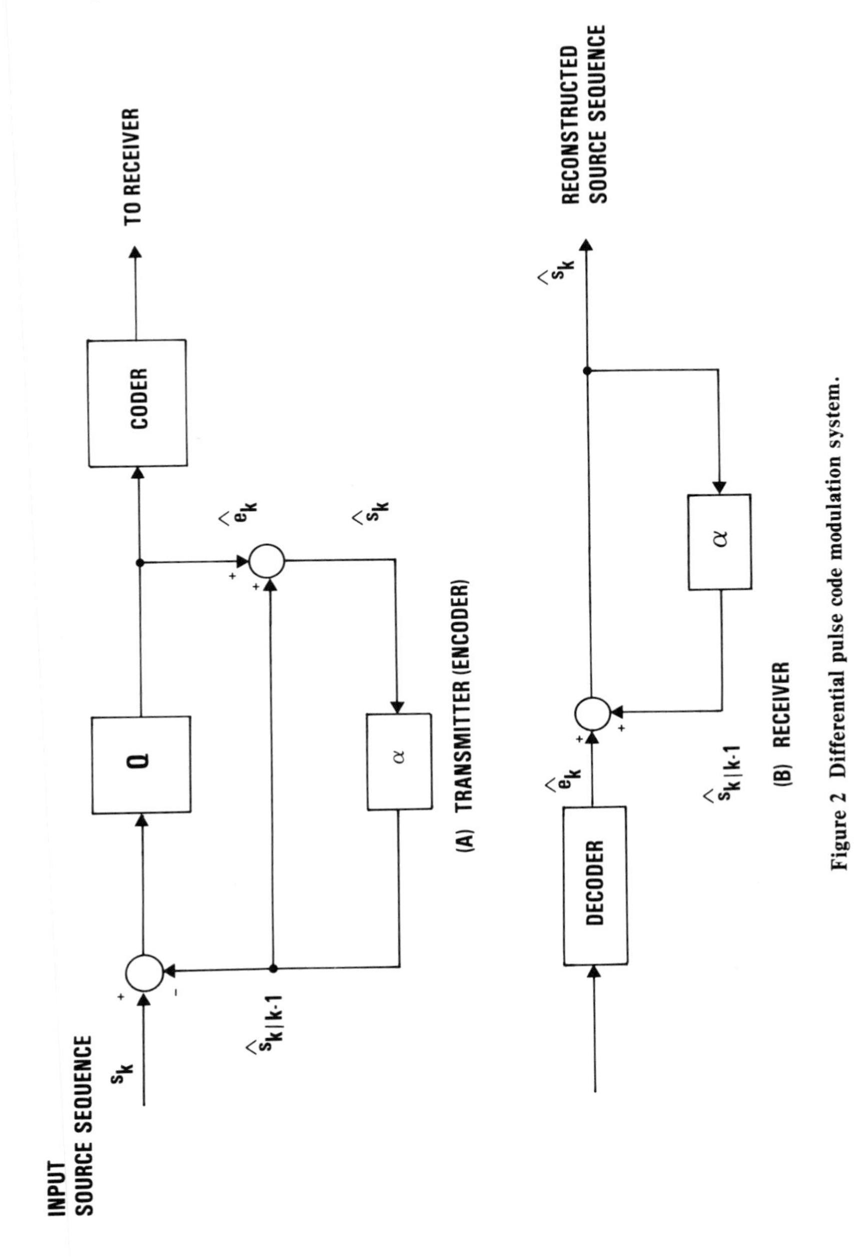

Figure 2 Differential pulse code modulation system.

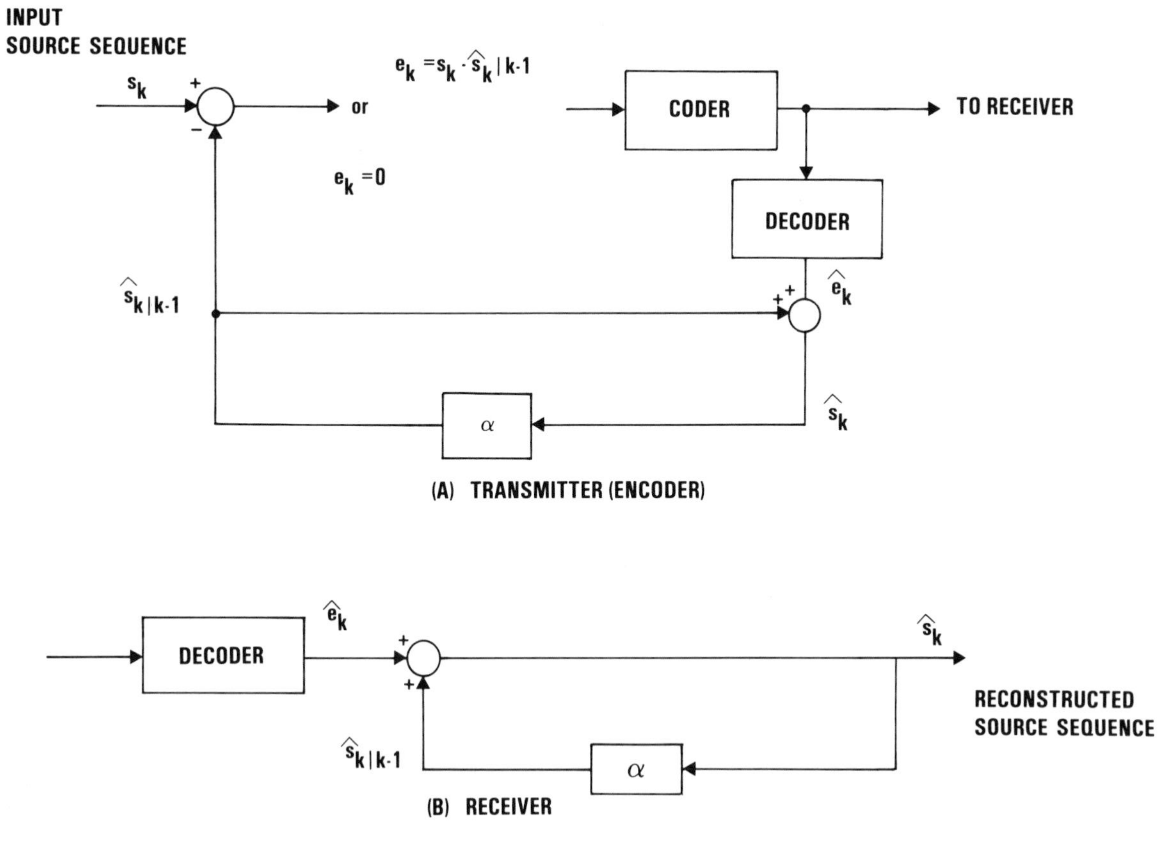

Figure 1 Data compression system for an IEAR source.

We can also use Eqs. (15) and (18) to find

$$\begin{aligned} E\{e_k^2\} &= E\{e_k^2 | s_k = \alpha s_{k-1}\} p + \\ &\quad E\{e_k^2 | s_k = \alpha s_{k-1} + w_k\}(1 - p) \\ &= \alpha^2 E\{q_{k-1}^2\} + (1 - p) E\{w_k^2\}. \end{aligned} \tag{20}$$

Assuming that all processes are covariance stationary or wide sense stationary (WSS) and letting $E\{s_k^2\} = \sigma_s^2$, $E\{w_k^2\} = \sigma_w^2$, $E\{e^2\} = \sigma_e^2$, and $E\{q_k^2\} = \sigma_q^2$, Eqs. (19) and (20) become, respectively,

$$\sigma_s^2 = \alpha^2 \sigma_s^2 + (1 - p)\sigma_w^2 \tag{21}$$

and

$$\sigma_e^2 = (1 - p)\sigma_w^2 + \alpha^2 \sigma_q^2. \tag{22}$$

Solving Eq. (21) for σ_w^2, substituting the result into Eq. (22), and manipulating yields

$$SNR = \frac{\sigma_s^2}{\sigma_q^2} = \frac{1}{1 - \alpha^2} \left[\frac{\sigma_e^2}{\sigma_q^2} - \alpha^2 \right]. \tag{23}$$

An important data compression system for AR processes is differential pulse code modulation (DPCM) illustrated in Fig. 2. While Figs. 1 and 2 are visually quite similar, there are substantive differences in their mode of operation. DPCM injects quantization error for each and every input sample, even if $e_k \equiv 0$, while the system in Fig. 1 encodes $e_k = 0$ perfectly. This difference is critical for sources of the form in Eq. (8) with $p > 0.5$. To see that this is true, recall the rate calculation in Eq. (12) for the source in Eq. (13) with $\alpha = 0.9$. There we determined that we could quantize nonzero values of the prediction error with 8 bits/sample for the system in Fig. 1 with a total transmitted data rate of 2 bits/sample. For DPCM at 2 bits/sample, every sample is represented with 4 levels. At 8 bits/sample, σ_q^2 is on the order of 10^{-5}, while for 2 bits/sample, σ_q^2 is about 0.1.

The SNR expression for a DPCM system with an input sequence given by Eq. (8) is exactly the same as Eq. (23). However, as just noted, the value of σ_q^2 is different for the two systems (Figs. 1 and 2), and thus the first term within brackets in Eq. (23) is much larger for the Fig. 1 system. It must be mentioned before proceeding that the differences between σ_q^2 values will not be as large as indicated if "entropy" coding is employed in conjunction with the DPCM system. Details of this procedure are left to the references [10,11].

Another indicator of differential encoding system perfomance is the mean squared prediction error. From Eq. (20), we can develop an expression for $E\{e_k^2\}$ given by

$$\begin{aligned} E\{e_k^2\} &= (1 - p) \sum_{j=1}^{k} a^{k-j} \alpha^{2(k-j)} E\{w_j^2\} \\ &\quad + a^k \alpha^{2k} E\{e^2(0)\}. \end{aligned} \tag{24}$$

To derive Eq. (24), we have assumed the relationship

$$E\{q_k^2\} = aE\{e_k^2\} \tag{25}$$

for all k. This expression is approximately true for both DPCM and the system in Fig. 1, however, it breaks down for DPCM when $e_k = 0$, but it still holds for Fig. 1. If we assume that the sequence $\{w_k\}$ is WSS then Eq. (24) can be rewritten as

$$E\{e_k^2\} = \sigma_w^2(1 - p)\left\{1 + \frac{a\alpha^2[1 - (a\alpha^2)^{k-1}]}{1 - a\alpha^2}\right\} + a^k\alpha^{2k}E\{e_0^2\}. \tag{26}$$

Since, in general, $a\alpha^2 << 1$, we find that

$$\lim_{k\to\infty} E\{e_k^2\} = (1 - p)\sigma_w^2\left\{1 + \frac{a\alpha^2}{1 - a\alpha^2}\right\} = \frac{(1 - p)\sigma_w^2}{1 - a\alpha^2}. \tag{27}$$

If p is near one, then the asymptotic mean squared prediction error is much less than σ_w^2. Although surprising initially, this result is intuitive since the ensemble averaging includes the highly probable samples for which $e_k = 0$, as well as the lower probability cases when $e_k \geqslant w_k$.

4 PREDICTOR COEFFICIENT ESTIMATION

In all of the preceding analyses, we have assumed that the parameter α in Eq. (11) and Fig. 1 is known. It is often true in data compression problems that this coefficient is unknown or varies for different input sequences and must be estimated. Due to its location in Fig. 1, α is usually called the predictor coefficient. We discuss two methods for determining the predictor coefficient of an IEAR(1) source.

The first method relies on what Gaver and Lewis [7] call the "zero-defect" of their positive valued gamma sequence. To estimate α, we monitor the incoming source sequence $\{s_k\}$. We assume $\alpha > 0$. The random variable Y_k is defined according to $Y_k = \dfrac{|s_k|}{|s_{k-1}|}$, $k = 1, 2, \ldots$, so

$$Y_k = \begin{cases} \dfrac{|s_k|}{|s_{k-1}|} = \alpha, & \text{w.p. } p \\[2ex] \dfrac{|\alpha s_{k-1} + w_k|}{|s_{k-1}|}, & \text{w.p. } 1 - p. \end{cases} \tag{28}$$

We monitor the series of Y_k's for tie values. If we get a tie, that tie value is α. Unlike Gaver and Lewis [7], it is not possible to simply observe the minimum value of Y_k here since the driving sequence $\{w_k\}$ can take on both positive and negative values. Further, s_{k-1} can be zero, and hence, provisions must be made for discarding overflows and exceptionally large values of Y_k.

The waiting time (in number of input samples) before obtaining β ties is $Z + \beta$, where Z has a negative binomial distribution with $E[Z] = \beta(1 - p)/p$ and var $(Z) = \beta(1 - p)/p^2$. If we are satisfied with one tie, $\beta = 1$, Z has a geometric distribution, and the waiting time to obtaining one tie has a "geometric plus one" distribution [12].

Another way to view this estimation problem is to compute how many samples are needed such that the probability of getting one tie (two or more "successes") is, say, 0.99. Thus, for example, if we have $p = 0.8$ in Eq. (11), and thus in Eq. (28), then we find that only 5 or more input samples are needed to achieve a probability of a tie of 0.99.

A second method for estimating the predictor coefficient is feasible if the input sequence has the form in Eq. (13) so that $p = \alpha^2$. In this method, one simply counts the number of ties obtained in a suitably long sequence and uses this number to compute the proportion of ties which is α^2. The analysis of this method is not included here since it seems more difficult than the previously-described approach.

Irrespective of which technique is used to estimate α, the resulting estimate will have to be transmitted to the receiver. This is necessary since both techniques employ the source sequence $\{s_k\}$, which is unavailable at the receiver, and because the transmitter and receiver must use the same predictor coefficient to achieve good performance. Computing the predictor based upon the input and transmitting the value to the receiver is called forward adaptive prediction.

5 RELATED SYSTEMS

In addition to DPCM, another system similar to that in Fig. 1 is the predictive-comparison system previously studied by Davisson [13,14]. In his structure, the present input sample is compared to a predicted value. If the prediction error is sufficiently small, "nothing" is transmitted to the receiver and the reconstructed output is the predicted value. Actually, the transmission of some binary word is required to inform the receiver of this decision. If the magnitude of the prediction error exceeds a threshold, the input sample itself is transmitted to a very high degree of accuracy. This is clearly wasteful relative to the Figure 1 system which, since the predicted value is available at the receiver, transmits only the prediction error. Of course, the prediction error should have a narrower dynamic range than the source input sample, and thus, a better representation is possible with a fixed number of bits. Davisson analyzed his system only for classical Gaussian AR process sources.

6 CONCLUSIONS

The use of intermittently excited autoregressive processes for image source modeling in a data compression task is proposed. A Laplacian IEAR(1) process is defined and some of its properties are examined. A new differential encoding structure that is well-matched to IEAR sources is presented and investigated. The new system is shown to offer better performance than classical DPCM at the same transmitted data rate. Two new, simple techniques for estimating the predictor coefficient are described. It is shown that the predictor coefficient can be determined within a waiting time of five samples with a probability of 0.99 for an IEAR(1) source with $p = 0.8$.

REFERENCES

[1] B.M. Oliver, "Efficient coding," *Bell Syst. Tech. J.*, Vol. 31, pp. 724-750, July 1952.

[2] C.L. Mallows, "Linear processes are nearly Gaussian," *J. App. Prob.*, Vol. 4, pp. 313-329, 1967.

[3] P.A. Jacobs and P.A.W. Lewis, "A Mixed Autoregressive-Moving Average Exponential Sequence and Point Process (EARMA1, 1)," *Adv. Appl. Prob.*, Vol. 9, pp. 87-104, 1977.

[4] A.J. Lawrance, "An Exponential Moving-Average Sequence and Point Process (EMA1)," *J. Appl. Prob.*, Vol. 14, pp. 98-113, 1977.

[5] A.J. Lawrance and P.A.W. Lewis, "The Exponential Autoregressive-Moving Average EARMA(p,q) Process," *J. R. Statist. Soc. B*, Vol. 42, No. 2, pp. 150-161, 1980.

[6] A.J. Lawrance, "The Mixed Exponential Solution to the First-Order Autoregressive Model," *J. Appl. Prob.*, Vol. 17, pp. 546-552, 1980.

[7] D.P. Gaver and P.A.W. Lewis, "First-Order Autoregressive Gamma Sequences and Point Processes," *Adv. Appl. Prob.* Vol. 12, pp. 727-745, 1980.

[8] P.A.W. Lewis, "Simple Models for Positive-Valued and Discrete-Valued Time Series with ARMA Correlation Structure," in *Multivariate Analysis-v*, P. R. Krishnaiah, ed., North Holland Publishing Co., 1980, pp. 151-166.

[9] E. McKenzie, "Extending the Correlation Structure of Exponential Autoregressive-Moving-Average Processes," *J. Appl. Prob.*, Vol. 18, pp. 181-189, 1981.

[10] A.D. Wyner, "Fundamental Limits in Information Theory," *Proc. IEEE*, Vol. 69, pp. 239-251, Feb. 1981.

[11] J.D. Gibson, "Adaptive Prediction in Speech Differential Encoding Systems," *Proc. IEEE*, Vol. 68, pp. 488-525, April 1980.

[12] W. Feller, *An Introduction to Probability Theory and Its Applications: Vol. I*, third edition, John Wiley, New York, 1968.

A Modular Software for Image Information Systems

Shi-Kuo Chang[1]

Department of Electrical and Computer Engineering
Illinois Institute of Technology
Chicago, IL

Chung-Chun Yang[2]

Digital Image Processing Laboratory
Naval Research Laboratory
Washington, DC

ABSTRACT

Images are natural means for man-machine communication. Recent advances in hardware technology have made feasible the design of sophisticated image information systems. Software requirements for such systems have been identified, although the problem of flexible integration of various software modules is not yet fully solved. There are needs for image information systems in many application areas. On the other hand, an image information system should be considered as a subsystem of the integrated information system. A survey of R&D efforts indicate that many sophisticated techniques and concepts have been developed, but lack of integration again prevents their immediate applications. Based upon these considerations, an approach of software design for image information systems to achieve system integration is suggested, which emphasizes self-descriptive data structures, program modularity, unified image processing language and programming environment, and flexible user interface. A scenario of user interface is presented. Interactive commands for picture tree restructuring are then described.

1 IMAGE INFORMATION SYSTEMS

An *image information system* (IIS) provides an integrated collection of image data for easy access by a large number of users. An *image database* is a collec-

[1]The research was supported by the National Science Foundation under Grant ECS-8005953 and by the Office of Naval Research under Contract N00014-82-C-2156.
[2]The research was supported by the Office of Naval Research under Work Request No. NR 609-007.

tion of sharable image data encoded in various formats. The image database is the core of an image information system.

Advances in image information systems are technology-driven. New hardware devices include the following:

(a) Image input device: High-resolution scanners, with resolution of several hundred to a couple of thousand lines per inch are now available [1].

(b) Image processor: Stand-alone image processors incorporating microprocessors and array processors can be interfaced with most major mini/micro-computers. Their processing functions include pixel-by-pixel processing, spatial convolution, zoom, rotation, floating-point vector/matrix processing, smoothing, image enhancement, etc. The availability of image processors imply that many image processing and signal processing functions previously requiring large CPU time can now be off-loaded to these processors and computed efficiently. Highly parallel systems can also be configured to further improve processing speed [2].

(c) Image output device: High resolution color graphics raster displays, typically with display resolution of up to 1024 by 1024 pixels and multiple colors, offer high quality image and graphics output.

(d) Image storage system: In addition to the traditional mass storage systems, there is renewed interest in using microfiche for storage/retrieval of large quantities of images. A minicomputer stores the microfilm address and key descriptors, and controls the retrieval of specific document images. Such electronic filing system can be used to create and maintain an image database [3]. The video-disk technology offers another means of creating and maintaining image database [4].

In addition to these hardware components/systems, the recent advances in data communications, especially broad-band local area networks, provide the means to interconnect many workstations for *multimedia communication*, including voice, data, and image.

These technology advances are exciting and stimulate the design of advanced image information systems. From the software viewpoint, system-wide compatible software modules must be designed for: (a) image input, (b) image editing, (c) image processing, (d) image storage, (e) image retrieval, (f) image output, and (g) image communication.

2 IMAGE DATABASE TECHNOLOGY

There are two basic tasks that face designers of image information systems. The first is the storage, retrieval and processing of a large number of images; the second is the storage, retrieval and processing of a very large image or images of great complexity. Also to be considered is the intended usage of an image information system—whether it is intended mainly for retrieval of images, or processing and manipulation of pictures is the main purpose. These considerations can lead to the design of entirely different image database systems to support the intended application. However, common to these design problems, are two central issues: (1) how to provide a unified approach to retrieve/manipulate image information, and (2) how to utilize data structures to improve/develop algorithms for image information retrieval/manipulation [8].

In what follows, we describe recent approaches in applying the relational database technology in image database design, intelligent user interface design, image data encoding techniques, and special-purpose data structures.

2.1 Image Database Design

The relational approach for image database system is first proposed by Kunii. A relational database schema for describing complex images with color and texture is presented. Structure independence, modularity, associativity, and machine independence for complex image description are discussed.

To motivate the relational database approach for image data management, we will use geographic information system as an example. Much data have been collected according to regional or geographic units for analysis. For example, the USA Census Bureau collects innumerable descriptive statistics about population and housing by various political units. Marketing companies collect market data by zip codes to determine which areas to target their advertising campaign. Land use planning organization requires information on soil type for each land parcel. Storing data as attributes for each geographic unit lends itself to the development of a relational database. This database organization allows the user to algebraically manipulate the data to answer complex questions and to generate new information which could be converted to images. Several researchers have suggested the desirability of the relational database as the underlying structure for a geographic information system with cartographic output. The structural power of a relational database requires data aggregated by and associated with geographic units. Theoretically, the image can be called a *map*, and the image relational database can be called a d-map. The process of converting a map into a d-map is called abstraction, and the inverse process of converting a d-map into a map is called materialization.

Early work in the relational model approach to designing an integrated database for tabular data, graphical data and image data is described by Kunii of Tokyo University. At IIT's Information Systems Laboratory headed by Prof. Chang, extensive results for integrated relational database design are presented, and the traditional relational algebra is found insufficient to manipulate these data, for example, to *analyze spatial relationships* among various image objects. A set of picture operations, called image algebra, is designed to handle the spatial data for storage, retrieval, manipulation and transformation. An integrated image database management system is designed by Prof. Chang. This system combines a relational database management system, RAIN, with an image manipulation system to enable the user to perform various image information retrieval by using zooming, panning, and spatial relation analysis operations to manipulate image database. This image database contains image data in two forms: relational table form and raster image form. The user can specify his image query in GRAIN language which is imbedded in the DIMAP (Distributed Image Management And Projection) system. A hierarchically structured collection of image objects is organized in order to allow logical zooming which is used to navigate into the detailed description or to browse through the image objects similar to the interested image. This concept can be applied to the storage and manipulation of a large number of images, some of which can be very complex. A query language is developed for image information retrieval, which utilizes an *image database skeleton* as the *knowledge model.*

2.2 User Interface Design

In order to facilitate man-machine interaction, information retrieval requests, or queries, for retrieval images from an image database can be specified in several ways.

The simplest way is by means of an index or a key. A command language can be designed to retrieve images by keys from an image storage system.

Many researchers have worked on *picture languages* to specify user's queries. The GRAIN system described above is an example. For example, to retrieve the image containing major highways through Tokyo, the GRAIN query is:

Sketch highway; through (cityname is 'tokyo').

As another approach to image query translation, Chang and Fu at Purdue University describe a relational database system which is interfaced with an image understanding system, and the design of a flexible query language for image information retrieval based upon Zloof's query-by-example (QBE) approach. An algorithm to convert a relational graph into a relational database is described and the importance of spatial operations is also emphasized as the clue to retrieve images from an image database.

2.3 Compaction and Encoding of Image Data

There are two commonly used formats in representing image information. Computer cartography, topography and spatial analysis in the geographic information system have traditionally used a vector format. However, more and more data are becoming available via new image capture devices which generate data in raster format, such as LANDSAT imagery and drum scanner output. Advanced technologies in computer graphics such as video frame buffer memories and improvement in CRT resolution have made feasible the increasing use of raster-scan displays as interactive graphics terminals. Algorithms and data structures for manipulating raster-formatted spatial data has also been explored. Recursive description and processing of hierarchically structured image data in a raster-format is proposed, and its implementation is reported. The logical basis of raster-format processing is established. These recent advances have led to the development of image database systems capable of managing image data encoded in various formats.

2.4 Spatial Data Structures

Generalized spatial data structure which can be used to represent spatial objects has been proposed by many researchers. We will survey a few interesting approaches.

Shapiro's structure consists of a set of N-ary relations, often including an attribute-value table. The entries in the table and the objects on which the relations are defined may also be spatial data structures. In this approach, the description of spatial structure is emphasized. Retrieval processing is through the attributes of spatial data. A semantic table is set up for the spatial relation analysis in order to speed up the retrieval process. Database updating is somewhat complicated.

Sties described an image information system in which an object is described in terms of shape, symbolic description, and relations with the other objects. Objects in an image can be retrieved in terms of symbolic description and/or objects relationships. A multisensor image database system (MIDAS) which can perform image understanding and knowledge acquisition is described by McKeown. A hierarchical data structure is chosen primarily to store symbolic representation in the image database system. Images are stored hierarchically with different resolutions, which is called an iconic data structure. Partial image description is stored in a relational database which is used for performance evaluation purpose.

Klinger has proposed a hierarchical data structure scheme for storing image by regular decomposition of images into adjacent quadrants with different resolutions. Each quadrant corresponds to a node in a tree called W-tree. In keeping with the theories on database design, this image data structure facilitates accessibility of image data in storage, flexibility in image data storage, image data independence, redundancy reduction in multilevel structured data, and analysis of small portions of an image with efficiency. A further generalization of quadrant tree structure to iconic/symbolic tree structure has been proposed by Tanimoto.

3 SOFTWARE DESIGN APPROACH

Based upon the above considerations, an approach of software design to achieve system integration is suggested.

3.1 Modularity

It is clear that image communications will be an important aspect of future image information systems. Standard image formats and standard image communication protocols are needed for image communications, either across the boundary of two systems, or within the same system. Self-descriptive data structures, therefore, becomes an important concern.

In Appendix I, the standard image file header is presented. Since all image files are self-descriptive, all program modules can be implemented separately. It should be noted that the relational files are described using the same header format. Therefore, the software can be extended easily.

In software development, *self-descriptive program modules* can also be enforced to ensure system extendibility. An example of a C program to create the standard image file header is given in Appendix II. In each program module, a concise user manual is included as the first comment. The user help command, therefore, can access the program module to display the user manual. This approach also makes program maintenance easy.

3.2 Image Processing Language and Programming Environment

We need a unified image processing language IPL, and a programming environment supporting the language primitives for an image information system. The main characteristics of IPL is that it should be flexible and extendible. It also allows the user to navigate through the image database using image icons.

The image processing language IPL consists of three subsets: the logical image processing language LIPL, the interactive image processing language IIPL, and the physical image processing language PIPL.

The user interacts with the image information system using the interactive image processing language IIPL. The logical image processing language is used to retrieve/manipulate logical images stored in the image database. The user can also employ the physical image processing language to process/manipulate physical images stored in the image store.

A concise manual for the current modular software system is given in Appendix III.

3.3 Flexible User Interface

The user interface should be customized to local needs. Yet it should be capable of translating the local image manipulation commands into the unified IPL language. User should be able to navigate in the image database with ease by manipulating user-defined icons, windows, and ports.

An image database generally has a complex hierarchical structure. Since at any given time the user will always focus his attention on a small portion of the image database. We could maintain a user workspace WS which is a temporary work area to store the current entities of interest. Thus all data which are used for a specific application can normally be handled in a workspace by means of predefined meta-data structures and user commands. A WS can be saved, loaded and deleted, new WS can be created when needed. Communication between a WS and a database of any external device can be made by associating data in WS with ports. Implementations of workspaces can easily be done in, for instance, LISP, APL or FORTH, because these languages support already workspaces in their programming environment. However, it could be implemented in other programming languages as well.

An experimental software package following the above design approach is currently being implemented jointly by Information Systems Laboratory, IIT, and Digital Image Processing Laboratory of Naval Research Laboratory. In what follows, we describe a scenario to clarify the concepts.

4 A SCENARIO

The following scenario illustrates how a user may interact with an image information system using IIPL.

(1) Create icon:

The following command creates an icon, whose name is "b25", corresponding to all the bridge objects. In other words, the icon "b25" is a virtual object set.

```
icon b25 "bridge"
```

No icon sketch is associated with this icon, so that it will be displayed as an alphanumeric string "b25".

(2) Create icon and icon sketch:

The following command creates an icon, whose name is "b26", corresponding to bridges over the river Kwei. The icon "b26" is a virtual object set corresponding to some, but not all, bridge objects.

```
icon b26 "bridge:bridge over river; river.name = Kwei"
```

The next command illustrates how an icon sketch might be created.

```
icon camera sketch (chain,0000022442464466)
```

In this command, the icon "camera" has no corresponding database objects. Later on, it will be associated with an input port for taking video camera input. The icon sketch for camera is specified by chain codes. An interpretation is given below:

```
      * 4 *
      6   2
* 4 4 *   * 4 4 *
6               2
6               2
^ 0 0 * 0 * 0 0 *
```

The origin is indicated by the special symbol "^". The symbols "*" are fillers, which do not appear in the actual chain code.

Instead of using chain codes, the system may ask the user to interactively create an icon sketch. The details of sketch creation is implementation-specific, and therefore left to the implementor.

(3) Create ports:

The following command creates a disk input/output port associated with icon b25. The port name is port2.

```
port port2 (disk,rk05,inout) b25
```

The following command creates an input port for camera unit c01. The port name is port1. It is associated with icon "camera".

port port1 (camera,c01,in) camera

(4) Logical zoom using icons:

The following command loads icon b25 into the current workspace.

load b25

or equivalently

load b25 via port2

Since b25 has been associated with port2, the clause "via port2" is optional. The following command then materializes an image, image1, from the icon b25.

sketch b25 into image1

We can also load an image from the camera port, and then creates the image using the paint command.

load camera

paint camera into image2

(5) Show image:

The show command is used to display an image. The images must already exist in raster format.

show image1

show image2

(6) Create window and pan:

The following command creates a window over an image whose name is image1. The dimension of the window is (10,10).

window image1 100 100

The following command moves the window to a different location (45,45) and shows the image through the display window.

pan -a image1 45 45

We can now zoom into the area where the window currently is located for details.

zoomi image1

(7) Database query:

The following command retrieves the descriptions of all the bridges over the river Kwei within 10km of location (100,30).

get bridge: bridge over river; river.name = Kwei;
distance (((bridge.x), (bridge.y)), (100,30)) =< 10.

(8) Save workspace:

The following command saves the current workspace into ws10.

unload ws 10

(9) End session:

This is the usual quit command.

quit

5 PICTURE TREE OPERATIONS

We have discussed the concept of picture trees, picture queries, and iconic indexing techniques. In this section, we describe a set of *picture tree operations* for the retrieval and manipulation of picture trees.

Given a picture object set V = {v1, ..., vn}, we can construct a picture tree PT = {H1, ..., Hm}, where the Hi's are subsets of V, satisfying the following conditions:

(A1) H1 = V.

(A2) For any two sets Hi and Hj, either they are disjoint, or Hi contains Hj, or Hj contains Hi.

The picture tree is constructed as follows: If Hi $\supset$ Hj, and there is no Hk such that Hi $\supset$ Hk $\supset$ Hj, then there is a directed arc from Hi to Hj, written as Hi -> Hj.

Example 1: The Hi's are:

H1 = {v1, v2, v3, v4, v5}
H2 = {v1, v2}

H3 = {v3, v4, v5}
H4 = {v1}
H5 = {v2}
H6 = {v3, v4}
H7 = {v3}
H8 = {v4}

The picture tree is as illustrated in Figure 1(a).

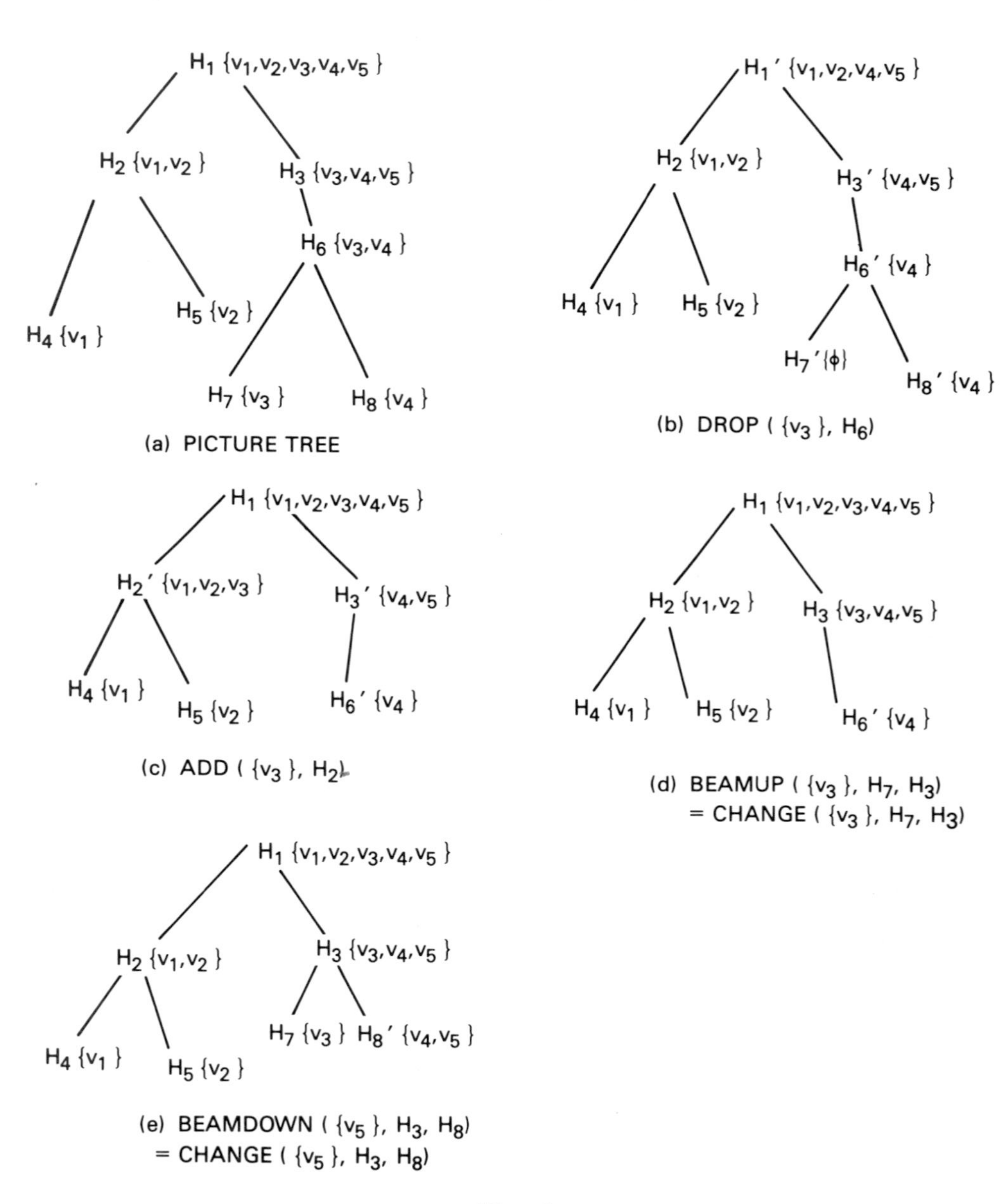

Figure 1

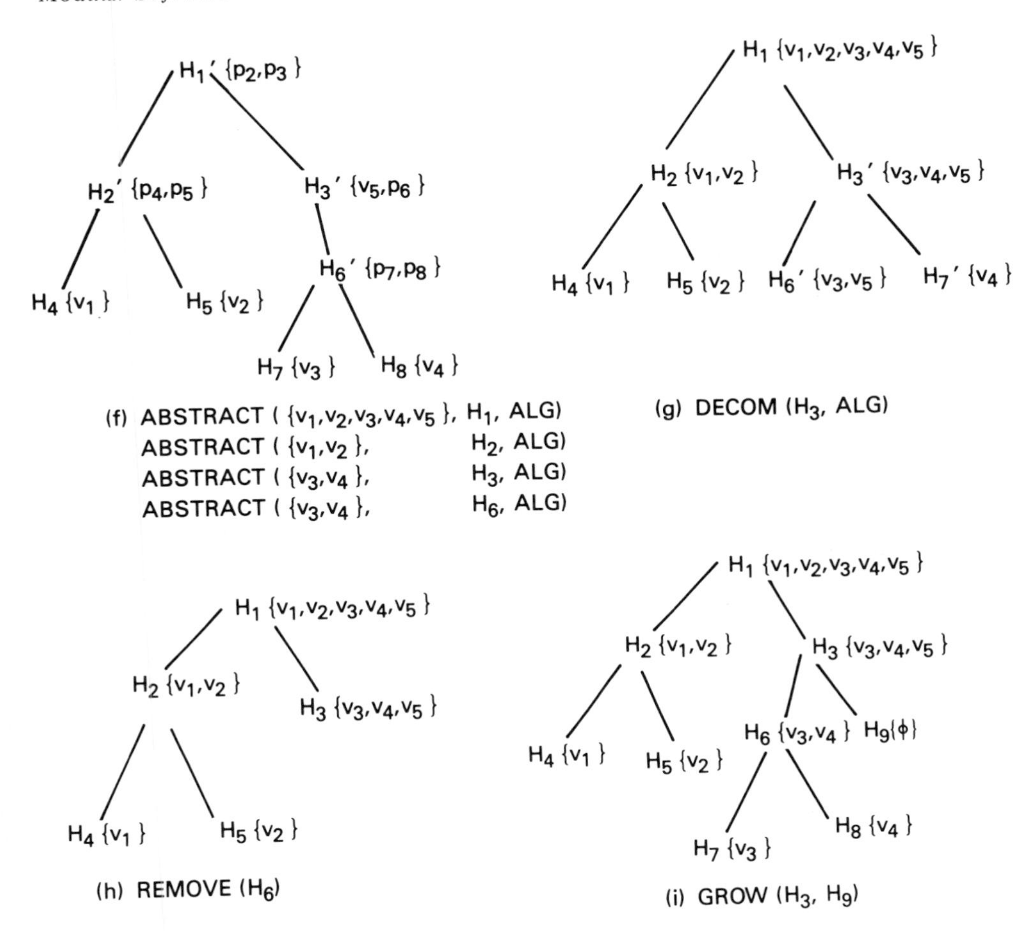

Figure 1 (Continued)

Each node Hi in a picture tree is associated with a *depth* D(Hi), which is the distance from the root node H1. For example, in Figure 1(a), D(H2) is 1, and D(H8) is 3.

In the picture tree induced by PT, a node Hi is a *leaf node*, if there is no Hj such that Hi > Hj. We define the *ancestor set* of Hk as A(Hk) = {Hi: Hi >= Hk}. We define the *follower set* of Hk as F(Hk) = {Hi: Hk > Hi}.

In Example 1, A(H6) = {H1, H3, H6}, and F(H6) = {H7, H8}.

To restructure a picture tree, the following operations on the picture tree H are introduced.

(a) DROP(Y, Hk)

The set Y is contained in Hk. We want to drop Y from Hk. To preserve (A1) and (A2), DROP(Y,Hk) has the following effects: Every Hi in A(Hk) ∪ F(Hk) is changed to H'i = Hi − Y = {v: v in Hi and not in Y}. An example of performing DROP({v3}, H6) on the tree of Figure 1(a) is illustrated in Figure 1(b).

(b) ADD(Y, Hk)

We want to add Y to Hk. To preserve (A1) and (A2), ADD(Y,Hk) has the following effects: Every Hi in A(Hk) is changed to H'i = Hi ∪ Y. An example of performing ADD({v3}, H2) on the tree of Figure 1(b) is illustrated in Figure 1(c).

(c) CHANGE(Y, Hi, Hj)

This operation is equivalent to a DROP(Y, Hi), followed by an ADD(Y, Hj). In other words, it drops Y from one node Hi, and then adds Y to another node Hj. For example, the tree of Figure 1(c) can be obtained from the tree of Figure 1(a) by CHANGE({v3}, H6, H2).

In performing DROP, ADD, and CHANGE operations, some node Hi may become the empty set, and some Hi may become identical to another node Hj. Such nodes should be removed.

Suppose Hj is in F(Hi), and H'j and H'i become identical after a DROP operation, then H'k becomes empty for all other Hk in F(Hi). To prove this, suppose Hi = A ∪ B ∪ C, Hj = A ∪ B, and Y = B ∪ C is dropped from Hi and Hj. By (A2), all other Hk in F(Hi) must be disjoint from Hj = A ∪ B. Therefore, Hk is contained in C. When Y is dropped from Hk, it becomes empty.

The CHANGE(Y, Hi, Hj) operation can be used to restructure a picture tree. It can be seen that CHANGE(Y, Hi, Hj) only affects nodes in F(Hi) ∪ (A(xi) ∪ A(Hj) − A(Hi) ∩ A(Hj)). In other words, CHANGE(Y, Hi, Hj) affects only nodes in a subtree having the common ancestor of Hi and Hj as the root node. The common ancestor node itself is not affected. Therefore, CHANGE is a *local operation* on the picture tree.

In a picture tree, information (picture objects) carried by a node is also always carried by its ancestor nodes. The converse is not true. As illustrated in Figure 1(a), an ancestor node (such as H3) may carry some information (such as v5) not carried by its follower nodes (such as H6, H7, and H8).

Let B(Hk) be the set of picture objects that can be found only at Hk, and C(Hk) the set of picture objects that can be found at follower nodes. Formally,

$$C(Hk) = \bigcup_{i:Hk \supset Hi} Hi$$

and

$$B(Hk) = Hk - C(Hk)$$

Hk is called the *host node* of the picture object set B(Hk) or any picture object v in B(Hk).

In Figure 1(a), H4 is host node of v1, and H3 is host node of v5. It can be seen that all leaf nodes in a picture tree are host nodes. If only the leaf nodes are host nodes, then the picture tree is called a *skirting tree*. The classical Quad-tree can be regarded as a skirting tree, if we consider the pixels to be the set V, and the partitions to be the Hi's.

The host node of a picture object set Y is where information concerning Y can be retrieved. We can change the host node of Y by performing the following operations, both being special cases of the CHANGE operation.

(d) BEAMUP(Y, Hi, Hj)

If Hj contains Hi, CHANGE(Y, Hi, Hj) will move Y from a node Hi to its ancestor node Hj. Hj becomes the host node for Y. An example of performing BEAMUP({v3}, H7, H3) on the tree of Figure 1(a) is illustrated in Figure 1(d). This example also illustrates the elimination of empty nodes (such as H7) and identical nodes (such as H8).

(e) BEAMDOWN(Y, Hi, Hj)

If Hi contains Hj and Y is contained in B(Hi), CHANGE(Y, Hi, Hj) will move Y from a node Hi to its follower node Hj. Hj becomes the host node for Y. An example of performing BEAMDOWN({v5}, H3, H8) on the tree of Figure 1(a) is illustrated in Figure 1(e).

Intuitively, the BEAMUP and BEAMDOWN operations move information up and down a picture tree. In a picture tree, only host nodes actually store information. Information (picture objects) in nonhost nodes is *abstracted* and *condensed* into picture indexes.

(f) ABSTRACT(Y, Hk, ALG)

Y is in C(Hk), i.e., Hk is not the host node of Y. The picture objects in Y are abstracted and condensed into picture indexes using algorithm ALG. An example of performing ABSTRACT operations on the tree of Figure 1(a) is illustrated in Figure 1(f). We note that since Hk is not the host node of Y, condensed abstraction of Y can be allowed, so that picture indexes to other host nodes can be constructed.

Finally, to perform complete restructuring of a picture subtree, we have the decomposition operation.

(g) DECOM(Hi, ALG)

DECOM(Hi, ALG) replaces the subtree with root node Hi by another subtree. The new subtree is constructed by decomposition of Hi, using decomposition algorithm ALG. An example of performing DECOM(H3, ALG) on the tree of Figure 1(a) is illustrated in Figure 1(g).

In addition to the above picture tree operations, there are two usual tree operations, REMOVE and GROW.

(h) REMOVE(Hi)

REMOVE(Hi) removes node Hi and its follower nodes from the picture tree. An example of performing REMOVE(H6) on the tree of Figure 1(a) is illustrated in Figure 1(h).

(i) GROW(Hi, Hj)

GROW(Hi, Hj) adds a new node Hj as the follower node of Hi, where Hj must be a new node name, and Hj is initially empty. An example of performing GROW(H3, H9) on the tree of Figure 1(a) is illustrated in Figure 1(i).

Given a picture tree PT and query Q, the pictorial retrieval problem is to find QS, the picture query set, by searching the picture tree PT. To facilitate pictorial retrieval, we describe the technique of *structured retrieval* as follows.

Given a picture query Q and a node Hi in picture tree PT, it is assumed that we can compute the following:

We can derive (TS′,Q′,S′) from (Q,Hi), where TS′ is the *partial picture query set* obtainable from (Q,Hi), Q′ is the *modified query* specifying the remaining query after the processing of (Q,Hi), and S′ is the set of descendent nodes Hj of Hi to be visited next.

We can now describe informally the structured retrieval technique as follows. We start from node H1 and compute (TS′,Q′,S′). If the query Q cannot be fully answered, we will have nonempty Q′. We then retrieve all nodes Hi in S′. We now use (Q′,H′) to compute (TS″,Q″,S″), where H′ is a descendent node in Sk. If Q′ cannot be fully answered, we will have nonempty Q″, and we should retrieve all nodes Hi in S″. We proceed as above recursively, and QS is Q(TS), where TS is the union of all the TS′s. We have the following RETRIEVE operation.

(j) RETRIEVE(Q,QS)

RETRIEVE(Q,QS) is the structured retrieval operation, which provides a query set QS for a given query Q, computed as follows:

STEP 1: LIST <-{(Q,H1)}, TS <- empty set.

STEP 2: Remove one tuple (Q,Hi) from LIST. Compute (TS′,Q′,S′) from (Q,Hi). TS <- TS ∪ TS′.

STEP 3: For each successor node Hj of Hi, if {Hj} ∩ S′ is nonempty add (Q′,Hj) to LIST.

STEP 4: If LIST is nonempty, go to STEP 2, else go to STEP 5.

STEP 5: QS <- Q(TS).

For the example illustrated in Figure 1, the transformation (Q,Hi) -> (TS′,Q′,S′) is as follows:

$$(1)\ TS' \begin{cases} = PSi \text{ if } mi = 0 \\ \varnothing \text{ otherwise} \end{cases}$$

(2) Q' = Q

(3) S' = {Hj: Hj is a descendent of Hi and (Pkj Δ Q) is nonempty}

The transformation (Q,Hi) -> (TS′,Q′,S′) is a way to incorporate a *knowledge model* into the RETRIEVE operation. We can then study the effects of different knowledge models on such transformation rules.

6 INTERCHANGE RESTRUCTURING TECHNIQUE

In a picture tree, the CHANGE(Y, Hi, Hj) operation may not be applicable to all node pairs. There may be geometric constraints (such as in the Quad-tree), orr semantic constraints. We can incorporate such knowledge-model into a test function, CTEST(Y, Hi, Hj), which is logical 1 if CHANGE(Y, Hi, Hj) is allowed, and 0 otherwise. A picture tree is *homogeneous*, if CTEST(Y, Hi, Hj) is always 1. Otherwise, it is a *nonhomogeneous* tree.

The CHANGE operations can be used to dynamically restructure a picture tree, based upon user's queries or information retrieval requests. Given a query set QS, let HB denote the set of host nodes with nonempty intersection with QS, i.e., HB = {Hi: B(Hi) ∩ QS ≠ 0}. Let HC be its complement, i.e., HC = H − HB.

We can investigate two types of CHANGE operations:

(1) *Neighborhood Oriented*: We select Hi in HB with fewest neighbors in HB, and Hj in HC with largest number of neighbors in HB, and let Y be Hi in QS. CHANGE(Y, Hi, Hj) is then performed.

(2) *Depth Oriented*: We select Hi in HB with largest D(Hi), and Hj in HC with smallest D(Hj), and let Y be Hi ∩ QS. Again, CHANGE(Y, Hi, Hj) is performed.

The cost function, COST(PT, QS), depends on the tree structure PT, and the query set QS. The cost function could be an extension of those used in [9], [10]. The following cost function can be used:

$$\mathrm{COST(PT,\ QS)} = \mathrm{C1} \times \sum_{\mathrm{Hi\ in\ HB}} |\mathrm{B(Hi)}| + \mathrm{C2} \times \begin{array}{l}\text{number of arcs in}\\ \text{the subtree containing}\\ \text{all Hi in HB}\end{array}$$

The first term represents amount of data retrieved in answering user's query, and the second term represents size of the subtree searched. In interchange restructuring technique, we can perform CHANGE operations iteratively, so that COST(PT, QS) is decreased.

The effects of the two heuristic CHANGE operations need to be investigated. These operations are related to, for example, the rotation operations discussed in [11], [12] for local tree optimization.

The interchange restructuring technique performs local tree reorganization by transferring information from one node to another node. Like dynamic tree balancing, the interchange technique has its limitations. The DECOM (Hi, ALG) operation, on the other hand, performs local tree reorganization by clustering the objects in Hi, thus generating a new subtree.

If V represents pixels, the paging algorithms and decomposition algorithms developed in [13], [14], [15] can be used in DECOM operation to restructure a picture tree. These decomposition algorithms take geometric constraints into consideration.

If V represents local picture objects, we could use *picture information measures* [16] to measure the similarity of objects to perform *objective clustering* on Hi.

We could also use other similarity measures of semantic relatedness, such as fuzzy measures [17] to perform *subjective clustering* on Hi.

An interactive raster graphics terminal can be used to allow the interactive selection of decomposition techniques and changing of thresholds of abstraction. Moreover, CHANGE and DECOM can be applied iteratively, until a picture tree with adequate COST(PT, QS) and adequate structure is generated.

REFERENCES

1. G. Nagy, "Optical Scanning Digitizers," Computer, Vol. 16, No. 5, May, 1983, 13-24.

2. Kai Hwang (ed.), Special Issue on Computer Architectures for Image Processing, *Computer*, January 1983.

3. Toshiba Corporation, "Toshiba Document Filing System: Specification Manual," 1982.

4. H. Tan, "High Density Optical Storage," C&IT Report Number 9, 1982.

5. P.C. Treleaven and I.G. Lima, "Japan's Fifth-Generation Computer Systems," *Computer*, Vol. 15, No. 8, August 1982, 79-88.

6. Robert A. Myers, "Trends in Office Automation Technology," Technical Report RC9321, IBM Watson Research Center, Yorktown Heights, New York, April 1982.

7. Spectrum Special Issue on Data Driven Automation, *Spectrum*, May 1983.

8. S.K. Chang (ed.), Special Issue on Pictorial Information Systems, *Computer*, November 1981, 13-21.

9. V.K. Vaishnavi, H.P. Kiegeland, D. Wood "Optimum Multiway Search Trees," Acta Informatica, Vol. 14, 1980, 119-133.

10. L. Gotlieb, "Optimal Multi-Way Search Trees," SIAM Journal on Computing, Vol. 10, No. 3, August 1981, 422.

11. James R. Bitner, "Heuristics That Dynamically Organize Data Structures," SIAM Journal on Computing, Vol. 8, No. 1, February 1979, 104-107.

12. Mark R. Brown, "A Partial Analysis of Random Height-Balanced Trees," SIAM Journal on Computing, Vol. 8, No. 1, February 1979, 37-39.

13. S.K. Chang, "A Methodology for Picture Indexing and Encoding," in *Picture Engineering*, (K.S. Fu and K.L. Kunii, Eds.), Springer Verlag, Berlin 1982, 33-53.

14. J.L. Reuss and S.K. Chang, "Picture Paging for Efficient Image Processing," Proceedings of IEEE Computer Society Conference on Pattern Recognition and Image Processing, IEEE Computer Society, 69-74, May 1978.

15. S.H. Liu and S.K. Chang "Picture Covering by 2-D AH Encoding," Proceedings of IEEE Workshop on Computer Architecture for Pattern Analysis and Image Database Management, Hot Springs, Virginia, November 11-13, 1981.

16. S.K. Chang and S.H. Liu, "Indexing and Abstraction Techniques for a Pictorial Database," Proceedings of International Conference on Pattern Recognition and Image Processing, June 14-17, 1982, Las Vegas, 422-431.

17. L.A. Zadeh, "A Theory of Approximate Reasoning," Machine Intelligence, Vol. 9, Halsted Press, 172-177, 11979.

A Space-Efficient Hough Transform Implementation for Object Detection[1]

Christopher M. Brown

Computer Science Department
University of Rochester
Rochester, NY

ABSTRACT

The Hough Transform (HT) is a well-known method for detecting objects in images, and can be generalized to recognition of abstract situations characterized by perceptual features. Due to its mode-based nature, HT is robust in situations that would defeat least squared error methods. Implementation of HT involve a histogram that can be large, and previous solutions to the space problem have sacrificed continuity properties useful in HT analysis. The result here is to achieve space efficiency and maintain continuity properties in the histogram. The results are encouraging, and the basic mechanism has been implemented in hardware.

1 BACKGROUND AND OVERVIEW

The Hough Transform (HT) is a parameter estimation strategy based on the statistical mode (Duda and Hart 1972; Ballard 1981). More common strategies such as least squared error fitting (e.g. linear regression) are based on the statistical mean. The HT has achieved engineering importance in several areas of image understanding. It is an efficient implementation of generalized matched filter (template matching) detection, and its mode-based nature makes it highly noise-resistant. Outliers do not affect it, whereas they always have some effect on simple regression (of course, much research in statistics is concerned with rejecting outliers). HT is an important technique in some massively parallel computing architectures (Feldman and Ballard 1982).

[1] This work was supported by the Office of Naval Research under contract N00014-82-K-0193.

In the HT, features (in the transform space) produce votes for parameters with which the features are compatible. After the voting process, the cell with the most accumulated votes indicates the parameters explaining (consistent with) the most input evidence. HT implementations usually use an N-dimesional array that accumulates votes in discrete cells of N-dimensional parameter space. Then HT interpretation usually consists of finding the winning cell (or cells) by searching the array for local or global maxima using more or less complex algorithms. If the array is considered as a histogram, then the maxima (peaks) correspond to its modes.

Figure 1 shows HT in action, detecting circles of various radii in the somewhat confused image shown. The figure makes two points.

(1) The HT is good at detecting phenomena in cluttered signals. Mode-based detection inherently ignores outliers that confuse other detection methods.

(2) The HT can be expensive in space if naive methods are used to collect the histogram.

In an effort to avoid the space requirements inherent in an N-dimensional accumulator array, we at the University of Rochester (Brown and Sher 1982; Brown 1983) proposed to implement vote accumulation and peak-finding

Figure 1 **The Hough Transform detects circles of different radii in a complex input image (upper left). In the upper right is an edge-enhanced image that was used to generate votes into the histogram arrays shown in the bottom two images. Each edge element votes for the center locations of circles that would explain the edge as part of the circle. These votes are collected in a three-dimensional array indexed by x and y position of the center and the radius of the circle. The lower images are two slices through this array at two different radius values, showing vote count as intensity. The bright spots correspond to modes in the histogram and therefore to good candidate circle locations. The results in this paper show how to implement the histogram array with a small amount of storage. (Figure courtesy of Laura Sanchis, U. Rochester).**

(mode estimation) in HT with a content-addressable cache (a hash-table is a software equivalent). The cache is smaller than the full accumulator array, but it must be flushed when its capacity is filled to make room for more votes. The hope is that in a space sparsely filled with votes the cache reliably finds the mode using much less storage then the array. Preliminary experiments with a simulated cache and both simulated and real data were undertaken, and are reported in (Brown and Sher 1982), and summarized in Section 3.4.

I here propose a cache-flushing technique and associated architecture that may significantly improve the performance of cache-based sequential mode-estimation schemes. The basic strategy is inspired by an iterative mode-estimation scheme in the statistical literature (Robertson and Cryer 1974), which uses spatial contiguity (intervals on the real line). The Robertson and Cryer algorithm is similar to the "converging squares" mode-finding algorithm (O'Gorman et al. 1983; O'Gorman and Sanderson 1983). The new flushing algorithm takes into account contiguity in parameter space; the old flushing algorithms did (indeed, could) not. The new scheme needs more cache complexity than the old, and its logic is more complex. However, it seems the added complexities are not ruinous, are perhaps amenable to hardware implementation. The new scheme generalizes most mode-finding algorithms in that it can find multiple modes.

Section 2 outlines the problem and the proposed solution. Section 3 is a list of ideas that are more or less related, and an assessment of their utility. Section 4 explores the proposed scheme in more detail. Some technical details appear in Section 5.

2 THE PROBLEM AND PROPOSED APPROACH

This section is meant to provide context for the basically bottom-up organization of the remainder of this report. One mode-estimation strategy takes samples from a one-dimensional density function, sorts them, and then iteratively constructs ever smaller intervals, each the smallest in the last containing some number of samples determined by the sample size (Robertson and Cryer 1974, and Section 3.3). This basic converging strategy easily extends to many dimensions, though this may affect the technical results on convergence, consistency, etc. The iterative convergence behavior is achieved also by the converging squares technique of (O'Gorman and Sanderson 1983).

A convergence algorithm for sequential sampling was proposed by Hall (1982). The work in this report is an implementation inspired by Hall's suggested approach, filtered through some current computer science ideas and existing implementations, such as Cache-based implementations of the Hough transform.

A cache-based Hough scheme (Brown and Sher 1982; Brown, Curtis, and Sher 1983; Section 3.4) maintains a content-addressable cache of vote tallies for parameter vectors. Content-addressability means that geometric relations (e.g. contiguity) between vectors are not necessarily mirrored in the cache data structure. The algorithm of Robertson and Cryer thus differs from current Cache-Hough schemes in two important ways (and some unimportant ones such as requiring absolutely continuous pdfs for convergence proofs).

1) Their algorithm assumes a fixed sample size, and that all the sampling is completed before it runs.

2) It is based on spatial contiguity (intervals), over which it constructs density measurements.

In both these respects the convergence algorithm has an advantage over Cache-Hough, which must deal with samples collected sequentially and which (so far) has no notion of geometric contiguity and hence of vote "density" over a finite area. The first difference is fundamental. It is the purpose of this report to address the second difference and endow cache-Hough with a flushing strategy guided by geometrical contiguity. The resulting cache-management algorithms resemble existing convergence algorithms to an extent, especially as regards the use of intervals converging on the mode. They are much closer to the suggestion of Hall (1982).

The idea is to use a cache version of quad (oct,...2^d) trees to guide flushing of the highest-resolution tally cache (HRC). In the lower resolution caches (LRC's), each tally entry corresponds to a $2 \times 2 \times \ldots \times 2$, d-dimensional hypercube of the next higher-resolution cells or parameter vectors, and contains the total number of votes in the corresponding higher-resolution hypercube. In the cache version, only cells and vectors with non-zero counts are explicitly represented. Since the vector components are quantized and thus discrete, I shall usually refer to vector (HRC) entries as cells. The hope is that flushing (and perhaps thereafter barring) votes from contiguous hypervolumes of low voting strength renders mode-capturing more robust. (This hope is so far unsubstantiated by experiment, and arises from intuition only).

In the proposed scheme, a cache version of a 2^d tree is constructed as votes come in to the HRC. The LRCs implementing lower-resolution levels are updated to be consistent with HRC. Flushing is triggered by a full HRC, but emanates from some LRC, say FlushCell, whose vote count is low, say FlushCount. Vote counts in LRCs of lower resolution than FlushCell are decremented by FlushCount, and higher-resolution cells in FlushCell's hypervolume are completely removed from the cache. Thus the flush proceeds in both directions, decrementing in the lower-resolution direction and removing in the higher-resolution direction.

Current Cache-Hough schemes only have the HRC, and flush on the basis of low individual tallies (optionally using random mercy to preserve a fraction of low tallies). Single-resolution, content-addressable caches are limited to such "pointwise" flushing strategies. The hierarchical cache captures geometrical structure in the data for use be cache-maintenance algorithms. It provides some concept of vote density, can support a version of accumulator space smoothing, and can implement a flushing algorithm that captures the advantages of convergence schemes without accepting all their limitations. The known convergence algorithms concentrate on finding a single peak. This algorithm penalizes low counts but allows for multiple peaks to survive in the cache, thus implementing multiple mode-finding.

3 RELATED WORK

The ideas in Section 2 suggest various results from statistics and computer science. This section mentions several, and passes judgement on their utility in this context. Briefly we conclude the following.

1) Convergence algorithms for mode estimation seem at this writing to be the most directly relevant statistical methods, though they are most at home in unimodal situations. Other problems, such as PDF estimation and the maximum of a sequence problem, are not as relevant.

2) No technical results about the exact problem of mode estimation with finite memory and a discrete sample space have been found in the literature.

3) Cache Hough methods have promising performance, but suffer from severe myopia when flushing. The incorporation of geometrical contiguity information may help.

4) The usual management of multi-resolution data structures (such as Quad (oct...) trees, Dynamically Quantized (DQ) spaces, and DQ pyramids) is not suited for this application.

3.1 Nonparametric Multivariate PDF Estimation

One natural idea is to estimate not just the mode of the histogram embodied in the accumulator array, but to estimate the corresponding PDF (Wegman 1972, 1982). PDF estimation is inherently harder than mode estimation, because there is more information to be gleaned from the same input. In fact, some PDF estimation assumes the mode is known. N-dimensional (i.e., multivariate) PDF estimation is thought in the statistical community to quite difficult and to require many samples because N-dimensional spaces are (hyper) voluminous. Another difference is that usually a PDF is to be estimated from a fixed-size, completely gathered sample, rather than from the sequential samples that arise in cache methods. Because of the greater inherent difficulty and the seeming lack of relevant methods, nonparametric PDF estimation does not seem to be an attractive alternative to mode-finding.

Often data is neither purely parametric nor completely non-parametric, and partial information about the form of the underlying PDF can be used to advantage (Sager 1985). Translating to HT terms, Sager advocates ordering N-dimensional cells by their counts and then applying conventional (one-dimensional) rank statistics. In other words, he would use vote count contours in n dimensions to help describe the PDF. This suggestion appeals to me because sometimes we know the shape of the vote count surface (Brown 1983) and it could be used to help locate peaks. However, again, the resulting algorithm is not for sequential data.

3.2 The Maximum of a Sequence

The sequential nature of cache-based mode finding leads one to associate it with the Maximum of a Sequence (MaxSeq, Secretary, Beauty Contest, etc.) Problem. The pure version of MaxSeq is: you are presented n face-down slips of paper, on each of which is written an integer, and you are to turn them up sequentially until you decide that the current slip has the maximum integer. What strategy should you use to maximum the probabilty of choosing the slip with the maximum integer? It is surprising that even under such seemingly underconstrained conditions an optimal strategy exists and gives you the respectable and elegant winning probability of 1/e. MaxSeq has been extended in several ways, including allowing a finite buffer of candidates, searching costs, call backs, and so forth (Gilbert and Mosteller 1966; Smith and Deely 1975; Lorenzen 1979; 1981).

Loui (1983) investigated the application of these results to cache-based mode-finding: the conclusion is that MaxSeq is a different problem. In HT mode-finding terms, MaxSeq assumes that all the votes are in and saved, that the final tallies are sequentially presented, and that there is no recalling a previously dismissed tally. The non-recall constraint is especially stringent and non-intuitive in the HT context, and again there is the basic difference that in cacheing the votes come in sequentially, whereas in MaxSeq the totals are in and they (not individual votes) are presented sequentially.

3.3 Mode Estimation

A common method of estimating the mode of a continuous unimodal distribution from n samples is (in 1-D) to sort them and then find the shortest interval containing some number $h(n)$ of them. Then the mode is taken to be some point in that interval (e.g. its midpoint, or the mean or median of the samples it contains) (Venter 1967). The convergence scheme of Robertson and Cryer is designed to lend robustness to finding the mode of "contaminated" distributions. It refines the interval iteratively, at each stage finding the shortest subinterval containing $c^{(t)}(n)$ samples, for $t = 1,2,\ldots s(n)$ stages (Robertson and Cryer 1974). Thus the interval is cut down by a fraction of $[(c^{(t-1)}(n)-c^{(t)}(n)]/[c^{(t-1)}(n)]$ on iteration t. In all these cases, the technical statistical results have to do with choices of $h(n)$, $c(n)$, and $s(n)$ that yield consistent and quick convergence, and with asymptotic distributions. Experimental results of Robertson and Cryer using a $c(n)$ of approximately $2n/3$ to $3n/4$ (here n is the number of samples in the interval being refined, not the total number of samples), indicate that with outliers (noise) or contaminated data (multimodal or in their case a mixture of two distributions) the intervals must start smaller and converge faster to avoid getting confused by local maxima. They recommend that $c(n)/n$ be significantly smaller than $1-p$, p the fraction of contaminated data.

The mathematical restrictions on these methods that must be imposed to allow analytic results are fairly severe, but the strategies are clear and appealing and extend to multiple dimensions. They inspired the approach proposed in this paper.

Finding the shortest interval containing a given number of samples requires search. In statistical models, the samples are from a continuum, and hence will be duplicated only with probability zero. In the accumulator array search, the samples are discrete, and the cells can have more than a unit count. Thus the search will have to keep an updated sum of counts in a (d-dimensional) rectangular volume, which requires slightly more computational effort than merely counting single samples. Fast techniques exist for running rectangular averages, however.

The iterative search of convergence methods is not more expensive than one-time search. To compare Robertson and Cryer's approach to that of Venter, both presume n $1-D$ sorted real data. The search for the smallest interval containing $h(n) = c^{(t)}(n)$ of them requires comparing the lengths of $n - h(n) + 1$ intervals (thus $n-h(n)$ comparisons). Robertson and Cryer point out that in the iterative scheme, the number of intervals to compare is

$$\sum_{i=1}^{t} [c^{(i-1)}(n) - c^{(i)}(n)] = n-c^{(t)}(n) = n-h(n).$$

If the fraction by which the interval is diminished on each iteration is constant, write it as r. Then the above result is derivable from geometric series summation, by which it generalizes to d dimensions. Surprisingly, the iterative search in d dimensions takes fewer comparisons than the one-time search in d dimensions. (Remember, this is the number of hypervolume densities (or total occupancies) that must be computed to find the densest interval. It does not include the number of operations need to compute each total. For that operation, fast running total algorithms exist (Rosenfeld-Thurston 1971; Narendra 1978). In the one-time search, the number of hypervolume densities in d dimensional $M \times M \times ...M$ space that must be computed to find the densest $h \times h ...\times h$ hypervolume, $h = r^t M$ is

$$C_1 = (M-r^t M)^d = M^d \ (1-r^t)^d.$$

(Here t is an honest exponent, not an index.) In an iterative search, the number of hypervolumes to be considered is

$$\begin{aligned} C_2 &= (M - r\ M)^d + (r\ M - r^2 M)^d + (r^2 M - r^3 M)^d + \cdots \\ &= (1 - r)^d \ (M^d + r^d M^d + \cdots + r^{(t-1)d} M^d) \\ &= M^d (1 - r)^d \ (1 + r^d + r^{2d} + \cdots + r^{(t-1)d}) \\ &= M^d \ (1 - r)^d (1 - r^{td}) \ / \ (1 - r^d). \end{aligned}$$

The ratio C_2 / C_1 is the fraction of comparisons that the iterative search must make compared on the one-time search. The behavior of this ratio is not obvious from the formulae, although for small r it is approximated (from below) by

$$(1 - dr)/(1 - dr^t).$$

Table 1 gives values of the ratio for relevant r, d, and t. It shows that as dimension, the size of the ratio, and the number of iterations go up, the number of density comparisons falls off.

d	t: 2	4	8	t: 2	4	8	t: 2	4	8
2	0.680	0.605	0.600	0.556	0.378	0.336	0.510	0.275	0.175
4	0.411	0.323	0.318	0.556	0.086	0.068	0.140	0.026	0.009
8	0.168	0.103	0.100	0.039	0.007	0.004	0.013	0.000	0.000
		r: .25			r: .5			r: .75	

In the converging squares algorithm (O'Gorman et al. 1983; O'Gorman and Sanderson 1983), a space of size $n \times n$ (in two dimensions) yields four smaller overlapping spaces. In 2-D, the new spaces are the $(k-1) \times (k-1)$ squares in the four corners of the old space. The single $(k-1) \times (k-1)$ square of maximum density is chosen for expansion at the next cycle. The common area between the overlapping spaces allows it to be disregarded in computing the differences in density, resulting in substantial computational savings.

Comparing the computations needed for converging squares to two other simple mode-finding algorithms:

maximum value: $C(n^2-1)$

smoothing (four points) then maximum value: $3An^2 + C(n^2-1)$

converging squares: $A(n^2+7n-22) + C(5n-7)$

where C is a conditional operation, A is an addition operation. Typically a C takes twice as long as an A, and an implementation of converging squares is in fact about three times faster than maximum value and six times faster than smoothing on a VAX 11/780.

Finally, some results on mode-estimation with small sample sizes in unordered bins (equivalent to the cache-based method without the 2^d-tree structure) appears in (Brown 1984).

3.4 Cache Hough Implementation

The performance of Cache Hough schemes under a variety of conditions (noise, cache length, length of vote bursts, image scanning order, flushing strategy) is tested in (Brown and Sher 1982). These experimental studies are not backed by formal analysis.

The cache model is that of a single-resolution tally cache (the HRC of Section 2), flushed by either of two strategies. In "Slaughter of Innocents" flusing, all tallies below a threshold are flushed. In "Draft Lottery" or "Random Mercy" flushing, a fraction of all tallies below a threshold is selected at random and flushed. The performance of a cacheing scheme is measured by the ratio

$$SNR\,3 = \frac{(\text{votes for the vector known to be correct})}{(\text{maximum votes for any incorrect vector})}.$$

There are few qualitative surprises in this work. Performance improves with increasing cache length and falls off with increasing noise and fraction of incorrect votes in a vote burst. Scanning strategies seem equally matched except for random with replacement, which may have been prejudiced by relatively small sample size (400 samples (vote bursts) from a 20 × 20 array of features). The lottery flushing strategy works better than the slaughter strategy. Figure 2 shows some sample results.

After it fills (which can be after a few features cause vote bursts), the cache is continuously flushing. Since the content-addressable cache does not maintain contiguity information, a low tally from an active (dense) region of parameter space is as likely to be flushed as a low tally from an inactive (sparse) area. The noise modeled in the experiments is additive noise that does not "spread the peak" in parameter space as does quantization noise. In fact quantization noise is important, and is sometimes taken to be the only important noise effect (Shapiro and Iannino 1979). It is usually combated by smoothing the accumulator array before searching for modes. Such contiguity-based techniques are difficult and unnatural using only the content-addressable HRC, but become possible in the proposed scheme which leads to an "urban renewal" flushing strategy in which good neighborhoods are preserved. The analysis of (Brown 1983) shows that when multiple votes are produced for each feature, neighborhoods of high voting strength arise around peaks. Thus the "urban renewal" strategy offered hierarchical caches seems a promising approach for all known voting schemes.

3.5 Quad Trees and DQ Methods

Multiresolution approaches to image understanding and processing have been popular and useful for a long time (Kelly 1971; Warnock 1969). Keeping the multiple resolutions explicitly in a pyramid data structure has also proven quite useful (Samet 1980; Tanimoto and Pavlidis 1975). When the resolution pryamid is made of predefined cells, they are usually split symmetrically along each dimension, and the resulting structure is called a quad (oct,...) tree. I call them 2^d trees here. When the cells are split asymmetrically, or the density of resolution varies over a pyramid level, especially when the splitting varies as the contents of the data structure arrive sequentially, a Dynamically Quantized (DQ) space or pyramid results (O'Rourke 1981; Sloan 1981).

A control program usually adapts these data structures to high-resolution data by generating new cells where data is densest. In 2^d trees, this is done by splitting the lower-resolution cells. In DQ spaces, the data structure is a $k-d$ tree (Bentley 1975), and cells are split and merged as data arrives sequentially. In DQ Pyramids, the number of cells at a level is fixed but their extent is varied by moving the d-dimensional "crosshairs" that split each level into 2^d. Usually the management of the data structures has the goal of producing cells with equal complexity, or numbers of counts. The density of the data is thus mirrored in the data structure (dense data in a region produces a tree that is deeper for that region, for instance). This approach is natural in a sense — it does not require search if enough information is kept in the cells to allow splitting and merging. The data structure and some ancillary information is sufficient to reconstruct the approximate density of the original data, which is inversely proportional to the size of the cells.

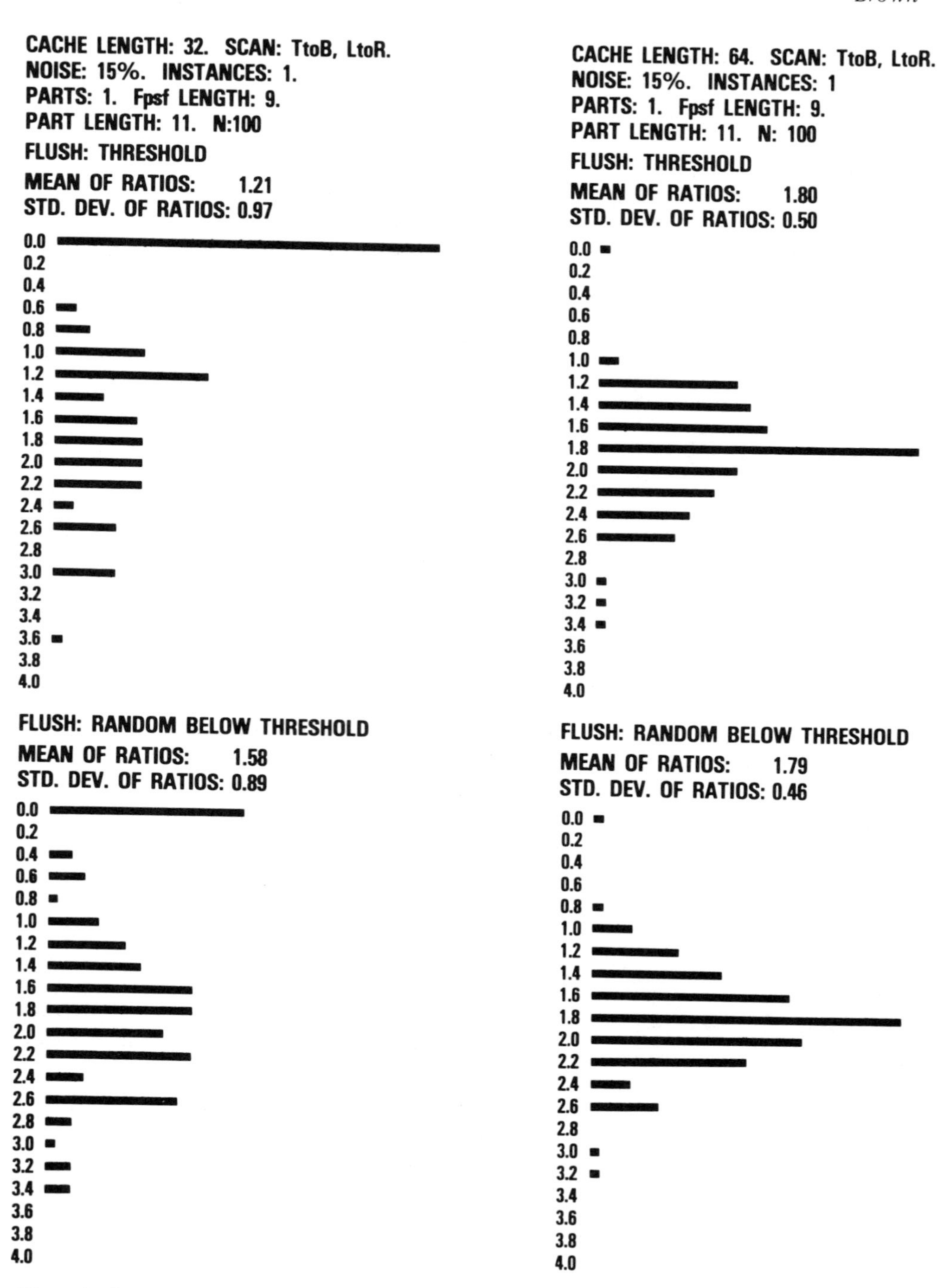

Figure 2 Sample SNR3 histograms for four configurations of cache HT, showing the beneficial effects of increased cache length and flushing with random mercy. SNR3 > 1 means the correct parameter vector received the most votes. The bimodality of the distributions is unexplained—the peak at SNR3 = 0 represents trails in which the correct vector was not even in the cache after the HT.

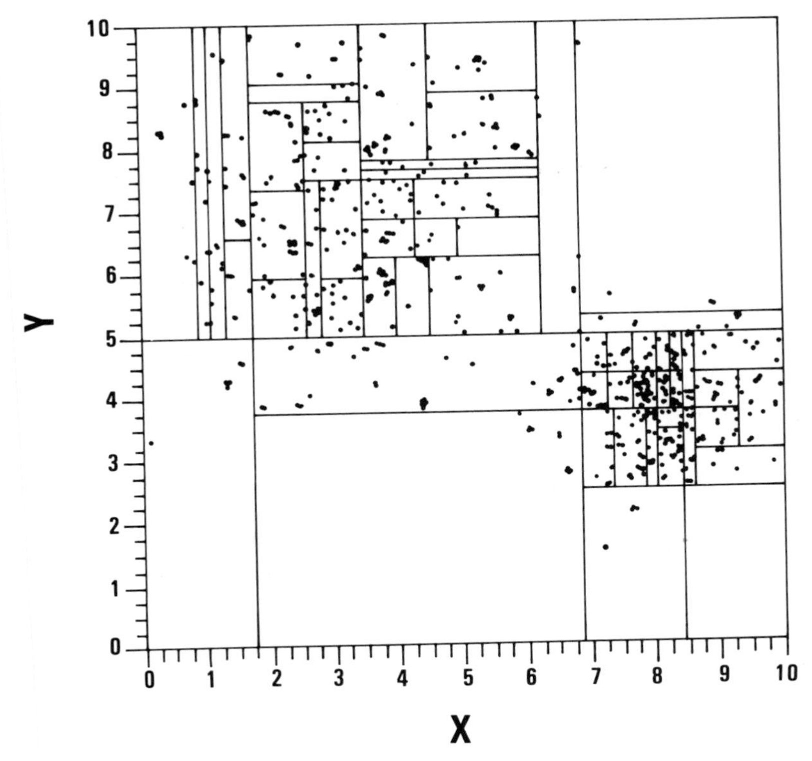

Figure 3 A DQ Space with two modes from (O'Rourke 1981). The cells in the space result from splitting and merging as data arrive

DQ structures in the literature suffer from a few difficulties. The cells in DQ spaces can be unintuitive (Figure 3). The splitting and merging algorithms are complex. The cells contain only approximately correct counts after splitting and merging, operations which are controlled by total counts and by count gradient information within a cell. The high-resolution parameters (locations) of counts are lost. DQ Pyramids, since the number of cells is fixed, have simpler algorithms, but again produce wrong counts as the crosshairs are moved away from their original positions by adaptive warping as data comes in. Despite all this, the DQ structures seem to be practically usable for some applications, including HT accumulation.

The usual count- (or complexity-) equalizing control strategies for hierarchical data structures have a dual sort of effect from the one we desire, although it is possible to imagine working with (i.e., around) them. In the cache mode-estimation application there is one cache entry per cell, and it would be best if modes were captured inside single cells instead of distributed across several. Also the possibility of flushing everything but one cell (at some level) is attractive. Thus the data structures of dynamically quantized structures are useful, but the management algorithms are inherently difficult and in any case can be modified to match our purposes better.

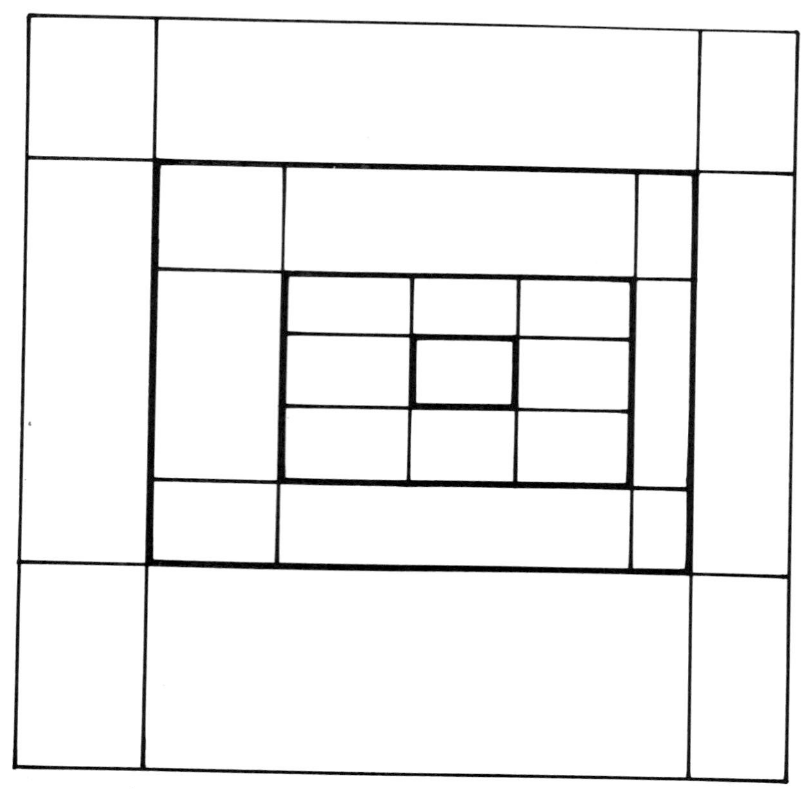

Figure 4 The DQ-3^d Tree. In two dimensions, the DQ non-tree. Central cells contain and converge upon dense areas of data. Non-central cells are candidates for flushing from a cache. Multiple modes simply require splitting a non-central cell at some level.

4 THE DATA STRUCTURE

4.1 DQ 3^d -Tree

The DQ 3^d Tree is a DQ 2^d Tree (Pyramid) modified to be more useful for our purposes. It is a natural data structure to associate with the convergence mode-finding algorithms. The idea is simple, and shown in Figure 4 in the 3^2 (two-dimensional) case: Construct cells that contain and converge on dense areas, rather than splitting dense cells. The search necessary to create these structures is related to that analyzed in Section 3.3. I do not seriously propose this structure for implementation since adaptive warping would raise all the problems encountered in standard DQ Pyramids, but offer an approximate and presumably simpler version in the next section.

4.2 Unphased 2^d Tree

As an approximation to the DQ 3^d Tree, consider the Unphased 2^d Tree (Figure 4). It is simply a 2^d tree augmented at each level by a phase-shifted version of the cells. For definiteness, call the usual cells the U cells, and the shifted ones the S cells.

This compromise alleviates (but does not cure, unfortunately) the problem of dense areas that lie on predetermined boundaries in the tree, at a considerable computational saving over the (better) solution offered by DQ 3^d Trees. A vote for parameter vector increments the count of a sequence of cells, namely all those cells containing the vector. Each such U cell may be addressed by the highest-order componentwise bits of the last. This produces the normal 2^d tree structure. The construction of the out-of-phase cells is treated in some detail in Section 5.1.

The shifted cells are some insurance against splitting a peak over several cells, as is guaranteed to happen in traditional multi-resolution schemes. We do not want this to happen: it loses a resolution level. Worse, it can make the new cells (each with only about 2^{-d} of the votes for the mode) vulnerable to flushing. Thus, both non-traditional tree management and phase-shifted cells may have a more important effect than just a gain of resolution in the context of cache-based schemes.

5 TECHNICAL DETAILS

5.1 Vector Addresses and Arithmetic

The 2^d tree is implemented as a set of separate but communicating caches, one per level for each of S and U cells. The Vector addresses have a number of bits that increases by d with each level of increasing resolution. I propose to use a straightforward translation for the address of U (natural) cells in the 2^d tree, based on their Cartesian coordinates in d-space. The S cell locations in natural, Cartesian coordinates do not have the elegant leading-bits relation with their underlying cells, and so are transformed into T addresses that do. U and T addresses must be differentiated.

The following discussion relates to U cells. In order for a parameter vector (address, d-dimensional location) to be related to its 2^d tree address, it is represented as follows. If x is a d-vector of m-bit quantities ($m = \log_2 M$)

$$\begin{array}{lllll} x = & x_{11} & x_{12} & x_{13} \cdots & x_{1m} \\ & x_{21} & x_{22} & \cdots & x_{2m} \\ & x_{d1} & x_{d2} & \cdots & x_{dm}, \end{array}$$

where the x_{ij} are bits. The write x as the single bit string (dm vector)

$$Y = x_{11}\, x_{21} \cdots x_{d1}\, x_{12}\, x_{22} \cdots x_{d2} \cdots x_{dm}.$$

In other words, read the above array of bits out columnwise. Thus the d high-order bits come first, and last come the d lower-order bits. We shall need one bit to distinguish U LRC addresses from T LRC addresses. The final form of address is

$$\text{address} = \{U/T\}\; y.$$

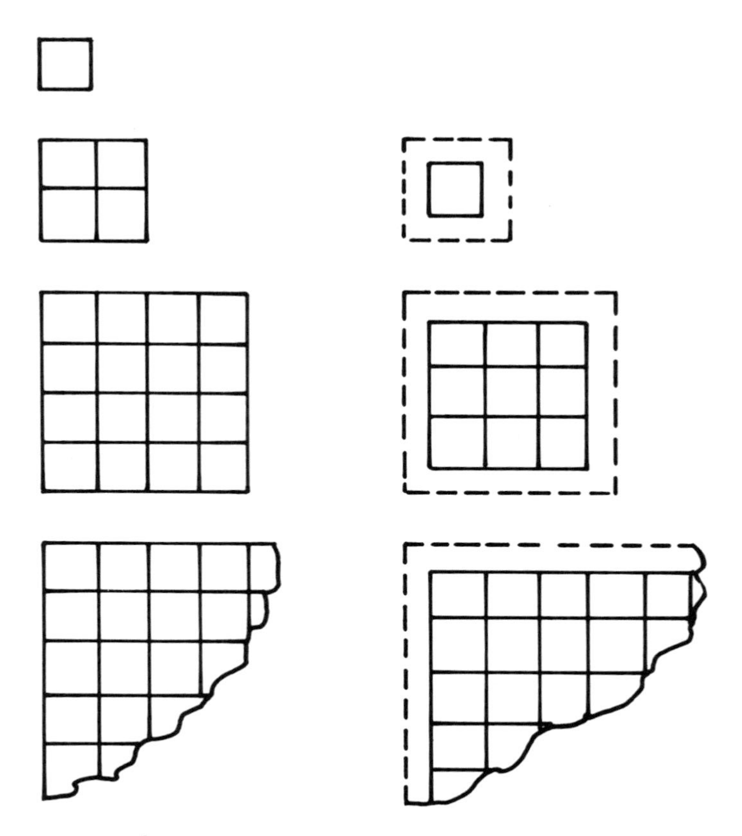

Figure 5 The unphased 2^d tree. In 2-D, the unphased quad tree. The kth layer consists of 2^{kd} U (usual or unshifted) cells and $(2^k-1)^d$ S (shifted) cells. The usual cells are augmented by those shifted half a cell size. The size of shift depends on the level. Their sub-cell inclusion rules are more complicated than for U cells.

Let the HRC be assigned level m and the lowest resolution cache entry (a single entry counting the total votes in each cache) have level 0. Then at level $k, 0<k<m$, $x's$ parameter vector (address) is the bit string of length kd (interpreted as a d-vector of k bit quantities)

$$\text{Right Shift}(y,\ d(m-k)).$$

Now consider S cells (Figure 5), which introduce considerable complication. They are shifted by a different amount on every layer. The kth layer of the 2^d tree has 2^k U cells in it. That layer has $(2^k-1)^d$ S cells of the same size. To generate a unique vector address (the T address) for the S cells, I subtract half their linear dimension from their cartisian (U) addresses. (Think of sliding the 3×3 S cells in layer 2 of a quad tree down so they cover the "lower left" 3×3 square in the 4×4 array of U cells). This is to generate a unique address for the S cell — all its members will have T addresses with identical leading bits (kd of them at level k), just like U addresses.

In a natural way, each U cell on any level k has associated cells on all other levels. They are the cells of higher k whose (hyper) volume it contains

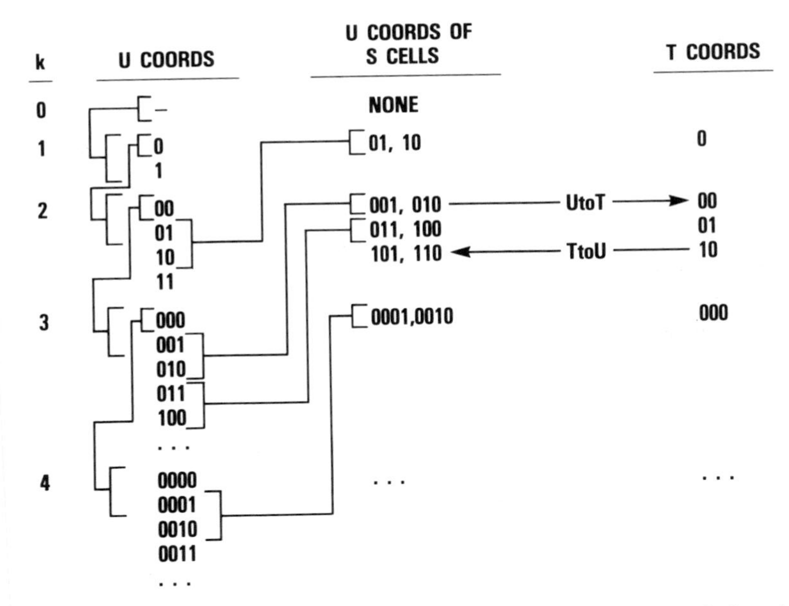

Figure 6 U, S, and T coordinates. The connected brackets show association relations between U cells and other U and S cells. These 1-D coordinates and transformations extend, componentwise, to d dimensions. U to T and T to U transformations are given in the text.

and the cells of lower k that contain it. The rule associating U cell addresses at level k_1 with a cell address y at level k is the following.

(R1) If $k_1 > k$, all cells at level k_1 having addresses whose higher-order kd bits are the same as y are associated with (lie within) the k-level cell at y.

(R2) If $k_1 < k$, the single cell at level k_1 whose address is the first $k_1 d$ bits of y is associated with (includes) the k-level cell at y.

S cells are not simply associated with each other, since the amount of offset varies from layer to layer. However, an S cell on the k layer is made up of 2^d U cells on the k + 1st layer, and it is this correspondence that is used in flushing strategy.

Figure 6 shows the U, S, and T coordinates of cells (and the association between U and S cells) in a "two-tree," where $d = 1$.

Figure 6 shows that two (in d dimensions, 2^d) cells at level $k + 1$ are included in a cell at level k. The transformation U to T maps U addresses at level $k + 1$ to T addresses at level k. T to U maps T addresses at level k back to U addresses at level $k + 1$.

U to T (k,d): Subtract $2^{(d-k-1)}$ (k leading 0's and *a 1*) from the k + 1-bit U address. The leading k bits are the T address.

T to S (k,d): Take the k bits of T address, append to them the two (in d dimensions, 2^d) possible configurations of one (d) bit (s). Then add

U	XXXXXXX	XXXXXXX	. . .	XXXXXXX	1010110	1011010
BORROWS			. . .	0100001	0100101	
SUBTRACT COMPONENTWISE						1111111
DISCARD						0100101
T			. . .		1110011	
T	XXXXXXX	XXXXXXX	. . .	XXXXXXX	1110011	
APPEND, AS ONE OF 2**D CHOICES						0100101
CARRIES			. . .	0100001	0100101	
ADD COMPONENTWISE			. . .			1111111
U			. . .		1010110	1011010

X	Y	X − Y AND X + Y	BORROW	CARRY
0	0	0	0	0
0	1	1	1	0
1	0	1	0	0
1	1	0	0	1

Figure 7 Addition and subtraction of address vectors. Simply a carry-ripple operation done bitwise (the ith bit in the each block belongs to the ith dimension's coordinate). The carries and borrows alone are added and subtracted since the leading bits of the addend and subtrahend are 0. "Overlow" here results from using illegal operands (trying to compute T addresses for U addresses that do not have them, or vice versa).

$2^{(d-k-l)}$ (k leading 0's and *a l*). The leading $k + l$ bits of the resulting two (2^{d}) addresses are the U addresses.

U to T and T to U are easily extended to operate on the linearized dM-vectors in d dimensions. Use the usual truth tables for addition and subtraction (Figure 7).

A more elegant scheme for accessing the shifted cells might result from a more sophisticated coding scheme. Other space-filling addressing schemes are possible and are potentially useful. The Generalized Balanced Ternary scheme is one such, using hexagonal cells in 2-D, truncated octahedra in 3-D, and in general $n + 1$ permutahedra in n-space (Gibson and Lucas 1982). We hope to pursue these topics as time allows.

5.2 Flushing Algorithms

Two strategies for flushing the caches seem useful. The first is called static, and seems primarily useful for unimodal accumulators and data that comes in from a static situation (for example, a random scan of a single image). The second is called dynamic, and seems more suited for multimodal data or data from changing sources (time-varying images or raster scans). Both flushes are initiated by conditions in the HRC, usually that it is close to filling up. In both flushing strategies, S or U LRCs with low counts are found and flushed.

To flush a U cell at level k, use UFlush(ACell): remove all ACell's entries. Decrement the count of the cells including ACell in lower-k U cells by ACell's count. Remove entries in higher-resolution (higher-k) cells that are included in ACell—whose leading address bits agree. Every time a U cell at level $k + l$ is flushed, decrement its associated S cell at level k.

To flush an S cell at level k, use SFlush(ACell): remove its entries and UFlush its associated U cells at level $k + 1$. This latter flush works its way up and down the hierarchy, flushing and keeping S and U counts consistent.

Flushing could trigger other flushing, as lower-k cells may become flushable through higher-k flushes. In the dynamic algorithm, flushing is all that happens. In static flushing, information is recorded about which cells were flushed, and no new votes in those areas are accepted. This can be implemented several ways, using filter registers to check on the addresses of incoming votes. The registers can contain acceptable or unacceptable ranges of addresses to be checked before votes are inserted in the HRC.

5.3 Number of Lower Resolution Cells

How many lower resolution cells will there be? We can easily put upper and lower bounds on their number, and can appeal to statistics for some more intuitions.

If the HRC is 2^m on a side for d dimensions, there are 2^{md} possible HRC cells. There are $2^{(m-1)d}$ possible LRC U cells on level $m-1$, $(2^{md}-1)/(2^d-1)$ LRC U cells, and a total of $(2^{(m+1)d}-1)/(2^d-1)$ potential cells in all caches. U cells in $m+1$ levels (down to the single cell at $k = 0$). With increasing d the number of LRC cells approaches the number in the first LRC, or $2^{(m-l)d}$. For example, in the four-level quad tree with 64 HRC cells, there are 85 total U cells and $64 + 16 = 80$ of these are in the HRC and the first LRC. The S cells approximately double the size of the LRC cache. The entire set of U and S LRCs thus is at worst only about $2^{(m-d)}$ as big as the HRC.

If all HRC votes are in a small area (a single HRC cell), then only $k + 1$ cells are allocated in the U cache, and k in the S cache, for a minimum of some $2k$ cells.

We may wonder how the cache will look between these two extremes. How likely are we to get empty cells that do not appear in the caches? This question is addressed by occupancy statistics (Johnson and Kotz 1977) and urn models (Cohen 1978). Say we have C cells and v votes. Then there are C^v ways to vote into the cells. Suppose we wish to count the number of ways to vote into C cells so that exactly $C-p$ of them are empty (p of them contain votes). For p chosen cells, the number of district ways to vote is

$$p!\,\{v,p\}$$

where $\{v,p\}$ is the Stirling number of the second kind (see below). There are (C,P) ways to choose the p full cells, where (C,p) is the binomial coefficient (see below). Thus the fraction of voting trails in which exactly p cells is filled is

$$F = \frac{p!\,(C,p)\,\{v,p\}}{C^v}$$

This quantity may be interpreted as a probability if each cell is equally likely to receive votes. If X is the number of occupied cells, $Pr[X = p] = F$ is known as the classic occupancy distribution. If cells are not equally likely to receive votes, the expression becomes extremely complex (Johnson and Kotz 1977, eq. 3.5). The equal-probability situation minimizes the expected number of empty cells, and so is a worst case (ibid).

The binomial coefficient (C,p) is $C!\,/\,(C-p)!\,p!$

The Stirling number of the second kind $\{v,p\}$ counts the number of ways of partitioning a set of v elements into exactly p subsets, none empty. We have

$$\{v,0\} = 0\ \{v,1\} = 1,\ \{v,2\} = 2^{(p-1)}-1,\ \{v,v-1\} = (v,2),\ \{v,v\} = 1,$$

and the recurrence

$$\{v,p\} = p\{v-1,p\} + \{v-1,p-1\},$$

which leads to a Pascal's triangle-like construction.

v \ p	1	2	3	4	5	6
1	1					
2	1	1				
3	1	3	1			
4	1	7	6	1		
5	1	15	25	10	1	
6	1	31	90	65	15	1
. . .						

There is a helpful identity for computing occupancy numbers:

$$p!\{v,p\} = (p,p)p^v - (p,p-1)(p-1)^v + (p,p-2)(p-2)^v - \cdots .$$

As an example, the distribution of full bins after 8 votes into 8 bins (there are 2^{24} possibilities) is approximately (in percent)

Occupied bins	1		2	3	4	5	6	7	8
% of trials	.0	...	.04	2	17	42	32	6	.2

Occupancy distributions have been thoroughly studied in the literature (Johnson and Kotz 1977). They would be expected to occur in studies of cacheing, and in fact were used to formalize the behavior of working sets (Denning and Schwartz 1972). For the classical occupancy distribution of interest here, a normal approximation is quite good (Vantilborgh 1974).

Certain limit theorems are known for the classical occupancy distribution. For a fixed number of votes v, as the number of cells p goes to infinity, the expected number of empty cells goes to infinity, the expected number of cells with one vote goes to v, and the expected number of cells with more than one vote goes to zero. As v and p go to infinity, with $pe^{(-v/p)} \rightarrow w$, then the probability that there are t empty cells approaches a limit

$$\lim_{v \rightarrow \infty} Pr[M_0 = t] = w^t / e^w t!$$

Also, if v and p go to infinity, with $v/p \rightarrow w < \infty$, then the limit standarized distribution of M_r (the number of cells with r votes) is unit normal, and

$$E[M_r] = p(v,r)p^{-r} \ (1-1/p)^{(v-r)} \ \ w^r/(e^w r!)$$

$$\mathrm{var}(M_r)/\ E[M_r] \simeq 1 - w^r \ [1 + ((r-w)^2)/w](r!e^w)\ .$$

The last two paragraphs deal with limits of expected values in occupancy distributions, not with the distributions themselves. The results in this section may be useful in calculating expected cache occupancy, or at least in lending some basis of order of magnitude calculations should they be desired. It appears that as a practical matter, the allocation of adequate space for the LRC caches might pay for itself in simplification of the management algorithms

6 CONCLUSIONS AND FUTURE WORK

Real HT data includes the effects of quantization error as well as inherent sidelobes surrounding peaks. In practice, accumulator arrays are usually smoothed to gather local evidence into a point. With only a few modes in the accumulator array, large volumes of it will be subject only to votes from noise. In traditional cache HT, spatial contiguity is lost, and the above observations do us no good. A vote flushing and filtering strategy that makes use of spatial contiguity seems likely to improve cache-HT performance, and this report proposes an architecture and algorithms for that purpose. The strategy is based on statistical mode-estimation algorithms, and the data structure is an augmented version of quad (oct,...2^d) trees. The management of the data structure differs from the usual in that the goal is to keep votes in the fewest cells possible, rather than to spread them out evenly between cells.

For a small investment in space, a hierarchical data structure that keeps track of geometric contiguity can be implemented in a cache environment, with vector addresses encoding the inclusion between multiresolution cells. The flushing algorithms for this structure are simple. Matters become more complicated when an ancillary data structure of shifted cells is added to cope with phasing problems (peaks being split across predetermined cell boundaries). Some aspects of the resulting structure are subject to analytic treatment.

One desirable analytic problem that might be feasible is the treatment of discrete sample space mode-estimation with finite memory, and in particular some properties of an iterative technique such as the one proposed here. How often, say, will it fail to find the mode of an analytically tractable distribution? Continuous approximations (say a continuous version of the whole problem) begin to resemble known convergence algorithms.

Dave Sher implemented a software simulator for HRC-only caches (Brown and Sher 1982). He has completed a VLSI implementation of a content-addressable tally cache (Sher and Tevonian 1984). Neither of these implementations incorporates the hierarchical structure discussed here. We have extended the software simulation to hierarchical flushing algorithms. The relation of the complex flushing algorithms to hardware is under study.

The next step is to run experiments with the hierarchical cache (initially only with U cells). Methods for vote filtering should be developed and tested. Static and dynamic flushing should be tried with various scanning strategies. If hierarchical caches perform significantly better than single-resolution caches, we must investigate the interaction of hierarchical structure with hardware caches under development.

7 ACKNOWLEDGMENTS

This work was carried out under NSF Grant MCS-8302038, ONR (DARPA) Grant N00014-82-K-0193, and CNA Grant SUB N00014-76-C-0001. R. Gabriel, T. Therneou, J. Hall, and J. Wellner have provided encouragement and guidance, and are in no way responsible for technical inadequacies. P. Meeker helped greatly in document preparation.

REFERENCES

Ballard, D. H., "Generalizing the Hough transform to detect arbitrary shapes," *Pattern Recognition 13*, 2, 111-122, 1981.

Bentley, J. L., "Multidimensional search trees used for associative searching," *CACM 18*, 509-517, September 1975.

Brown, C. M., M. B. Curtiss, and D. B. Sher, "Advanced Hough transform implementations," TR 113, Computer Science Dept., U. Rochester, March 1983; *Proc.*, 8th IJCAI, Karlsruhe, West Germany, August 1983.

Brown, C. M. and D. B. Sher, "Hough transformation into cache accumulators: Considerations and simulations", TR 114, Computer Science Dept., U. Rochester, August 1982; *Proc.*, DARPA Image Understanding Workshop, Palo Alto, CA, September 1982.

Brown, C. M., "Bias and noise in the Hough transform I: Theory," TR 105, Computer Science Dept., U. Rochester, June 1982; to appear, *IEEE Trans. PAMI*, 1983.

Brown, C. M., "Mode estimation with small samples and unordered bias," TR 138, Computer Science Dept., U. Rochester, June 1984.

Cohen, I. A. *Basic Techniques of Combinatorial Theory.* Wiley and Sons, 1978, p. 118ff.

Denning, P. J. and S. C. Schwartz, "Properties of the working set modal," *CACM 15, 3*, 191-198, March 1972.

Duda, R. O. and P. E. Hart, "Use of the Hough transform to detect lines and curves in pictures," *CACM 15*, *1*, 11-15, 1972.

Feldman, J. A. and D. H. Ballard, "Connectionist models and their properties," *Cognitive Science 6*, 205-254, 1982.

Gibson, L. and D. Lucas, "Spatial data processing using Generalized Balanced Ternary," *Proc.*, IEEE Conference on Pattern Recognition and Image Processing, 566-571, Las Vegas, June 1982.

Gilbert, J. P. and F. Mosteller, "Rocognizing the maximum of a sequence," *J. Amer. Stat. Assoc. 61*, 1966.

Hall, W. J., "Estimating the mode of a multivariate density, based on sequential sampling and with finite storage," Unpublished Note, Statistics Dept., U. Rochester, October 1982.

Johnson, N. L. and S. Kotz. *Urn Models and Their Application.* Wiley and Sons, 1977, p. 107ff, 315ff.

Kelly, M. D., "Edge detection by computer using planning," in Meltzer, B. and D. Michie (Eds). *Machine Intelligence 6.* Edinburgh University Press, 1971.

Kong, A., "The Beauty Contest with a searching cost," Unpublished Working Paper, Dept. of Statistics, Harvard U., 1982.

Lorenzen, T. J., "Generalizing the secretary problem," *Adv. Appl. Prob. 11*, 384-396, 1979.

Lorenzen, T. J., "Optimal stopping with sampling cost: The secretary problem," *Annals of Prob. 9*, 1981.

Loui, R.P., "How fast Hough?," Internal Working Paper, Computer Science Dept., U. Rochester, May 1983.

Narendra, P. M., "A separable median filter for image noise smoothing," *Proc.*, PRIP *78*, 137-141, 1978.

O'Gorman, L., A. C. Sanderson, K. Preston, Jr., and A. Dekker, "Image segmentation and nucleus classification for automated tissue section analysis," *Proc.*, IEEE Conference on Computer Vision and Image Processing, 89-94, Washington, DC, June 1983.

O'Gorman, L. and A. C. Sanderson, "The converging squares algorithm: An efficient multidimensional peak picking method," *Proc.*, IEEE Int'l. Conference on Acoustics, Speech, and Signal Processing, 112-115, Boston, MA, April 1983.

O'Rourke, J., "Dynamically quantized spaces: A technique for focusing the Hough transform," *Proc.*, 7th IJCAI, 737-739, Vancouver, B.C., August 1981.

Robertson, T. and J. D. Cryer, "An iterative procedure for estimating the mode," *J. Amer. Stat. Assoc. 69, 348*, 1012-1016, December 1974.

Rosenfeld, A. and M. Thurston, "Edge and curve detection for visual scene analysis," *IEEE TC-20*, 562-569, 1971.

Sager, T., "Dimensionality reduction in density estimation," This volume (1985).

Samet, H., "Region representation: Quadtrees from boundary codes," *CACM 23, 3*, 163-170, March 1980.

Shapiro, S. D. and A. Iannino, Geometric constructions for predicting Hough transform performance," *IEEE Trans. PAMI-1, 3*, July 1979.

Sher, D. and A. Tevanian, "The vote tallying chip: A custom integrated circuit," *Proc.* Custom Integrated Circuit Conference, Rochester, N.Y., May 1984.

Sloan, K. R., Jr., "Dynamically quantized pyramids," *Proc.*, 7th IJCAI, 734-736, Vancouver, B. C., August 1981.

Smith, M. H. and J. J. Deely, "A secretary problem with finite memory," *J. Amer. Stat. Assoc. 70*, 1975.

Tanimoto, S. and T. Pavlidis, "A hierarchical data structure of picture processing," CGIP *4, 2*, 104-119, June 1975.

Vantilborgh, H., "On the working set size and its normal approximation," *BIT 14*, 240-251, 1974.

Venter, J. H., "On estimation of the mode," *Annals of Mathematical Statistics 38*, 1446-55, October 1967.

Warnock, J. G., "A hidden-surface algorithm for computer-generated halftone pictures," TR 4-15, Computer Science Dept, E., U. Utah, June 1969.

Wegman, E. J., in S. Kotz and N. L. Johnson (Eds). *Encyclopedia of Statistical Sciences*, Vol IV. Wiley and Sons, 1982.

Wegman, E. J., "Nonparametric probability density estimation: I. A summary of available methods," *Technometrics 14, 3*, 533-546, August 1972.

New Computing Methods in Image Processing Displays

Cliff Reader

International Imaging Systems, Inc.
Milpitas, CA

ABSTRACT

The video rate processing hardware of image display-systems has traditionally used fixed point arithmetic and been limited to simple functions. The issue of computational noise has been ignored since the quality of the (analog) display monitor has been presumed to mask errors. If results are needed for other purposes than visual inspection, the arithmetic is repeated using the high precision floating point processing of the host computer. Now with the advent of bit slice chip sets there is an opportunity to perform high speed image processing at a low cost provided that adequate computational accuracy can be preserved. The paper will comprise a survey of past work on fixed point and block floating point arithmetic and discuss it relative to image processing algorithms. Architectures for fixed point image array processors will be presented together with discussions of specific applications.

1 INTRODUCTION

Image processing display systems have been available in the marketplace for fifteen years. During this time, there has been a consistent approach to the basic architecture of the systems, with an evolution that has seen functionality and performance steadily increase, and with variations on the basic theme being invented by the various manufacturers. These variations have tended to suit particular machines for specific applications, and also have resulted in a range of cost/performance ratios. Across the gamut of available machines, the range of available functions has been very consistent. However, the variations have meant that a given function will execute in a very different time from machine to machine.

The functionality and performance of these machines has been, and continues to be, constrained by the base technology that is fundamental to the designs. Two thresholds can be defined: "possibility" at which a given function becomes implementable at all, and "cost effectiveness" at which an implementation of a function can be accomplished at a cost that can be justified for some application. These thresholds are bound up in performance levels, and it is useful to define four classes of response time, to quantify these definitions.

Real-time is defined as that which seems instantaneous to the human. Most particularly, if an operator is manipulating a hand control to vary a parameter of the function, then the image should be varying smoothly and with no hysteresis in response. Real-time is a major goal since the human hand-eye feedback process is very effective for rapid optimization of a process. Near real-time is defined as that response which is perceivable, but rapid enough that the operator does not get fatigued or frustrated while interacting with the system. Non real-time is defined as that response which is inadequate to support sustained interactive operation of the system. An operator may sit through occasional submissions of such functions, but background (batch) processing is the correct environment. Beyond that is the region of impractical response times. Depending on the function, real-time is around 1/15th-1/30th seconds. Human factors testing has established near real-time as better than about 0.2 seconds, for there to be no degradation in operator performance, but a time of about 3 seconds is a good figure for the onset of frustration. Again depending on the application, impracticality starts after some hours or maybe tens of hours.

Within a class, base technology constraints have restricted the sophistication of algorithms that can be implemented, and hence the utility of applying image processing to given problems. Now however, advances in the base technology promise quantum jumps in the thresholds. Moreover, these advances are causing new system architectures to be proposed, and a radical advance seems imminent.

Critical to progress is the question of digital arithmetic, and the computational precision that can be maintained. Until recently, even the seemingly simple operation of multiplication had to be accomplished by look-up table techniques with consequent sacrifice of accuracy. Now, fixed point arithmetic is available, and the open question concerns the thresholds of technology in floating point arithmetic. Alternatively, as a compromise between the two types of arithmetic, the system designer has the option to use block floating point arithmetic, but the cost effectiveness issues are complex.

The ensuing discussions characterize the architecture of current processing displays, explain the constraints on functionality/performance by linking the base technology to system design, enumerate the key technology advances and forecast some of the new capabilities. The discussion revolves around digital arithmetic issues, and concludes with the example of design for a new geometric warper for the I^2S Model 75 processing display.

2 ARCHITECTURE OF IMAGE PROCESSING DISPLAYS

Current displays are characterized by having "channels" of refresh memory, and processing data at a frame synchronous rate tied immutably to refreshing a CRT. (A selection of papers describing the architectures of processing displays and also other image processors is contained in references [1] — [6]). Arithmetic operations are pipelined, and may reside in the path between the refresh memory and the monitor. A crucial capability is a feedback path from the end of the pipeline back into refresh memory. This path may also contain a pipelined sequence of arithmetic operators. Either or both of the paths may in fact

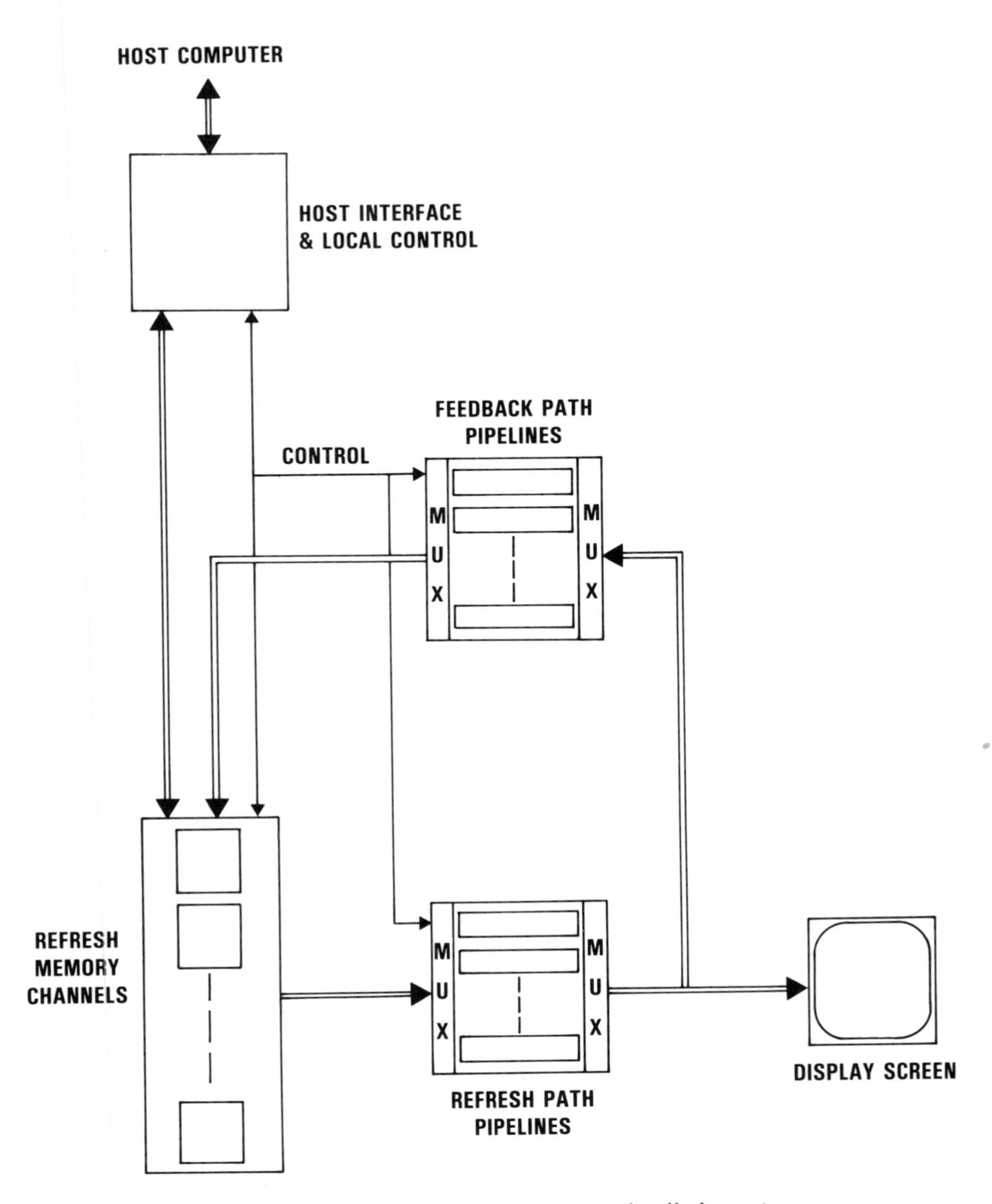

Figure 1 Conceptual design of the generic image processing display system.

comprise a set of (identical) parallel paths, with an interconnect network between them. Figure 1. Each pixel progresses down the pipes at the rate of about 100 ns per operation, this being the refresh time per pixel on the screen. (512 × 512 pixel, 30 Hz display). In one frame time (1/30th seconds) every pixel in the image has been processed (and /or displayed), leading to the interesting conceptual observation, that the system appears at this macro level

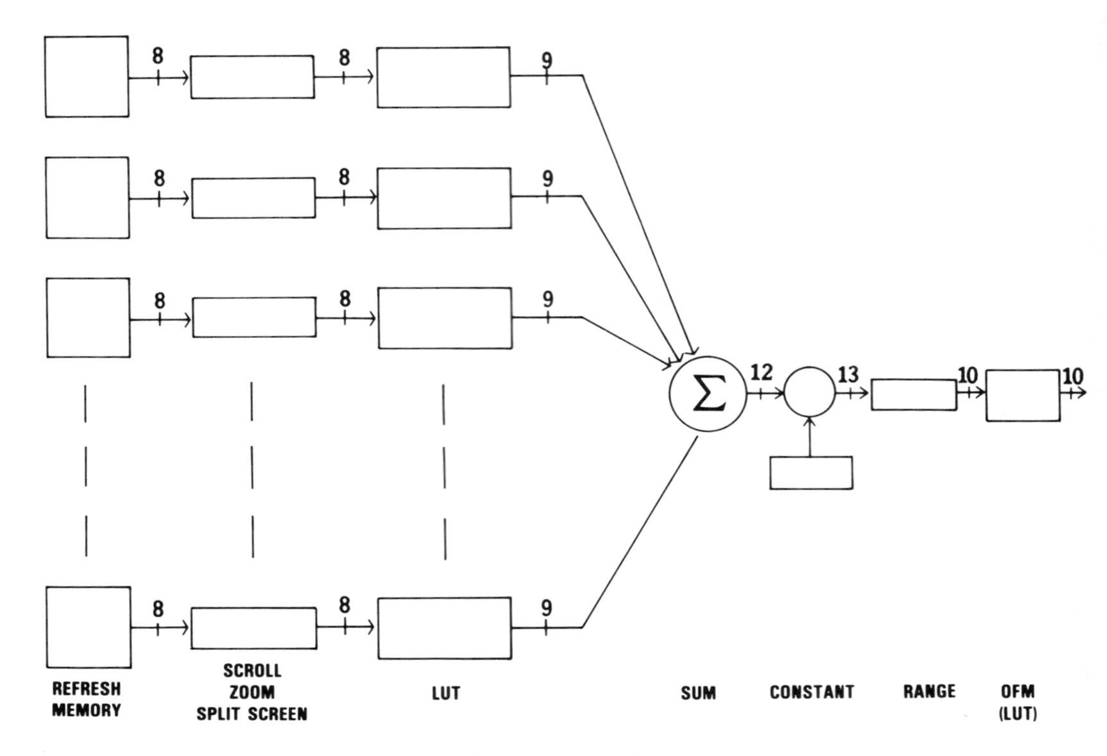

Figure 2 Refresh path parallel processor.

to be a parallel processor as wide as the number of pixels in the image, and as deep as the number of pipelined operations. The advantage of this architecture is simplicity in addressing and control. This yields benefits of ease in logic design, testing, maintenance, and software programming, all of which contribute to low cost. Notice however, that each logic element of the pipe must be capable of operation at the 100 ns rate.

An architecture for a refresh path parallel processor is shown in Figure 2. The pipelined sequence of operators is clear, and a summing tree allows the accumulation of results. In conjunction with a feedback path, iteration may be used if the number of sequential operations is longer than that actually implemented, or if the parallelism is wider. Note that the width of each path is limited, (cost constraint) but the parallelism is high (8 or 12 paths). The critical capability afforded by this architecture is to implement spatial operators, by the device of loading multiple copies of the input image into the refresh channels, and scrolling each of them separately to provide the spatial offset for each point of the kernel. This resolves the memory bandwidth problem at the expense of replicating the memory and provides the simplicity of addressing and control that yields a "well behaved system". Given that the memory may well have been installed for other purposes, this can be a very cost effective approach.

An architecture for a feedback loop processor is shown in Figure 3. This has less parallel width, but the paths themselves are wider. Therefore, arithmetic with higher precision can be performed, but the degree of iteration is int-

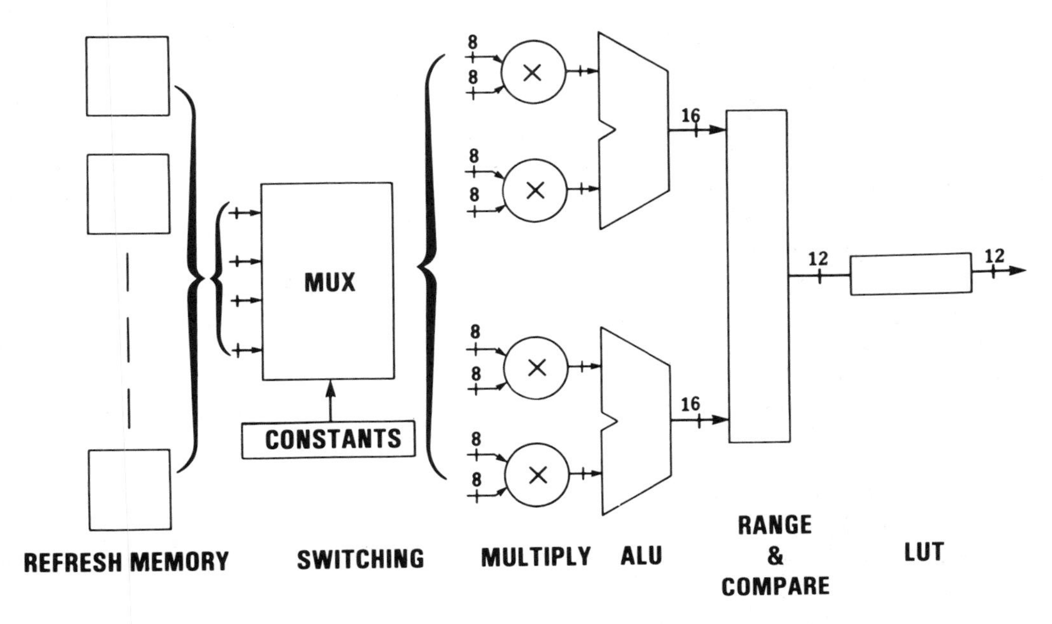

Figure 3 Feedback loop processor.

rinsically higher. This is disadvantageous in response time, but the processing is going on in the "background" thus the refresh path may be used concurrently. Note that the precision of the feedback path allows implementation of algorithms with many pipelined operations, notably recursive operators, without computational noise becoming unacceptable. Both of the architectures are effective for point operations and for spatial operations having limited width and depth. For more sophisticated spatial operators, the refresh path processor is better suited to real-time or near real-time applications whose output is display, since the CRT will mask digital noise effects, while the feedback path processor is better suited to near real-time and non real-time recursive operations.

These architectures are compatible with the particular characteristics of imagery data, and with an important subset of image processing algorithms. Imagery data is a uniform array of points each having a value that is integer, bounded and positive. Many algorithms are linear, or effectively so, meaning that intermediate and final ranges of computational precision and dynamic range are deterministic. Most importantly, many useful algorithms execute in a deterministic (i.e., non-data dependent) number of steps. These algorithms are well matched to synchronous, pipelined operation, including real-time implementation. Furthermore, the algorithms are highly structured, with symmetry between the coordinate axes, meaning that addressing and I/O between processor and memory are amenable to efficient, specialized implementation that exploits the structure.

The difficulty with increasingly sophisticated algorithms and implementation within a given class of response time, concerns the viability and cost of the intrinsic parallelism. This may immediately be broken into two components — the spatial "width" of the algorithm being the number of input pixels that contribute to an output pixel, and the pipelined "depth" being the number of concatenated operators required to produce an output point. This is a kernel function to be applied to every pixel, but in most cases it is an "outer" kernel in the sense that it may be decomposed into a small set of elemental "inner" kernels. In a simple example — spatial convolution with a 3 × 3 kernel, the width is nine and the depth is mathematically two, although in implementation, use of a summing tree would result in a depth of four. The inner kernel operator is multiply-add. From a base technology standpoint, there is a question of memory bandwidth to supply the parallel data, and of logic device functionality to perform any of the pipelined mathematical operations. From a system design standpoint, there are more obscure questions relating to basic timing along the pipeline, feasibility of wiring many parallel data lines, synchronizing those lines, dissipating heat, and so on.

The implicit assumption of the preceding discussion is that the architecture is exactly as wide and as deep as is necessary to perform the process at the desired rate. In actuality, variations from this ideal are practiced. In many cases, it is not cost effective or even practical to implement an algorithm in real-time. Then, an inner kernel of the process is implemented and the iterative capability of the feedback architecture is used to complete the algorithm, with consequent penalty in performance. At the other extreme, urgent requirements and availability of funds, may mandate preformance that is faster than the intrinsic design can accomplish. In this case, the logic may be replicated, with each identical section working on a different pixel of the output. The classic example of the former is iteration of spatial convolution through a single multiplier and adder, to accomplish say a 3 × 3 operation in nine frame times. A typical example of the latter is 1k × 1k pixel processing and display, in which four pipelines work on every fourth pixel to achieve an effective 25 ns per pixel rate. The key is availability of 100 ns logic, and there has been a gravitation of effort in the semiconductor industry toward this speed because it marks a significant threshold for video rate processing. (Presumably we will see a future striving toward 25 ns logic as 1k × 1k display becomes prevalent).

In the last few years, an alternate architecture has become viable due to the development of bit slice devices. Originally comprising programmable microsequencers and multi-function ALUs (arithmetic logic units) the addition of compatible multipliers and multiplier/accumulators set the direction for families of "building blocks". These devices operate at a uniform 100 ns rate, and are designed such that they can be integrated as a synchronous pipeline processor. (Principally this means that an input value may be latched on the same cycle that the previous value is being computed and made available to the succeeding device). In the last year, this approach has been considerably strengthened by the addition of devices that perform supporting roles within a complete system. ALUs with integral register files and multiple i/o ports, register file chips and address sequencers, provide the flexibility on a few chips to implement complex algorithms, with pipeline delays and attendant overlapped computations and i/o. The principal intent of these devices is to perform (one dimensional) signal processing. When applied to image processing, they do not contain adequate internal memory to support the second dimension, and moreover are too expensive, too bulky and consume too much power/produce too much heat for real-time processing. However, if the image processing algorithms are decomposed into a sequence of one dimensional (serial) operators, and the inner kernel of the process is implemented in bit slice logic as a pipeline, the processing rate of 100 ns per pipeline point is sufficiently fast to achieve a near real-time response. Moreover, the small number of chips used, results in a reasonably inexpensive implementation. The result is an optimum cost/performance ratio as compared with either a slow/cheap software implementation or a fast/expensive discrete hardware implementation. An additional degree of freedom for the designer is the length of the pipe i.e., the amount of implemented bit slice logic. As the example of the I^2S warper will show later, this enables the precise balance to be chosen between computation and i/o that yields full utilization of the hardware at maximum throughput rate.

At the present state of the art and cost of bit slice logic therefore, a single pipeline is installed, using iteration to cover the width of the algorithm, and perhaps also for the depth, if the pipeline length is insufficient. This pipe may have buffering at the input and output in order that computation and i/o are overlapped, and in this sense the architecture appears to be one of load-execute-store. However, the pipelining (and if necessary, separation of input and output data) guarantee a continuous flow.

3 VIDEO RATE COMPUTATION

Arithmetic can be performed by a number of techniques, choice among which is a question of tradeoff between cost, response time and precision. A selection of papers on computer arithmetic is contained in references [7] — [16]. Logic has been available for some years to perform the operations of add/subtract and Boolean algebra. Multiplication however, has been tackled a number of ways in image system designs. It is possible to implement "long multiplication" — shift and add in binary arithmetic. Obviously this is not timely, but is inexpensive and accurate. This capability usually comes "free" in the hardware that is implemented to do simple addition. If it is required to multiply an image with a constant (but not another image) this may be performed in look up tables (LUTs). This is cheap and quick, but accuracy is limited by the number of locations in the LUT. An alternate LUT approach, that also works for multiplying and ratioing two images is the use of concatenated LUTs and adders. Referring to Figure 2, if a log function is loaded in the LUTs and an antilog function is loaded in the OFM, the operation can be accomplished. This is also cheap, fast and limited in accuracy. The algorithm used to perform this in the I^2S system is:

$$\text{LUT}(1) = -255$$

$$\text{LUT (NLEVS)} = 255 \text{ (approx)}$$

$$\mathrm{LUT}(I) = \mathrm{BASE} \times \mathrm{ALOG}\ ([I-1)A]/[\mathrm{NLEVS}-1] + 1/A)$$

$$\mathrm{A} = \mathrm{EXP}\ (255/\mathrm{BASE})$$

$$\mathrm{NLEVS} = 256 (\#\ \text{of locations in the table})$$

$$\mathrm{BASE} = 200$$

$$\mathrm{OFM}\ (\mathrm{ONLEVS}) = \mathrm{MAXOFM}$$

$$\mathrm{OFM}(1) = 0\ (\text{approx})$$

$$\mathrm{OFM}(I) = \mathrm{MAXOFM} \times \mathrm{EXP}[(I - \mathrm{MAXINPT})/\ \mathrm{BASE}]$$

$$\mathrm{ONLEVS} = 1024 (\#\ \text{of locations in the table}).$$

BASE is an empirically derived constant, NLEVS and ONLEVS are determined by the hardware and MAXOFM is determined by the desired range of the output data.

The above three approaches were devised as compromises. They are rapidly being obsoleted by the multiplier and multiplier/accumulator chips. These are fast, accurate and becoming cheap. Over the last few years, there has been a tradeoff between speed and accuracy, however, 100 ns 16 × 16 bit devices are available, and progress is now forking into two directions — higher accuracy at the critical 100 ns rate, and sub-100 ns 16 × 16 bit devices. Moreover, the first floating point devices have appeared, and while very expensive at the moment, they are certain to become cheaper as the technology matures and volume of production increases. This raises the question of whether fixed point arithmetic will be obsolete.

Until now it has been mandatory to use fixed point arithmetic. For many algorithms this is acceptable due to the aforementioned deterministic nature of the computation. In fact, if it can be proven that overflow or underflow of the fixed point word length will not occur, fixed point arithmetic is preferable since no computational noise will be introduced. For typical problems though, the computational word length required to achieve this will be several times longer than the word length of the data itself, which is wasteful. More generally, if the occurrence and degree of overflow/underflow can be bounded, use of interim scaling will result in a controlled introduction of noise. Scaling may be actually coded into the implementation of an iterative (feedback) video rate process, or alternatively, special hardware may monitor the maximum and minimum values of the data on each iteration and activate scaling when necessary. In image processing, if all the data in the system is of the same precision (8-bits — 256 grey levels) and fixed size (512 × 512 pixels) and the algorithm is deterministic (convolution with 8-bit kernel coefficients), then the former can be practiced. The latter can apply if say the size is allowed to be a variable from task to task. The use of intra-process scaling is a type of block floating point arithmetic. Block floating point has been used extensively in dedicated hardware designs for which speed and cost considerations overide the generality requirement that mandates floating point arithmetic for general purpose computers. References [17] — [29] comprise a sample of papers that discuss the implementation of digital filters and references [30] — [40] are a similar selection of papers for the Fourier transform. Three categories of implementation for (automatic) block floating point can be defined: In the first, for every pipelined stage of arithmetic, the data can be scaled. In binary arithmetic this is a simple one-bit shift. A stage might typically comprise a multiply/add. (Which in most neoteric displays means one feedback iteration). This approach guarantees no overflow, but also guarantees that one bit of the implemented word length is not available as useful output. The implementation is of course very simple. At the other extreme, the hardware may iterate without scaling until an overflow is detected. At this point, the algorithm may be backed up a stage — it is necessary to store the inputs to each pass during that stage — the data is scaled and the process continues. It is also necessary to remember how many times the scaling is performed. This approach wastes no bits but is complex to control, and produces output at a variable rate. Between these approaches is a compromise in which the range of the data going into an iteration is examined — typically just the most significant bit (MSB) is examined. If overflow is possible (the MSB of any operand is set) then the data is prescaled. This implementation is simple to control and is synchronous, but on average half a bit of precision will be lost. A decision on this design must also include consideration of software costs. But it should be noted that the precoded approach to scaling is expensive if many algorithms will be coded.

4 TECHNOLOGY LIMITATIONS AND KEY TECHNOLOGY ADVANCES

There are two areas of critical technology — memory density and processing chip functionality. Memory density is important at two levels. Until now, the cost of systems has been dominated by refresh memory, which has been implemented using slow dynamic RAM devices, highly multiplexed to achieve adequate speed. The amount of memory has been limited as a result, and performance has been impacted by the I/O to more voluminous disk in a partitioned task environment. Less obvious has been the limited density of very fast registers and static RAM devices needed during processing for caching. The more subtle effects referred to earlier, can become significant when there are more chips required for a capability than can fit on a single board. There is a cost impact, but in addition it may not be possible to interconnect two boards in the middle of a function, either because of the sheer number of wires or because of signal timing. At the present time, however, it is apparent that some key thresholds are being passed in memory density, and a disproportionate, very welcome improvement in performance is anticipated with the imminent generation of chips.

The functionality of processing chips has been increasing very quickly. There are again practicality and cost benefits from integration of much logic into a physically small device. There is another key advantage in the computational accuracy that is now possible. A subtle benefit is the increased ease in programming the logic, but this is potentially theoretical, since the temptation to the hardware designer is to increase capability, ergo complexity, to a degree that outweights the gain. Of note has been the introduction of the single chip multipliers, and bit-slice processors. There is a multiplicatively improving situation as logic is integrated since the compactness itself increases speed and reduces interconnections. The current limitation is power consumption/heat dissipation. This will be overcome as CMOS technology advances sufficiently to match current bipolar device functionality at the 100 ns rate. At that point, opportunities for building very parallel processors, that are physically small enough to overcome the interconnect problem will increase rapidly. To illustrate this, consider a real-time 7 × 7 spatial convolver. Today this requires 49 multiplier accumulators (MACs). Problems of supplying the amperage on the 5 volt supply and dissipating the heat, make it prudent to implement them as seven boards, each comprising a 7 × 1 convolver (plus probably a summing board). Such a design is very expensive — the cost of a system is highly correlated with the number of boards. Now when it is possible to implement a 7 × 1 convolver on a single chip, the system will collapse to a single board with obvious positive cost impact. The feasibility is of course bound up in the base semiconductor technology (say 2-micron CMOS), but it is important to note that the structure of the algorithms favors the integration since the regularity of the 7 × 1 convolve allows pipelining with only the i/o lines for data in, data out, coefficient loading and a clock.

5 AN EXAMPLE — THE MODEL 75 WARPER

The rigid horizontal-vertical structure of refresh memories (or for that matter, the one dimensional structure of computer memory), has long presented an obstacle to even the seemingly simple warp function of rotation. Even if the i/o is solved with a very fast memory, there remains a problem of computational load since for every output point, it is necessary to evaluate a polynomial for each of the X- and Y- addresses, and then to interpolate the output point from a region of input points. This is illustrated in Figure 4, and fully described in Ref. [41]. It has been typical for warp to be implemented in software on a host computer and to execute in about five minutes for a 512 × 512 image. The goal of the Model 75 Warper was to find a cost/performance tradeoff point that was optimal. The design had to accommodate the existing architecture of the display, meaning that the i/o rate for random access to refresh memory was 800 ns per pixel. If no caching were attempted, then for a four point interpolator, the theoretical performance for a 512 × 512 pixel warp would be one second. (Four input accesses and one output access per pixel). This was considered to be acceptable if the cost were controlled by a single board implementation.

The addressing phase of generating an output point consists of evaluating two polynomials. For first order warp (translation, scale and rotation) these are:

$$x \cdot \Delta x = a_0 + a_1 u + a_2 v$$

$$y \cdot \Delta y = b_0 + b_1 u + b_2 v$$

Where u and v are the output space coordinates, and x and y are the input space coordinates, including a fractional pixel offset, (Δ).

To interpolate the output point, the fractional offsets can be used to look up a weighting coefficient in an interpolation kernel table for each of the neighboring points in input space.

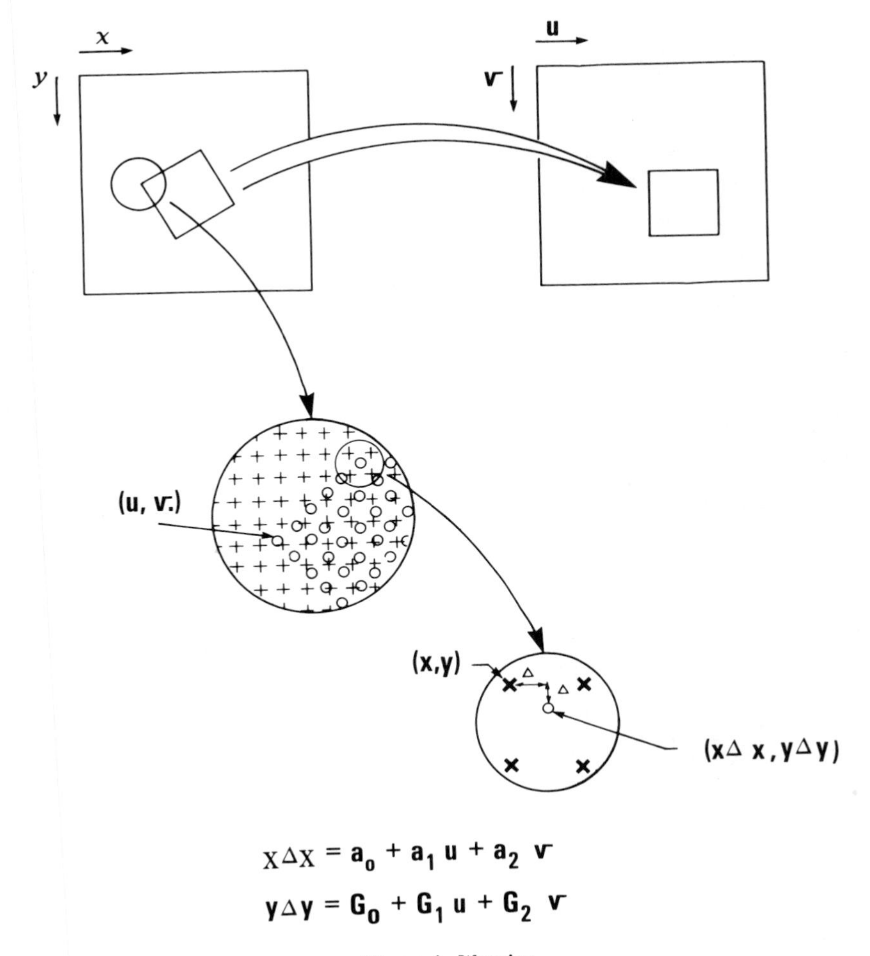

$$x\Delta x = a_0 + a_1 u + a_2 v$$
$$y\Delta y = G_0 + G_1 u + G_2 v$$

Figure 4 Warping.

$$\begin{aligned} f(u,v) = {} & f\ (x,y)\ k(\Delta\ x,\Delta y) \\ & + f(x+1,y)\ k\ (1-\Delta x,\ \Delta y) \\ & + f(x+1,y+1)\ k\ (1-\Delta x, 1-\Delta y) \\ & + f(x,y+1)\ k\ (\Delta x, 1-\Delta y). \end{aligned}$$

These algorithms were implemented in bit slice logic on a single board. Analysis of the computation showed that both could be accomplished in twenty eight cycles (per pixel), while the i/o takes forty cycles. Hence, a single arithmetic unit (ALU-MAC combination) was sufficient and the warper is in fact i/o bound for this operation. (For second order warp with more polynomial terms it becomes somewhat compute bound).

In implementation, it was required that the warper compute with sufficient precision to achieve better than 1/10th pixel accuracy on the output.

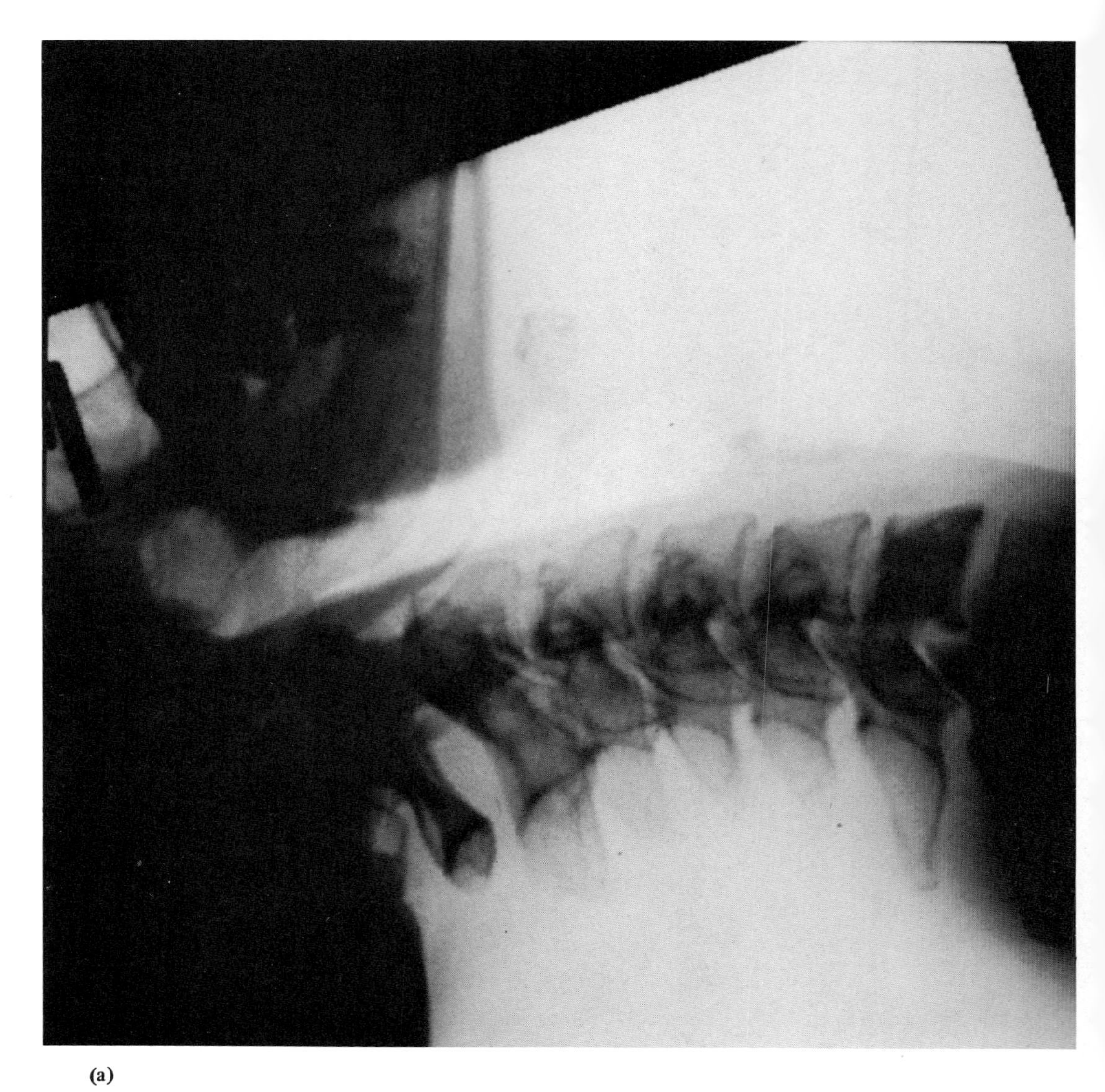

(a)

Figure 5 Effect of arithmetic precision on image rotation. (a) Original image rotated with floating-point arithmetic. (b) The difference image between (a) and image rotated with 8-bit logic. (c) The difference image between (a) and image rotated with 14-bit logic.

Therefore the addresses were computed to achieve four bits of fractional part accuracy. A constraint resulted from the particular design of mutipliers in that being "pin-out" limited, they multiplex the 32 bits resulting from a 16 × 16 multiply thereby taking two cycles. However, if only the top 16 bits are used, there is the option to procede at the full 100 ns per operation pipelined rate. Therefore the binary point was arranged as shown below to utilize full speed:

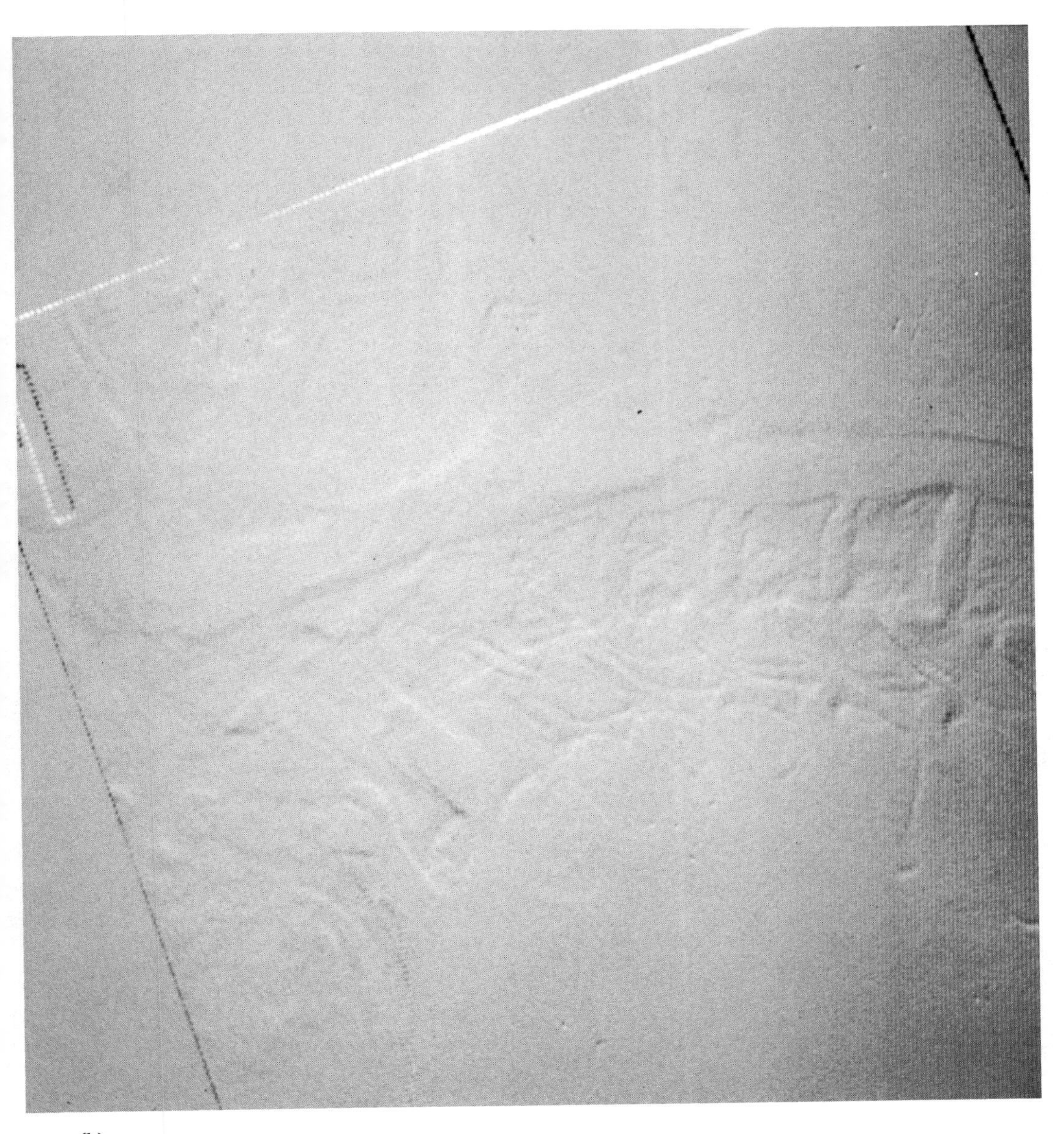

(b)

Figure 5 (Continued)

$$x \cdot \Delta x = 1 \times a_0 + a_1 x u + a_2 x v$$

	$1 \times a_0$	$a_1 x u$	$a_2 x v$	
	2.14 10.6	2.14 10.6	2.14 10.6	bits
	12.20	12.20	12.20	bits
		12.4		bits

(c)

Figure 5 (Continued)

The use of the unitary multiply with a_0 was a convenient, pipelined way to load the accumulator. The use of fourteen bits of accuracy in the logic, enabled a "dynamic range" of lk pixels and a "precision" of 1/16th pixel. The logic was simulated in software at various precisions and compared against a floating point implementation (24-bit mantissa). Figure 5 illustrates the results for 8-bit logic and 14-bit logic. With the exception of an aliasing effect on the image edge, the 14-bit result appears perfect. In fact there is a little statistical noise, however it is limited to errors of three grey levels or less for almost all pixels.

6 SUMMARY

The use of bit slice technology is now allowing complex image processing algorithms to be implemented economically with a near real-time response. Careful design, using fixed point and block floating point arithmetic allows the output images to be of full quality. At present, these designs employ a different architecture to that of video synchronous pipes, however, there is a clear trend to incorporation of bit slice logic into the pipes as the degree of chip-level integration increases and cost declines. We can look forward to cost effective real-time image processing with sophisticated and high quality results.

REFERENCES

Image System Design

1. Advances in Display Technology I, SPIE Vol 199 — Session 1, pp. 1-36 (1979).

2. Picture Data Description and Management, IEEE Computer Society (sponsored by the Technical Committee on Machine Intelligence and Pattern Analysis) 80CH1530-5 (1980).

3. Design of Digital Image Processing Systems, SPIE Vol 301 — Session 1 (1981).

4. Display Technology II, SPIE Vol 271 (1981).

5. Computer Architecture for Pattern Analysis and Image Database Management, IEEE Computer Society 81CH1697-2, Library of Congress 81-82808 (1981).

6. Duff, Levialdi; *Languages and Architectures for Image Processing*, Academic Press, (1981).

Computer Arithmetic

7. Dahlquist, "Convergence and Stability in the Numerical Integration of Ordinary Differential Equations," *Math. Scand.* **4**, pp. 33-53, (1956).

8. Wilkinson, *Rounding Errors in Algebraic Processes*, Prentice-Hall, Inc. Englewood Cliffs, N.J. (1963).

9. *IEEE Trans. on Computers*, Special Issue on Computer Arithmetic, **C-19**, No. 8, (1970).

10. Sikdar, "Determination of Multipliers Mapping an Integer into a Range of a Certain Type," *IEEE Transactions on Computers*, (1970).

11. "Stein, Munroe, "Scaling Machine Arithmetic," *IEEE Transactions on Computers*, (1971).

12. Morris, "Tapered Floating Point — A New Floating Point Representation," *IEEE Transactions on Computers*, (1971).

13. *IEEE Transactions on Computers*, Special Issue on Computer Arithmetic, **C-22**, No. 6, (1973).

14. Linnainmaa, "Taylor Expansion of the Accumulated Rounding Error," **BIT,16**, pp. 146-160, (1976).

15. *IEEE Trans. on Computers*, Special Issue on Computer Arithmetic, **C-26**, No. 7, (1977).

16. Bareiss, Barlow, "Roundoff Error Distribution in Fixed Point Multiplication," **BIT,20**, pp. 247-250, (1980).

Digital Filter Design

17. Oppenheim, "Realization of Digital Filters Using Block Floating Point Arithmetic," *IEEE Transactions on Audio and Electroacoustics*, **AU-18**, No. 2, (1969).

18. Jackson, "On the Interaction of Roundoff Noise and Dynamic Range in Digital Filters," *Bell System Tech. J.*, **49**, 159-184, (1970).

19. Jackson, "Roundoff Noise Analysis for Fixed Point Digital Filters Realized in Cascade or Parallel Form," *IEEE Trans. Audio Electroacoustic*, **AU-18**, pp. 107-122, (1970).

20. Liu, "Effect of Finite Word Length on the Accuracy of Digital Filters — A Review," *IEEE Transaction Circuit Theory*, **CT-18**, pp. 670-677, (1971).

21. Mitra, Hirano, Sakaguchi, "A Simple Method of Computing the Input Quantization and Multiplication Errors in a Digital Filter," *IEEE Transactions on Acoustics, Speech and Signal Processing*, **ASSP-22**, (1974).

22. Parker, Girard, "Correlated Noise Due to Roundoff in Fixed Point Digital Filters," *IEEE Transactions on Circuits and Systems*, 23, No. 4 (1976).

23. Maria, Fahmy, "Bounds for the Amplitude of Quantization Error in Forced First-Order 2-D Digital Filters," IEEE Proc. ISCAS/76 International Symposium on Circuits and Systems, (1976).

24. Heath, Nagle, Shiva, "Realization of Digital Filters Using Input Scaled Floating Point Arithmetic," *IEEE Transactions on Acoustics, Speech and Signal Processing*, **ASSP-27**, No. 5, (1979).

25. Chang, "Comparison of Roundoff Noise Variances in Several Low Roundoff Noise Digital Filter Structures," *Proceedings of the IEEE*, **68**, No. 1, (1980).

26. Dugre, Beex, Scharf, "Generating Covariance Sequences and the Calculation of Quantization and Rounding Error Variances in Digital Filters," *IEEE Transactions on Acoustics, Speech and Signal Processing*, **ASSP-28**, No. 1, (1980).

27. Arjmand, Roberts, "On Comparing Hardware Implementations of Fixed Point Digital Filters," *IEEE Circuits and Systems Magazine*, **3**, No. 2 (1981).

28. Haften, Chirlian, "An Analysis of Errors in Wave Digital Filters," *IEEE Transactions on Circuits and Systems*, **CAS-28**, No. 2, (1981).

29. Leuder, "Digital Signal Processing with Improved Accuracy," *European Conference on Circuit Theory and Design*, Delft University Press, (1981).

30. Welch, "A Fixed Point Fast Fourier Transform Error Analysis," *IEEE Transactions on Audio and Electroacoustics*, **AU-17**, No. 2, (1969).

31. James, "Quantization Errors in the Fast Fourier Transform," *IEEE Transactions on Acoustics, Speech and Signal Processing*, **ASSP-23**, No. 3, (1975).

32. Tran-Thong, Bede Liu, "Fixed Point Fast Fourier Transform Error Analysis," *IEEE Transactions on Acoustics, Speech and Signal Processing*, **ASSP-24**, No. 6, (1976).

33. Tran-Thong, Bede Liu, "Accumulation of Roundoff Errors in Floating Point FFT," *IEEE Transactions on Circuits and Systems*, **CAS-24**, No. 3, (1977).

34. Sundaramurthy, Reddy, "Some Results in Fixed Point Fast Fourier Transform Error Analysis," *IEEE Transactions on Computers*, (1977).

35. Patterson, McClellan, "Fixed Point Error Analysis of Winograd Fourier Transform Algorithms," *IEEE Transactions on Acoustics, Speech and Signal Processing*, **ASSP-26**, No. 5, (1978).

36. Fladung, Mergler, "High Performance Fast Fourier Transformer," *IEEE Transactions on Industrial Electronics and Control Instrument*, **IECI-25**, No. 4, (1978).

37. Knight, Kaiser, "A Simple Fixed Point Error Bound for the Fast Fourier Transform, *IEEE Transactions on Acoustics, Speech and Signal Processing*, **ASSP-27**, No. 6, (1979).

38. Reddy, Sundaramurthy, "Effect of Correlation Between Truncation Errors on Fixed Point Fast Fourier Transform Error Analysis," *IEEE Transactions on Circuits and Systems,* **CAS-27**, No. 8, (1980).

39. Trivedi, Lever, "Measured Performance of Block Floating Point Hardware FFT Processor for Real Time Speech Transform Coding," *IEEE Proceedings*, 128, Pt. F, No. 1, (1981).

40. Tran-Thong, "Algebraic Formulation of the Fast Fourier Transform," *IEEE Circuits and Systems Magazine*, **13**, No. 2, (1981).

Warping

41. Adams, Patton, Reader, Zamora, *Hardware for Geometric Warping,* Electronic Imaging Morgan-Grampian Publishing Company, (1984).

Part II
STATISTICAL GRAPHICS

Visualizing Two-Dimensional Phenomena in Four-Dimensional Space: A Computer Graphics Approach[1]

Thomas F. Banchoff

Department of Mathematics
Brown University
Providence, RI

ABSTRACT

Interactive Computer Graphics makes it possible for mathematicians to carry out exact visual investigations of geometric phenomena in dimensions four and higher, leading to insights and method which can be useful in analyzing data sets in the form of point clouds which cluster around two-dimensional surfaces. This article discusses techniques for interpreting such phenomena and applications to complex function graphs and dynamical systems.

1 INTRODUCTION

Visualization of point sets in higher dimensional spaces nearly always involves projection of these sets into planes or to 3-dimensional subspaces. The investigation of the resulting images gives us information about the original configuration only insofar as we are able to interpret the reduced data by comparison with familiar and well-understood behavior in our ordinary visual experience. Since ambiguities in the form of optical illusions already provide multiple hypothetical explanations of data sets viewed in 3-space we can expect that all these difficulties and more will arise when we attempt to use visual means to make direct investigations of sets of points in spaces of dimension four or higher.

[1]This research was supported by ONR contract N00014-83-K-0148.

For the past fifteen years, mathematicians and computer scientists at Brown University have been collaborating in research projects centering on the study of the geometry and topology of surfaces in 4-dimensional space. High-speed interactive computer graphics is providing entirely new ways of approaching this subject, and a number of classical and modern topics have been illuminated in the process. (See [3], [6], [7], and [8] for articles of the author which treat these results.) Since many of the problems to be expected for large-dimensional data sets are already present in the case of four dimensions, this ongoing project provides a good perspective on some of the ideas used in exploratory data analysis.

In this primarily expository paper, we indicate some of the phenomena which have occurred in our work that pertain to more general visualization projects. Many of the treatments of scatterplots in the plane concentrate on looking for linear correlations, approximating a configuration as closely as possible by a line. The higher-dimensional version of this study involves moving point clouds around in a 3-dimensional subspace determined by three of the variables, or by three linear combinations of a number of the variables. A situation where the projections of the data points cluster around a line on the 2-dimensional screen indicates a linear correlation in the original variables.

More subtle analysis is called for when the projections of the data points cluster around more complicated curves in the plane, such as parabolas or higher order polynomials in a Taylor approximation. The corresponding phenomena in 3-space are less well understood, and when we consider geometric loci in spaces of four or higher, experience in visualization is even more sparse. Over the years we have accumulated a certain amount of experience in dealing with families of 2-dimensional surfaces in 4-space which recur frequently in problems in geometry, topology, and more recently in dynamical systems. In these cases, familiarity with basic "test" configurations makes it possible to recognize similar patterns when the data are given only approximately. The first instance that we will examine in detail is a surface given as a complex function graph, the complex squaring function. The second involves flat tori embedded in a 3-dimensional sphere in 4-space.

In each case it is necessary to go beyond the straightforward device of orthogonal projection to 3-dimensional subspaces followed by rotation about an axis in such a space. The most effective means of studying these objects involve motion clues, incorporated either into real-time vector graphics displays or in films produced through computer animation. For a 2-dimensional exposition confined to the pages of a book, we resort to other devices to try to indicate the geometric information we obtain from the perception of the moving objects, primarily stereoscopic pairs and rendered diagrams with shading and highlighting. Such additional techniques are absolutely necessary when we deal with surfaces in 4-space since single wire-frame projections into 2-dimensional planes will very frequently be visually highly ambiguous.

All of the examples discussed here come from collaboration of the author and a number of students and faculty at Brown University in the departments of mathematics, applied mathematics, and computer science. Almost all of the early work was done in conjunction with Charles Strauss. Later collaboration has taken place primarily with students in the Computer Science department, many of whom are from the graphics group headed by Andries van Dam, including David Salesin, David Laidlaw, and Steven Feiner. The graphics research on the dynamical systems project is primarily due to Huseyin Kocak and Fred Bisshopp in the Division of Applied Mathematics, along with David Laidlaw and David Margolis. Particular responsibility for the computer programs used in the photographic process is due to David Margolis.

1 AMBIGUOUS VIEWS OF THE COMPLEX PARABOLA

In order to graph a complex-valued function of a complex variable, it is necessary to be able to display four real variables, two for the elements of the domain

$$z = x + iy$$

and two for the range

$$w = u + iv.$$

In the case of the complex squaring function,

$$w = z^2$$

we obtain

$$u = x^2 - y^2, \ v = 2xy.$$

Somewhat more convenient for computer graphics investigation is the representation of this function in polar coordinates, given by

$$x = r\cos(t), \ y = r\sin(t)$$

$$u = r^2\cos(2t), \ v = r^2\sin(2t)$$

where r goes from 0.1 to 1 and t goes from 0 to 2π.

We can then project into the 3-dimensional subspace of the first three variables to obtain the graph in 3-dimensional space of the real part of the complex function. Subsequent rotations about axes in 3-space bring this graph into an easily recognizable position (figure [1]). But it is precisely this sort of image in transparent or "wire-frame" mode which is most confusing in practice since it is visually identical with the image of a totally different view of this surface, namely the projection of the real part of its inverse function, the complex square root.

It is important to note here that the computer graphics approach to complex functions has the advantage of making it possible right from the beginning to treat any function and its inverse functions at the same time. Just as it is possible to transform the graph of a function of one variable into the graph of its inverse function, or the graph of its inverse relations if the function is not invertible, by the simple process of "picking up the graph and rotating it about the diagonal", we can use the graphics computer to manipulate the graph of a complex function in 4-space in an analogous manner. (Compare the discussion in [2].)

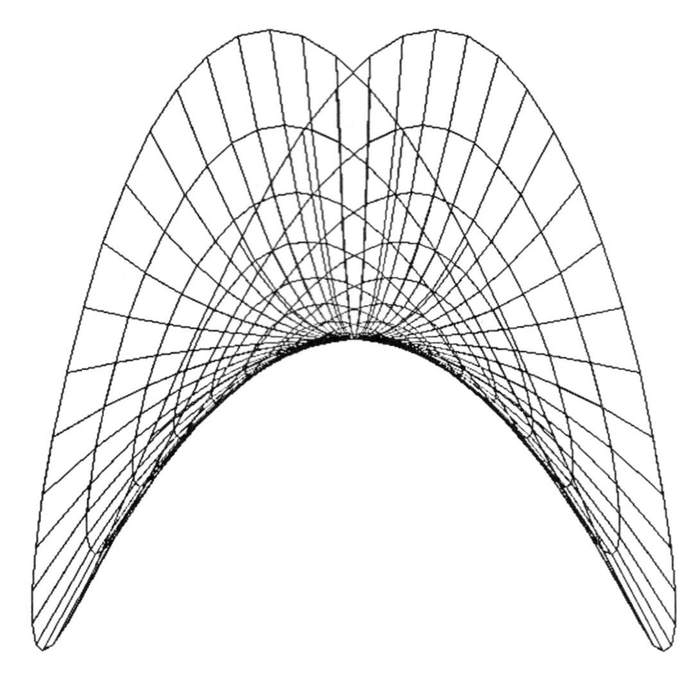

Figure 1

Projecting into the 3-space of the first, third and fourth variables gives us the graph of the real part of

$$z = \pm \sqrt{w},$$

giving the Riemann surface of the inverse function over this portion of the domain. The same pair of rotations in 3-space which positioned the real part of the squaring function yields a wire-frame image which is identical with the picture given above. It is precisely to avoid such ambiguities that we utilize additional graphic techniques such as motion clues, stereoscopic viewing, rendering techniques such as shading and highlighting, and color.

2 MOTION CLUES

For the study of wire-frame models, the use of real-time interactive motion is certainly the most effective manner of developing intuitions based on a kinesthetic appreciation of the geometry of the object. The experience of manipulating the dials of a machine which provides real-time response for rotations in 3- and 4-dimensional space to reposition wire-frame models is the most dramatic way to interact with such objects. It is clear that those who have this experience gain in their ability to interpret a situation in a short amount of time. The more time spent in such activity, the more quickly and surely someone can "untangle" a position on the screen and return to the start, even when the position presented involves rotations in 4-space. (The real-time vector graphics rotations in 3- and 4-space were an essential part of our project during the years 1970-1980 while we had the use of a hardware matrix multiplier and parallel processor developed by Harold Webber at Brown University. Unfortunately this equipment no longer functions and our real-time investigations are presently confined to 3-dimensional space.) A report of this early work is contained in [2].

3 EXPERIENCE WITH FILM

In order to give larger audiences and classes a direct visual experience of objects in 3- and 4-space, we have used the techniques of computer animation to make films based on our studies of complex function graphs and related loci in 4-dimensional space. Even though the experience is essentially passive for those who merely watch the films with no opportunity to interact with the objects of study, there is a definite sort of learning which takes place. The films in this series all depend on the same basic scenario, known as the "grand tour", and observers of some instances of this scenario appear to be able to interpret similar examples with far less difficulty than is common with first-time viewers. The primary evidence for such passive learning is the decrease in the incidence of perception of instantaneous reversal of direction, a phenomenon extremely common upon initial presentation of a film depicting rotations of objects in orthographic projection. Our experience indicates that after several viewings, this phenomenon becomes much less frequent. It appears that there is a strong predisposition to continuous rotation, so that the mind chooses some particular sense of rotation and continues to choose this as long as possible, even as the image passes through a point of high visual ambiguity. Repeated viewings may set up a sort of "warning system" so that such sources of ambiguity can be avoided. It is also possible that once the individual film or type of film becomes familiar, the observer is able to keep track of not one but a number of independent clues that make it possible to maintain a consistent rotation sense. When one of the key features "flattens out and turns inside out," there are still sufficient benchmarks to keep the sense of the rotation from reversing. After watching a number of films about complex function graphs over the course of a semester, a student generally experiences fewer such reversals, not only in films seen previously but also in new instances of the same grand tour scenario of a graph of a new complex function. Of course some of this increased facility in interpreting the images may be connected with an improvement in the ability to analyze the corresponding equations using techniques from algebra and the theory of complex variables, a combination of active and passive learning. Although there has been some preliminary interest in these phenomena from the cognitive viewpoint, as of yet there has been no systematic study of the psychology of perception in conjunction with our project.

Acquired familiarity with the behavior of complex function graphs under orthographic projections and rotations serves a useful purpose in debugging programs. If we study a new surface and if, upon rotation, it behaves in a way uncharacteristic of any previously studied surfaces, there is a fair chance that there is an error in the program and frequently the nature of the error is suggested by the unusual behavior of the rotating object.

4 STEREOSCOPIC VIEWING

Unfortunately it is impossible to present the total experience of motion clues in a book. We can suggest this experience by presenting a "story board" indicating the successive scenes that make up a film (a technique discussed in [5]), or by showing two nearby images as a stereoscopic pair. If the images represent closeby views of a figure rotating about a vertical axis, then the stereo view is identical in form with the conventional presentation of a stereo pair taken by a camera with slightly displaced lenses. If however the two nearby frames are the result of a rotation in 4-space which does not preserve any line in 4-space, the resulting stereo pair will not correspond to any standard view. (This last situation has been investigated in very deep ways by the late David Brisson in his construction of "hyperstereograms" without the use of computer graphics. Much work remains to be done in this new area of research, especially when Brisson's methods are combined with the techniques of computer-animated stereoscopic film.)

We present two examples to show various ways in which the ambiguous 2-dimensional image can be resolved, as the real part of the squaring function on one hand (figure [2]) and as the real part of the square root function on the other hand (figure [3]). [2]

In this article, we use red-green anaglyph stereograms to produce the effect of bringing the ambiguous image into some sort of resolution. In the laboratory this device is also possible, but even more effective is the technique of viewing stereoscopic images which are seen through polarizing filters by means of glasses with polarized lenses. The great advantage in this case is that we can use this approach in conjunction with the real-time capabilities of the vector graphics machine to see interactive 3-dimensional video output. As it happens, this technique is cumbersome enough that we have used it only in preliminary investigations of complicated objects. Once a sequence of geometrically illustrative views has been selected for presentation, the standard motion clues provide almost as much information without the device for a group of people as one or two observers can gain in front of the apparatus viewing a stereo pair.

It is also possible to put up two images in a split-screen video display and to allow observers to view single images or animated sequences by crossing or walling their eyes to fuse the images. Once this is done for a single image, it is not difficult to maintain fusion during an animation. The difficulties with such an approach include the strain on the viewer and the fact that each image is on half the screen, effectively reducing the area by four and reducing the resolution as well. The great advantage of using either of these non-anaglyph stereo modes is that we can incorporate the use of solid modelling techniques with shading and highlighting, as well as color coding.

---

[2] See Plate 1 for Figures 2 and 3.

It should be noted here that the experience of stereoscopic viewing varies quite a bit from individual to individual. A certain number of readers are functionally monocular, and for others the effect is very slight. On the other hand nearly everyone seems to appreciate the additional information afforded by motion clues.

5 POINT CLOUD ANALYSIS

In addition to the technique of wire-frame images, it is possible to define the shape of a surface in 3- or 4-dimensional space by using point clouds, randomly scattered over the surface. Once again a single image cannot distinguish between the views presented above of the real part of the squaring function and the real part of its inverse function, but some resolution of the ambiguity can be resolved by using red-green anaglyphs, as in figures [4] and [5].[3] The point cloud and the wire-frame images present some of the same geometric features, namely fold curves and apparent cusps. (In the example of the real part of the complex squaring function cusps will not occur since the locus is a quadric surface and any line not lying on the surface will intersect it in at most two points. This distinguishes the quadrics among all non-convex surfaces, since as soon as a higher order term appears in the Taylor approximation there will be lines meeting the surface locally at more than two points and projection perpendicular to such lines produces singularities more complicated than fold curves.)

6 RENDERED IMAGES

Research in solid modeling has provided a number of techniques which have been very useful in our computer graphics investigations of surfaces in 3- and 4-dimensional space, particularly in cases where wire-frame objects are inadequate for the display of certain important features. Specifically, the shading and highlighting procedures, which illuminate an object depending upon the position of surface normal vectors with respect to one or more light sources, are invaluable in the study of self-intersecting surfaces in 3-space, and for dealing with singularities of projections of surfaces in 4-space to 3-dimensional subspaces. We indicate in figures [6] and [7] the effectiveness of this method in distinguishing between the two ambiguous images of the complex function studied earlier. The first picture shows a saddle-shaped surface while the second clearly indicates a self-intersection line. At any point of 3-space which is the image of two separate points of the locus in 4-space, the two tangent planes are distinct so they receive different illuminations from appropriately chosen lights, clearly showing the self-intersection segment. The endpoint of this segment, situated at the origin of the picture, is a singular point of the projection and in any small neighborhood of this point there are double points for which the two tangent planes approach each other arbitrarily closely. This phenomenon presents a great deal of difficulty to routines for determining self-intersections of surfaces, leading to "feathering," where the two nearly identical parts of the

---

[3]See Plate 2 for Figures 4 and 5.

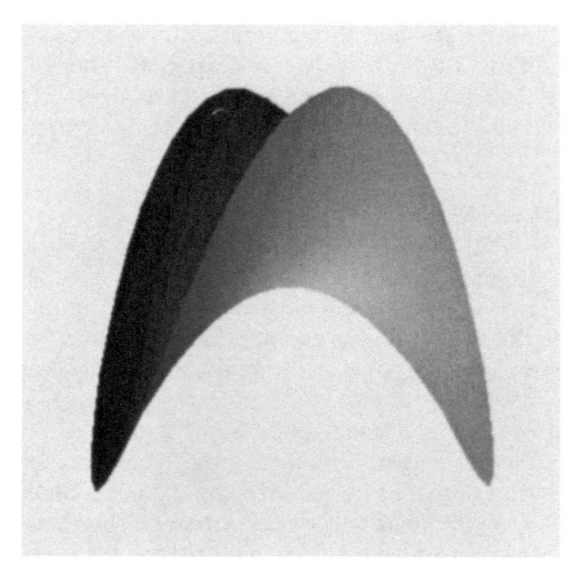

Figure 6

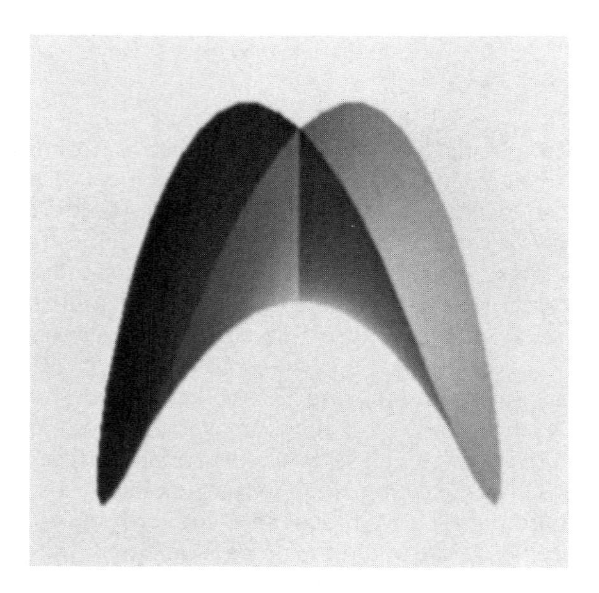

Figure 7

surface are both visible. Thus the self-intersecting surfaces provide a sort of "worst case" test for the algorithms, and the places where the algorithm has difficulty help to locate the singular points in the geometric situation.

Rendering algorithms which show highlighting suggest a number of geometrical projects concerning "specular points" i.e. critical points of the function measuring the angle between the surface normal and a fixed direction. Such points play an important role in the perception of shape of objects in 3-space and they can be expected to play some role as well in future studies of objects in spaces of dimension four and higher.

Recently our particular investigation has been aided by the ability to combine motion clues with rendering. We can now rotate an object in 3-space in real time in vector graphics mode until a particularly interesting position is reached and ship the information on the rotations to another machine which provides the rendered image.

7 COLOR CODING

Although it is usually not possible to present full-color illustrations in journal articles, this technique has great value in investigations in the laboratory. The most significant use of color in our work has been as a "tracer," identifying points on the domain of a function so that we can keep track of positions of the object as it rotates in 4-space.

This technique was used already in the final version of the film "The Hypercube: Projections and Slicing" by the author and Charles Strauss, described in [4]. Earlier versions had used rotating black and white wire-frame objects and these were often perceived in a highly ambiguous way. Due to the high symmetry of such an object, viewers frequently experienced not only the orientation reversal discussed above but also a discontinuous displacement effect where the object appeared to "snap back" and begin a half turn over again and again. When one of the 3-dimensional cubical faces is colored red and the opposite one green, the film is much easier to follow, without the disconcerting apparent discontinuities. We illustrate this with a pair of images, one without color, figure [8], and one with the red and green indicated, figure [9]. (This last image is not meant to be an anaglyph stereogram.)[4]

Much more elaborate effects can be produced when a full spectrum of colors is available. By coloring a disc in the plane so that the angular coordinate determines the position on a color wheel, we have an effective means of following a particular point during a series of rotations in 3- or 4-space. We can see easily at which points on the boundary a particular self-intersection is taking place, especially in the neighborhood of a singular point of a projection from 4-space to 3-space. In addition we may indicate the radial coordinate by changing the saturation or value so that we have even more clues as to the positions of various points in the course of the rotations.

This particular use of color has been especially valuable in our investigation of certain highly symmetric complex functions such as the complex reciprocal

$$w = 1/z$$

given in real coordinates by

---

[4]See Plate 3 for Figures 8 and 9.

$$\left(x, y, \frac{x}{x^2+y^2}, \frac{-y}{x^2+y^2} \right).$$

In this example the graphs of the real part and the imaginary part of the function are congruent and they are congruent as well to the real and imaginary parts of the inverse function. The same scenario that is useful in exploring the complex squaring in wire-frame mode is virtually useless for this example since rotations in 4-space are too easily interpreted wrongly as 3-space transformations. When color coding is introduced, this problem is avoided since it is apparent that the progression of colors around the boundary of the domain (in this case an annulus in the complex plane) is not merely the effect of an ordinary rigid motion in Euclidean 3-space.

8 STEREOGRAPHIC PROJECTION AND HOPF CIRCLES

Up to this time we have discussed examples in which a locus in 4-space is projected orthographically into 3-space and then rotated. This appears to be the most common method of dealing with sets of points in higher dimensional spaces but it is not always the most appropriate for a given set of points. An example which calls for a different way of reducing 4-dimensional information to 3-space is given by Hopf circles and flat tori on the 3-sphere. In studying this topic as part of an ongoing project in the geometry of dynamical systems, we have used all of the techniques of the previous paragraphs in conjunction with the device of stereographic projection in order to present the geometry with as little distortion as possible.

The Hopf mapping is a projection of the 3-sphere in 4-space to the 2-sphere in 3-space. We identify 4-space with the space of two complex variables (z,w) and we define the 3-sphere as the points at distance 1 from the origin:

$$S^3 = \left\{ (z,w) \epsilon C^2 \mid z\bar{z} + w\bar{w} = 1 \right\}.$$

The mapping is then given by

$$h(z,w) = (2z\bar{w},\ w\bar{w} - z\bar{z})$$

in 3-space expressed as

$$R^3 = \left\{ (z,t) \mid z \epsilon C,\ t \epsilon\ R \right\}.$$

The preimage of (0,1) under h is the great circle

$$w\bar{w} = 1$$

while the preimage of (0,-1) is

$$z\bar{z} = 1.$$

These two circles are linked on the 3-sphere in the sense that any disc lying on the 3-sphere which has one of these circles as boundary necessarily meets the other circle, in fact in an odd number of points if the second circle pierces through the disc at each intersection point. This linking can also be detected by the fact that the planar discs bounded by any two non-intersecting great circles intersect only at the origin in 4-space.

It is this last fact which already indicates that orthographic projection is not the best way of examining this linking, since the orthographic projection of any great circle is an ellipse centered at the origin, and no two such ellipses can be linked in 3-space.

An object of great significance in this subject is the preimage of the equator in the 2-sphere. This preimage will be a collection of mutually linked great circles on the 3-sphere whose union is the set of points with

$$w\bar{w} = z\bar{z}, \text{ i.e. } w\bar{w} = \frac{1}{2}, \ z\bar{z} = \frac{1}{2} \ .$$

Thus this preimage is topologically a torus given as a product of two circles in orthogonal planes in 4-space, a "flat torus" on the 3-sphere. If we project this torus perpendicular to a unit vector determined by two complex numbers a and b, then the image of the torus will have singularities whenever the tangent plane to the torus in 4-space contains the unit vector. This will happen at exactly four points, corresponding to the two points on the circle in the z-plane where the tangent line is parallel to the vector a and the two points in the w-plane where the tangent is parallel to b. The singularities are known as "Whitney pinch points" as illustrated in figures [10] and [11]. Notice the way that the extremely ambiguous wire-frame model is clarified by the rendered image. Such singularities are extremely significant in the singularity approach to normal characteristic classes, an ongoing research project in the geometry seminar at Brown University. In this case however we would like to avoid the singularities if possible since the torus is embedded in a symmetric way in the 3-sphere, in particular with no singular points.

The mapping which provides the best way of studying these phenomena is stereographic projection from the 3-sphere minus the North Pole into the equatorial 3-dimensional hyperplane, defined by

$$\pi(x + iy, u + iv) = \left(\frac{1}{1-v}\right)(x,y,z).$$

This mapping is one-to-one and it sends smooth surfaces in the 3-sphere which do not pass through the North Pole to surfaces without singularity in 3-space. One of the most important properties of stereographic projection is that it preserves circles. Linked circles on the 3-sphere not passing through the North Pole are sent to linked circles in 3-space, and if one of two linked circles does pass through the North Pole, its image will be a straight line which is linked by the image of the other circle.

Under this projection the flat torus discussed above is sent to a torus of revolution in 3-space such that the ratio of the radius of the circle of centers to the radius of the vertical circles is $\sqrt{2}$. If we rotate this torus in 4-space, keeping the point of projection fixed, then we obtain a one-parameter family of tori, all of which are covered by two one-parameter families of circles which are pair-

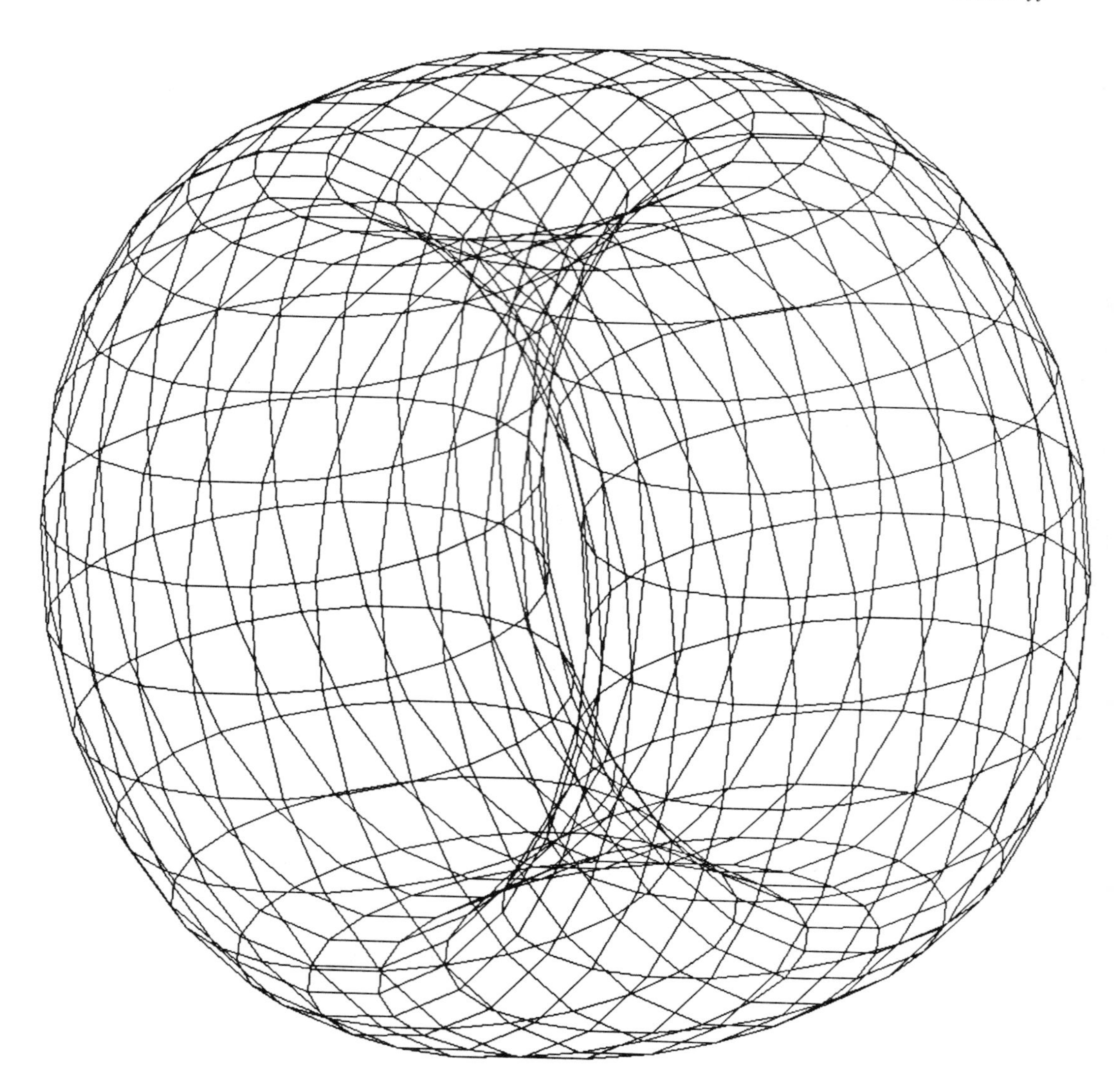

Figure 10

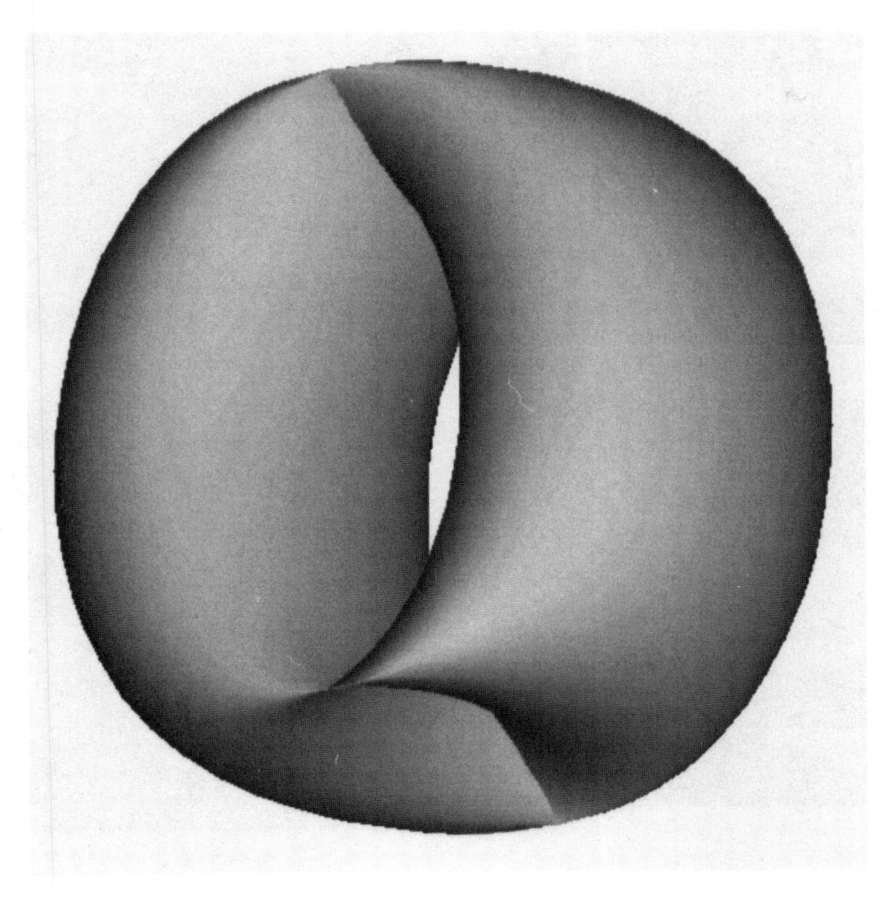

Figure 11

wise orthogonal. These tori have been studied classically under the name "cyclides of Dupin" and they appear in an essential way in modern investigations of minimal total absolute curvature. In fact it was the study of such cyclides which provided the subject for the first computer-animated film produced at Brown University in 1969 (compare [1] and [3]).

If we rotate the flat torus in such a way that it passes through the North Pole, then the images of the circles on the torus through this point will be straight lines and the image of the torus itself will be an unbounded surface which is also known as a cyclide of Dupin. In passing through this position, the image of the torus "turns inside out", illustrating that the 3-sphere can be expressed as a union of two solid tori bounded by the flat torus.

In our research in the topology of dynamical systems, such tori appear as unions of orbits of a integrable Hamiltonian systems, the Hopf circles corresponding to one of the most fundamental examples (compare [9].) In order to prepare for an investigation of more general systems we have carried

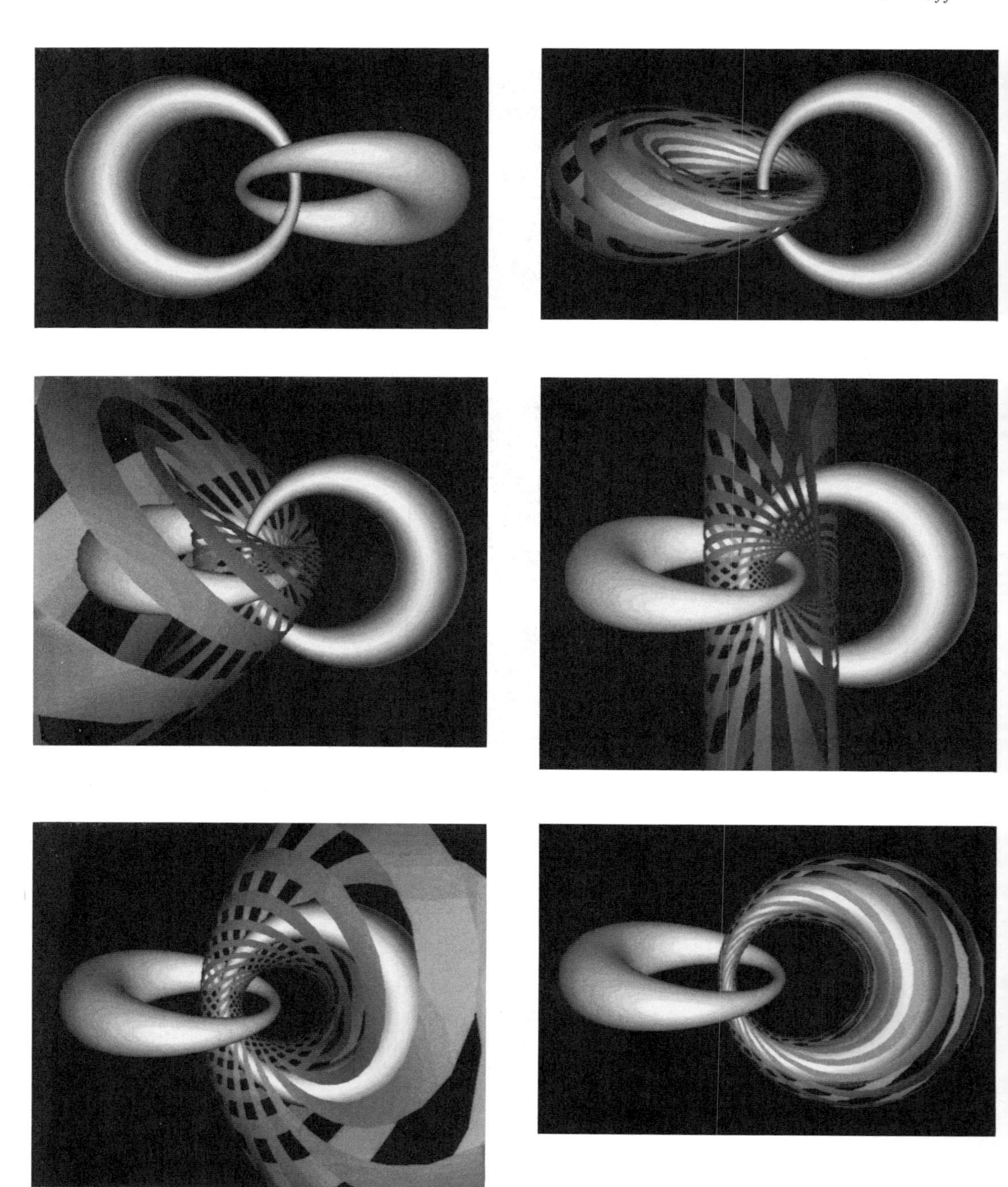

Figure 12

out an extensive study of the Hopf mapping using our computer graphics methods. We conclude this paper by showing a number of views of a scene from our film on the subject. We first rotate the 3-sphere so that the preimages under the Hopf mapping of the Tropics of Cancer and Capricorn are two tori whose images under stereographic projection are symmetrically situated linked cyclides. As we move from one of these circles to the other on the 2-sphere, we obtain a one-parameter family of cyclides, one of which is an unbounded cyclide, shown in Figure 12.

In order to illustrate this phenomenon in film, we used a device suggested by the way that the data points were presented in more complicated examples already studied by Huseyin Kocak and Fred Bisshopp. In plotting such data points it was possible to plot two nearby orbits connected by a "ribbon" so that it was possible to see the topology of the whole collection of orbits more clearly. This device suggested the representation of the intermediate cyclides not by solid objects but by a collection of half of the strips making up these surfaces. The resulting objects are sufficiently "transparent" that we can show the two fixed cyclides and the intermediate ones simultaneously, thus giving a good example of the way in which the abstract analysis of two-dimensional phenomena indicates new ways of visualizing their properties, and conversely, the way that enhanced visualization is playing a larger and larger role in research into the structure of such phenomena in spaces of higher dimension.

BIBLIOGRAPHY

1. Banchoff, T., "The spherical two piece property and tight surfaces in spheres," *J. Differential Geometry*, 4 (1970), 193-205.

2. Banchoff, T. and Strauss, C., "Real Time Computer Graphics Techniques in Geometry," Proceedings of Symposia in Applied Mathematics Vol. 20, The influence of computing on Mathematical Research and Education, Amer. Math. Soc. (1974), 105-111.

3. Banchoff, T., "Computer Animation and the Geometry of Surface in 3- and 4-Space," Proceedings of the International Congress of Mathematicians, Helsinki (1978) (Invited 45 minute address), 1005-1013.

4. Banchoff, T. and Strauss, C., "Real-Time Computer Graphics Analysis of Figures in Four-Space" *American Association of the Advancement of Science Selected Symposium* **24** (1979), Westview Press, Colo. pp. 159-168.

5. Banchoff, T., Feiner, S. and Salesin, D., "DIAL: A Diagrammatic Animation Language" *IEEE Computer Graphics and Applications*, Vol. 2, No. 7 (1982), 42-55.

6. Banchoff, T., "Computer Graphics in Geometric Research," Recent Trends in Mathematics (1982) Teubner-Verlag, Leipzig, 317-327.

7. Banchoff, T., "Differential Geometry and Computer Graphics Perspectives in Mathematics," Anniversary of Oberwolfach (1984), Birkhäuser Verlag, Basel, 43-60.

8. Banchoff, T., "Computer Animated Films in Geometry Teaching and Research," to appear in Proceedings of the International Conference on Film in Mathematics, Torino, 1984.

9. Banchoff, T., Kocak H., Bisshopp, F. and Laidlaw, D., "Topology and Mechanics with Computer Graphics, Linear Hamiltonian Systems in Four Dimensions," *Advances in Applied Mathematics*, June 1986

The Man-Machine-Graphics Interface for Statistical Data Analysis[1]

David L. Donoho,[2] Peter J. Huber, Ernesto Ramos, and Hans-Mathis Thoma

Department of Statistics
Harvard University
Cambridge, MA

ABSTRACT

We describe a system for the interactive analysis of multivariate data, using advanced computer graphics. We identify and discuss some crucial man-machine-graphics interface issues and describe the solutions we have chosen.

1 BACKGROUND

This paper reports on a project to explore the use of advanced computer graphics in statistics. The central objective is to engage the pattern discovery faculties of the human visual system to the fullest possible extent, that is, to utilize our instantaneous perception of structures in 3-d space, time (=motion) and color (hue, saturation, brightness) for the purpose of statistical data analysis.

We have access to a VAX-11/780 with an Evans & Sutherland Multi-Picture System attached to it. The latter is a black-and-white vector display with 3-d rotation in hardware, and with brightness depth cueing. We are currently porting the data analysis system we developed since 1979 on the above hardware to an Apollo DN600 (with a 1024 × 1024 color bitmap display).

Our system should enable the scientist to work with the raw data as far as appropriate and feasible, and so the graphical representations we use are inten-

[1] Prepared with the support of ONR contract N00014-79-C-0512.

[2] *Current affiliation*: Department of Statistics, University of California, Berkeley, California.

tionally simple; most of them are of the scatterplot type, but in three dimensions, and with various enhancements, and they are used in rather sophisticated ways. Also, these are about the only graphs that can be both instantly produced by our hardware and instantly processed by the human visual system.

This is a style of data analysis present-day statisticians are not used to. We may say that descriptive statistics — which flourished in the 19th century and which went into eclipse in the 1940's and 1950's — now is rising from the ashes in a new, rejuvenated form as computer assisted visual data analysis. See Huber (1983a) for further thoughts on this, and for a more detailed description of the steps through which such data analyses tend to progress.

Our system is an offspring of the PRIM family (see Fig. 1). This family is interconnected through some basic concepts, but the software of the different

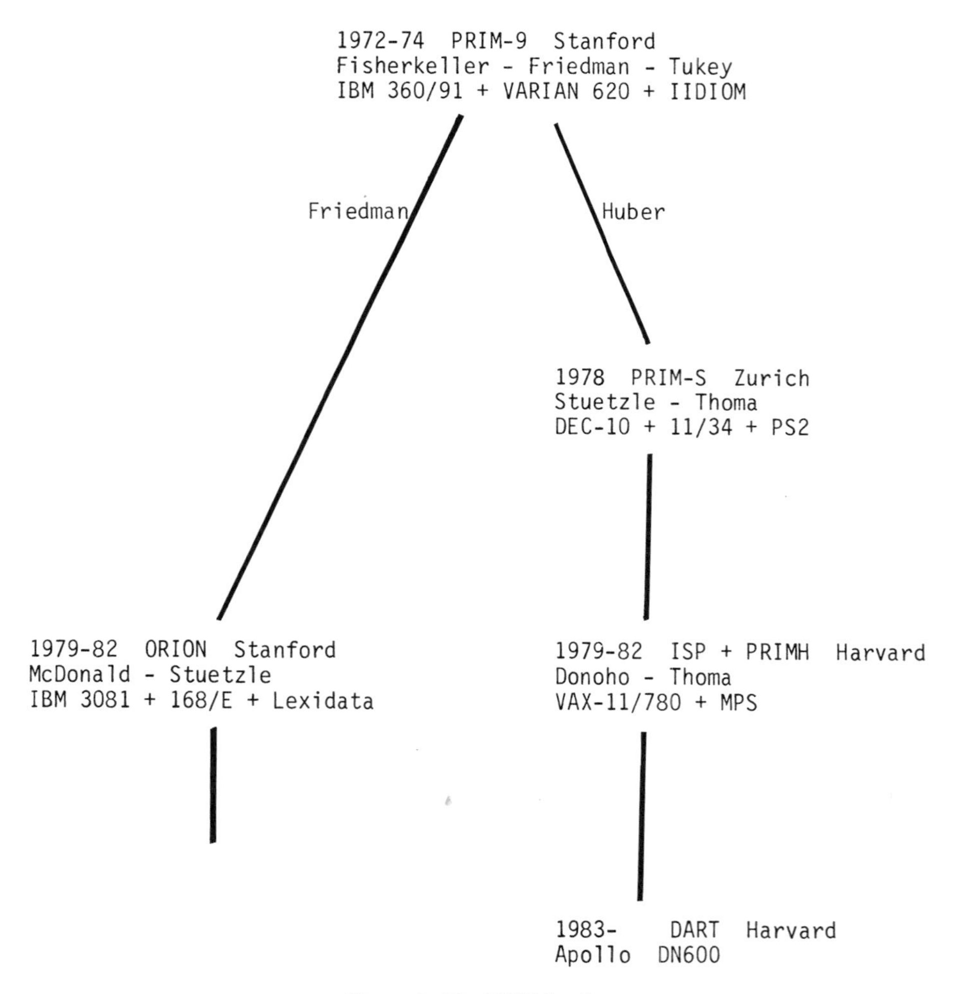

Figure 1 The PRIM family.

systems is disjoint, with the exception of our interactive statical processor (ISP), which runs in practically identical versions on the VAX and on the Apollo. This is so partly because of the dissimilarities of the hardware, partly because our views on what things to do, and how, have evolved.

2 THE FIRST INTERFACE ISSUE: THE NEED FOR A GENERAL PURPOSE INTERACTIVE DATA HANDLER

The earlier systems (up to and including the Zurich PRIM-S and the Stanford ORION) were naked systems for viewing raw data. However, each time we had inspected a real-life data set for 5 to 10 minutes, we invariably would want to modify the representation in some way. This modification might consist in a mere nonlinear transformation to better adapt the representation to the dynamical range of the display and of our eyes. Sometimes, we would want to identify some points, connect them by lines, or highlight some clusters. Perhaps we would want to fit a smooth curve or surface and then inspect the smooth and the rough part (i.e., the residuals) separately, and so on.

Since it is not possible to anticipate the exact nature of these wishes, any graphical data analysis system worth its name must be coupled with a general purpose interactive data handler.

Initially, when we started this project in 1979, we (i.e., PJH and HMT) contemplated to press MINITAB into service as our data handler, despite misgivings about its limitations. Other candidates, like APL, would not interface well, or were using the wrong operating system, or were simply too big. After DLD joined our group, we settled on a version of ISP (Interactive Statistical Processor), since he had been the principal designer of the Princeton ISP.

The next question which then arose was where to put the dividing line between the data handler and the graphics part: which tasks should be accomplished in which system? The crucial issue is that graphics is time-critical: we see more when we interact, but visual feedback must occur within a time span commensurate to our reaction time, if the man-machine feedback loop is to operate properly. To achieve the required speed, we must program the graphics part in a kind of hardwired, inflexible fashion.

The data handler on the other hand should be completely flexible and facilitate improvisation. This means that it should be interpretive, and this involves considerable system overhead.

We discussed for hours and never really resolved the problem, but it ultimately faded into the background when the total system became flexible enough so that essentially arbitrary graphical representations could be prototyped in a slow fashion through simple ISP programs ("macros"). Sections 3 and 5 below contain some more details on this.

In the end, we settled on the system structure diagrammed in Figure 2. Each of the connecting arrows in this diagram poses an interface problem. Most of them are crucial for the success of the integrated system — if one of them is solved badly, the total system will be difficult to use. We shall briefly discuss them in turn in the following sections.

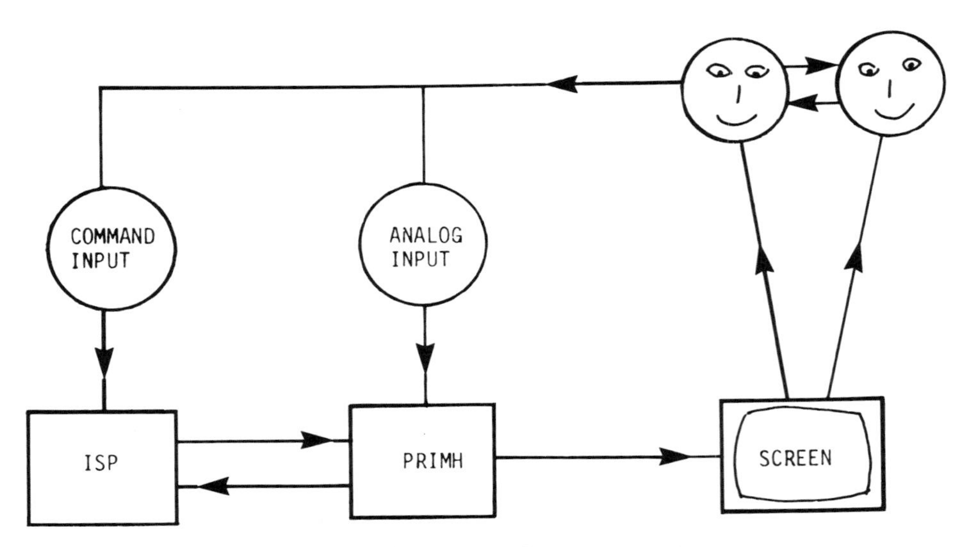

Figure 2 **Structure of the ISP and PRIMH data analysis system.**

3 THE COMMAND INTERFACE

Folklore has it that for any given command language, any given user, unless he is a full time hacker, only uses a very small fraction, maybe 10%, of the available commands. Rather than looking up a more powerful, but rarely used and therefore easily forgotten command, he will improvise around it by combining two or more commands in his ready repetory. Basically, this is a human memory problem. Well-chosen mnemonics help somewhat, but the problem seems to have to do more with the number of items and their hierarchical organization than with their names. We believe the classic paper by G.A. Miller (1956) on "The magical number seven, plus or minus two" bears some relevance also on this subject.

This remark implies that it may be counterproductive to provide too many commands, but that it pays to choose a few basic ones very carefully. The basic commands should enable the average user to do simple things simply, and the more sophisticated user to do everything which can be done by machine (but not necessarily in a very efficient way).

We tend to think in hierarchical structures: whenever a composite thing gets too unwieldy, we give it a name and treat it as a new, indivisible element.

A decent command language should mirror this structure of human thought. In order to solve a problem, the user will decompose it into its conceptual constituents, and each of these parts should ideally correspond to the execution of one command — no more and no less. In particular, looping should be implicit: if an operation is to be applied to all elements of a given structure, the user should not have to write it out in longhand. In other words, we need a full-scale interactive language, and the basic objects upon which this language operates should not be individual numbers, but arrays and more complex structures.

The following example may illustrate the idea; it also shows where our present version of ISP comes close to the ideal and where it falls short.

Suppose we should like to improvise comparative box-plots of a data set (see, e.g., Velleman and Hoaglin 1981 for a concise description of the notion), before a suitable hardwired command is available. At the moment of writing, the only hardwired ISP command for doing static graphics on the Apollo indeed is a simple x-y-scatterplot: the command

gscat/pch =* *x y*

plot the symbol * at the locations given by the column vectors x and y. If the plot type is set to *l*,

gscat/pty = *l x y*

these locations are connected by lines; if x and y are matrices with p colums, then p plots are superimposed. Lines to and from missing elements (symbolized by ?) are omitted.

The number of steps involved in the construction of a boxplot is such that we will afterwards want to preserve the sequence in the form of an ISP program or "macro". Figure 3 shows this macro, and Fig. 4 is a sample output.

The first and last few lines in Fig. 3 serve to organize the command sequence as a macro. For example, the line

getarg "data" > *x*

gets the first argument of the macro; if the user forgets to supply it when invoking the macro, the system asks for it by printing

data>

as a prompt.

The commands to construct the boxplots proper begin with *fivenum*, and on the whole they are pretty close to the ideal — one command line per conceptual element. A major deviation, with some unintuitive steps (e.g., the need to transform certain floating point numbers to strings through use of the function ftos) occurs in the calculation of the data window for the plot.

4 THE ANALOG INTERFACE

We decided on a complete, physical separation of the user interactions with the statistical processor and with the graphics subsystem. The former are through the keyboard, the latter through analog devices. This evades a constant source of confusion and reduces the number of user errors.

In data analysis, we find that the following actions need to be done in direct interaction with the picture:

- changing the viewing aspects;
- picking a point in a scatterplot, or an item in a menu;
- selecting a subset of points;

```
makmac > gboxplot
# makes p vertical boxplots on graphics device
# syntax:
#         gboxplot x title
# where x(n,p) is the data matrix
#
# declare local variables (ought to be automatic!):
   local fn h1 h2 md if1 if2 of1 of2 ad1 ad2
   local foo oo mi ma xma hw boxx boxy
# get the arguments:
   getarg "data" > x
   getarg "title" > title
# calculate extremes, hinges and median:
   fivenum x > fn
# extract hinges and median from fn:
   let h1 = trn(fn(*,3))
   let h2 = trn(fn(*,5))
   let md = trn(fn(*,4))
# calculate inner and outer fences:
   let if1 = h1 - (h2-h1)*1.5
   let if2 = h2 + (h2-h1)*1.5
   let of1 = h1 - (h2-h1)*3
   let of2 = h2 + (h2-h1)*3
# calculate adjacent values:
   reduce/axis=1/op=min (if x >= if1 then x else ?) > ad1
   reduce/axis=1/op=max (if x <= if2 then x else ?) > ad2
# get outside observations:
   let oo=if((x>=of1 & x<ad1)|(x<=of2 & x>ad2)) then x else ?
# get far outside observations:
   let foo=if(x<of1 | x>of2) then x else ?
# scale the plot properly (ought to be simpler!):
  # turn off the automatic scaling:
     set nice=n
  # find extreme values and set vertical plot limits:
     reduce/axis=1/op=min (fn(*,2)) > mi
     reduce/axis=1/op=max (fn(*,6)) > ma
     let ymin=ftos(mi)
     let ymax=ftos(ma)
  # set horizontal plot limits:
     let xma=max(10,sizeof(md)+1)
     let xmin='0'
     let xmax=ftos(xma)
# plot the far outside observations as *:
   gscat/init=y/out=n/pch=* (index(foo,2)) foo title "" ""
# plot the outside observations as o:
   gscat/init=n/out=n/pch=o (index(oo,2)) oo
# set half-width of box:
   let hw=0.2
# assemble the x-coordinates of a standard box+whiskers:
   glue/axis=1 (0)(0)(?)(hw)(-hw)(?)(0)(0)(?) \
               (hw)(hw)(-hw)(-hw)(hw) > boxx
# assemble the y-coordinates of boxes+whiskers:
   glue/axis=1 ad1 h1 md md md md h2 ad2 md \
               h1 h2 h2 h1 h1 > boxy
# plot the boxes+whiskers:
   gscat/init=n/out=y/pty=1 (boxx+index(boxy,2)) boxy
# that's all
   return
```

Figure 3 Macro for making comparative boxplots.

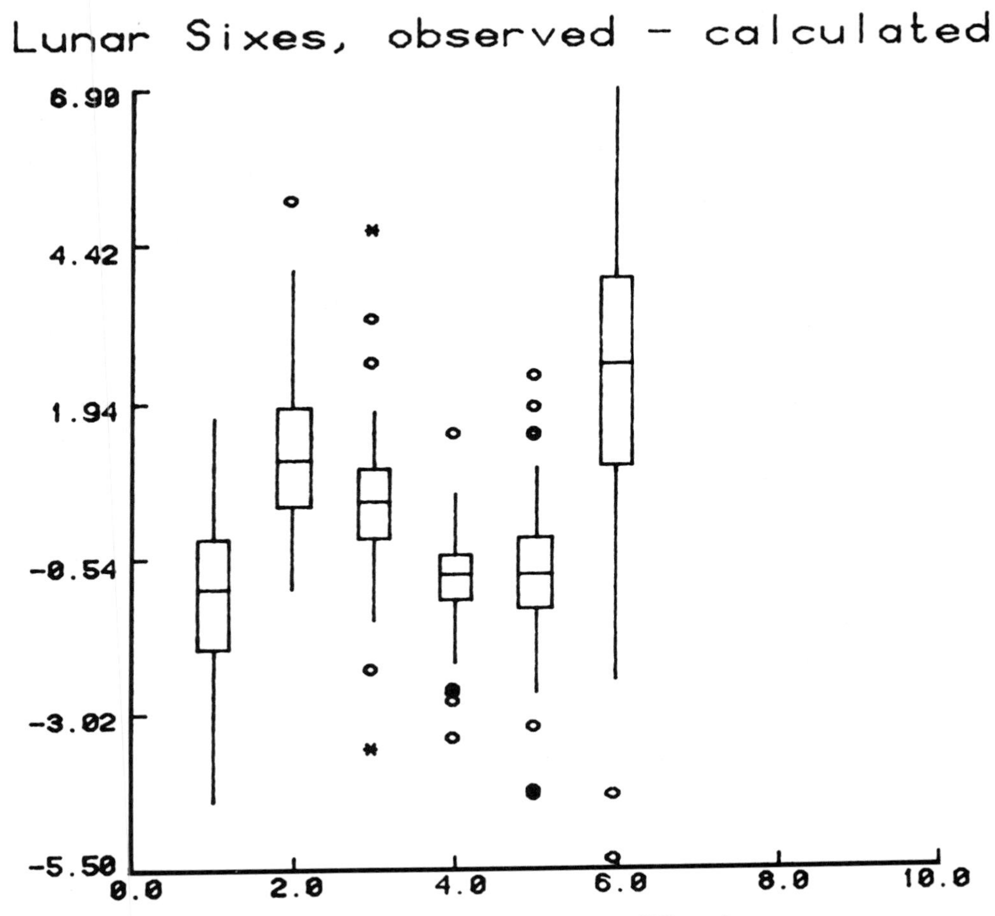

Figure 4 Sample output of the macro of Figure 3.

- changing internal parameters of the scene (e.g. the degree of smoothness);
- digitizing a photograph, or a geographical map.

In order to handle the viewing aspects, we need to change some 7 parameters in a more or less concurrent fashion:

- rotation around the 3 screen axes;
- overall scale;
- *x*-*y*-translations;
- relative scale (squashing or stretching the picture).

While we had perspective in the Zurich PRIM-S, we found little use for it in our kind of applications, and in PRIM-H we dropped this option altogether. Perspective is a strong monocular depth cue for scenes containing right angles and parallel straight lines, but with 3-d scatterplots we found it to be mostly useless, and sometimes even counterproductive.

Probably the best solution for changing the viewing aspects is a battery of at least 7 dials (knobs to be turned). With most other devices you either get trapped in modes (you have to switch modes if the device supports too few parameters, and you always seem to be in a mode different from the one you think you are), or you have to keep an eye on the position of a tracking cross instead of on the data themselves (because this position encodes which parameter you are modifying, plus the sign and speed of the modification). We found that the seemingly primitive touchpad offered by Apollo as the standard analog device (a 2.5" square rubber surface at the side of the keyboard) is surprisingly convenient: stroking with a finger along one of the 4 edges allows you to control 4 parameters (say rotation and scale), and the interior gives 2 more (say x-y-translation). By rolling the finger instead of stroking one can achieve very accurate small adjustments. Unfortunately, lifting the finger may cause a last-moment jump.

Picking can be done reasonably well with either a mouse or a tablet, but only the latter can be used for simple digitizing tasks (we often would want to enhance a scatterplot by a superimposed geographical map or the like). Selecting a subset of the data is most easily done by fencing it in by a polygonal line in a suitable 2-d projection (using a tablet or mouse).

It is surprising that the commercially available interactive devices still are as clumsy as they are — interacting through them is like unwrapping a bar of chocolate with fork and knife.

5 THE ISP-PRIMH INTERFACE

It may be of some interest to sketch the stages in the development of the interface between ISP and PRIMH. To simplify development and debugging, the two systems initially were kept completely separate. The user would massage his data in ISP, then save them in a file, quit ISP, start PRIMH, load the saved data, and then select the data set and the variables to be looked at. This would take at least 10-15 seconds.

After about a year of development we had a working demonstration system (i.e., one where the system designer — not a client — chooses the data and the method of analysis).

Only after a further one-and-a-half years of development did we dare to integrate the two systems more tightly: PRIMH became a command in ISP, and the intermediate saving of the data on disk was skipped. ISP now transmits the name of the data set, and optionally, subset descriptors, connections, row and column labels, and a status vector to PRIMH. Upon exit, PRIMH returns new subsets, new connections and a new status vector. This way, switching between ISP and PRIMH takes only about 1-2 seconds. One can set up the system such that PRIMH is invoked from inside an ISP macro.

Soon afterwards, in early summer 1982, we realized that the various improvements to the two systems had resulted in a prototype of a working data analysis system: for the first time, a sophisticated data analyst could take a scientist along to the workstation, and the two of them could improvise an analysis of the scientist's data on the spot, following the whims and hunches they developed while sitting in front of the screen.

Since then, we have had experiences with the analysis of about half a dozen major data sets. Perhaps the most striking experience was the intensely "conversational" workstyle that was emerging: the scientist and the data analyst would sit side-by-side in front of the screen, argue about the implications of what they saw, and mutually suggest the next actions to be taken.

Why did we suddenly have a working system for visual data analysis? There was no single decisive factor, except that the frequency of frustrating events — where one had to break off a session because one was unable to improvise a necessary or desirable next step of the analysis — had become tolerably low. The following items helped particularly:

- ISP had become an (almost) full-fledged language. At present, its weakest points are, like in ALGOL-60, its lack of character data type and poor *I/O* (points which we hope to improve soon). The fact that it is array based, and that it has macros (with loops and local variables) that look to the user like built-in commands, make it very powerful for interactive use.

- The system is fast: data sets and macros are objects in memory rather than disk files.

- There are various aids to the human memory:

  - Script files, containing an editable and executable record of what the user typed in;

  - monitor files, containing a complete session record;

  - help files;

  - adequate error messages;

  - the system prompts for forgotten arguments of a command or macro.

6 THE HUMAN VISION INTERFACE

The display devices finally begin to match the perceptual capabilities of the human visual system with regard to resolution, speed and color. This accentuates the psychological issues involved in the use of graphics, but their specific relevance still is largely unexplored. See in particular the recent book by Marr

Figure 5 Stereo pair showing a hollow ball.

(1982). Note that 3-dimensional scatterplots pose some unusual problems, so far hardly considered in the psychological literature.

While we easily perceive 3-d solid bodies, by reconstructing them in our mind from the 2-d surfaces we see, the perceptual task becomes much harder if such a body is represented by a random sample of points from its interior. From example, a hollow ball (that is, a sample from the standard normal distribution in 3-space, minus the points contained in the unit sphere) shows a faint annular structure in its 2-dimensional projections. But the underlying cause — the hole in the middle — is recognized only if we look at a stereo pair of the rotating ball. To our surprise, the spatial effect produced by rotation alone, or by stereo pairs alone (see Fig. 5), even with brightness depth cueing added for good measure, would not suffice for most of us to see the hole.

While there is evidence that the human mind — at least to some extent — can work with visual images of lower dimensional structures embedded in more than 3-dimensional spaces (e.g., 2-dimensional manifolds embedded in Euclidean 4-space, cf. Banchoff's contribution to this conference), we seem to encounter unsurmountable difficulties with genuine 4- and higher-dimensional structures.

Glyphs, like Chernoff faces, which sometimes are used to represent high dimensional data, unfortunately require to be looked at sequentially, and the human mind often does not process them faster, or more reliably, than a well-organized table. Moreover, the typical use of glyphs amounts to the representation not of a genuinely high-dimensional structure, but only to that of a 2-dimensional function surface in a high-dimensional space (two independent variables are encoded in the location of the glyph, and the dependent variables — 18 of them in the case of Chernoff faces — are encoded in the glyph itself).

Our conclusion is that in order to make higher dimensional data sets amenable to visual analysis, we need powerful dimension reducing techniques, of which principal component analysis, factor analysis, multidimensional scaling and projection pursuit (cf. Huber 1983c) are just a beginning. See Huber (1983b) for further remarks.

REFERENCES

P.J. Huber (1983a). Data analysis: in search of an identity. To appear in the *Proceedings of the Neyman-Kiefer Symposium*, Berkeley, June 20, 1983.

P.J. Huber (1983b). Statistical graphics: history and overview. *Proceedings of the NCGA Conference*, Chicago, pp. 667-676, June 26-30, 1983.

P.J. Huber (1983c). *Projection Pursuit.* Technical Report, Mathematical Sciences Research Institute, Berkeley, CA. To appear in *Ann. Statist.* (1985).

D. Marr (1982). *Vision.*W.H. Freeman, San Francisco.

G.A. Miller (1956). The magical number seven, plus or minus two. *Psychological Review 69*, 82-96. Reprinted in: G.A. Miller (1967), *The Psychology of Communication.* Basic Books Inc., New York.

P.F. Velleman and D.C. Hoaglin (1981). *Applications, Basics and Computing of Exploratory Data Analysis.* Duxbury Press.

Interactive Color Display Methods for Multivariate Data[1]

D. B. Carr, W. L. Nicholson
R. J. Littlefield, and D. L. Hall[2]

Pacific Northwest Laboratory
Richland, WA

ABSTRACT

Data analysts frequently find that multivariate data are difficult to understand. While graphics provide a set of tools for increased understanding there is no detailed theory for display in more than two dimensions. This chapter describes principles that can guide the selection and evaluation of display techniques. A wide variety of display techniques is discussed involving color, stereo, glyphs, sequential views, scatterplot matrices and rigid motion. These techniques are illustrated through the display of several multivariate data sets.

1 INTRODUCTION

This paper reviews statistical graphics research that is part of the Analysis of Large Data Sets (ALDS) Project, a United States Department of Energy, Applied Mathematical Sciences, sponsored research project at Pacific Northwest Laboratory (PNL). ALDS research is directed toward expediting the data analysis process on large, complex data sets. Thus, part of the ALDS focus is on multivariate statistical graphics. For further background on the ALDS project see Hall (1983).

Most interesting data sets are complex, due usually to the existence of many variables. The initial analysis phase for such data sets typically involves seeking insight into relationships among the variables. One approach, called

[1] Work sponsored by the U.S. Department of Energy, Applied Mathematical Sciences, under Contract DE-AC06-76RL0 1930.
[2] *Current affiliation*: Boeing Computer Services Company, Seattle, Washington.

exploratory data analysis (EDA), relies heavily on visualization as a method of finding structure and patterns. The more dimensions that can be displayed graphically, the better the chances of detecting patterns that lie in the higher dimensions. Thus, there is a continuing effort on the part of data analysts to develop methods of looking at data in higher-dimensional space.

Many techniques have been used to display more than two dimensions. Glyphs allow additional variables to be indicated with plotting symbols. Shading and color have been used to display additional variables. Stereo views and motion have been used to create a 3-D cue. Juxtaposed and sequential views have been used to convey additional dimensions. This paper discusses these and other methods as they have been adapted to the ALDS work.

2 GENERAL PRINCIPLES

The search for effective display techniques is itself a complex multivariate problem. Color, shading, glyphs, motion, stereo, transformation and aggregation can be combined almost without limit. Some combinations work, but many do not. Although there is no detailed theory of good statistical graphics, the literature does provide some general principles that can be used to guide the search for effective displays (Kruskal 1982; Tufte 1983; Chambers et al. 1983; and Cleveland and McGill 1984a).

Good graphs generally minimize the mental/visual processing required to draw conclusions. Methods for accomplishing this minimization involve the following principles:

- perceptual accuracy
- focus
- simple visual reference
- distributional reference
- proximity in space and time
- connectedness
- perceptual flexibility

These principles are described briefly in the following sections, along with guidelines for their application. Some principles are competitive; for example, perceptual flexibility often competes with focus. Thus, to follow one principle, others may have to be sacrificed. In later discussions of display techniques, only occasional reference is made to principles. The reader should keep the principles in mind to assess which techniques are effective for such perceptual tasks as looking for functional forms, changes in spread, outliers, and density variations.

2.1 Perceptual Accuracy

In many cases, interpreting a display requires quantitative comparisons among symbols. A moderate amount is known about the accuracy of elementary perceptual comparison tasks. Cleveland and McGill (1984a) reviewed psychophysics studies and established the following best-to-worst ordering for quantitative comparison tasks:

1. position along a common scale
2. position along nonaligned scales
3. length, direction, angle
4. area
5. volume, curvature
6. shading, color saturation

No preference is indicated for items in the same line. Our experience generally supports this order, except that for some purposes we prefer direction, or orientation, to length for conveying quantitative information.

2.2 Focus

Good graphics encourage focusing on relevant aspects of the data. For example, when the task is looking for functional relationships, adding smooths to plots can be helpful (Cleveland and McGill 1984b). Any graphic element is counterproductive if it does not contribute to understanding the data. In this vein, Tufte (1983) and Wainer (1984) have espoused the data-ink ratio criterion. That is, the ratio of ink used for data to the ink used for the whole display should be maximized. Note, however, that the usefulness of features can vary as the analysis evolves. For example, scales are helpful mainly at the beginning of an analysis, when they can be used to 1) verify that the right data have been plotted and have been plotted correctly, 2) aid in identifying individual points, and 3) draw simple conclusions based on extremes. Later in the analysis, when the main task is seeing interpoint relationships, the scale is not needed (Tukey 1977) and constitutes a distraction. This is particularly true if an interactive facility for identifying points is available.

2.3 Simple Visual Reference

The theme of simple visual references appears frequently in the literature. If the task is to look at residuals, plotting the data and the fitted values is poor; fitted values form a varying visual reference that requires the analyst to mentally obtain differences before making comparisons. Plotting the residuals directly is better because the implicit visual reference is a horizontal line, which is certainly a simple reference. Another example of simplifying the visual reference is supplied by Chambers et al. (1983). In the scatterplot matrix, they change the signs of variables to increase the number of positive correlations shown. Since correlation magnitude and functional shape are of primary interest, the changing of signs allows more plots to be viewed and compared with a single positive slope reference.

2.4 Distributional Reference

References to distributions help in understanding data. Many methods can be used to add distributional information. Tufte (1983) modifies the scatterplot frame to show the univariate medians and quartiles of the two variables represented by point position. When symbols are used to represent additional variables, they can be adapted to convey distributional information. In Section 4.3 a ray glyph is used to represent two variables using orientation for one variable and length for the other. The data are scaled so that a ray with vertical orientation corresponds to the median and a single horizontal orientation corresponds to the most extreme tail. Such scaling does an excellent job of conveying central tendency and direction of asymmetry. To show quartiles corresponding to length, four colors are used. Variations on using implicit visual references, such as horizontal and vertical, and color for adding distributional information have numerous applications.

2.5 Proximity in Space and Time

Juxtaposed plots such as those in Tukey and Tukey (1981) and the small multiples in Tufte (1983) illustrate proximity in space and time. However, the principle, especially the temporal aspect, does not seem to be amply discussed in the statistical graphics literature. Our experience tells us that it is very important. For example, we have had to walk down the hall to look at a time-series plot. It has taken a lot of looking to see patterns because the plot had to be committed to long-term memory. Memory is not always reliable, and space/time barriers inhibit checking and rechecking. Thus we strive to minimize the eye movements required for comparison and use film-loop techniques (Section 4.4) so that overlays and sequences can be viewed closely in time.

The decision as to what should be closest in space and time is important. Kleiner and Hartigan (1981) use a clustering algorithm to determine which variables will be displayed as adjacent elements in trees or castles. Similarly, the variables in correlation matrices and scatterplot matrices can be ordered so that sets of correlated variables appear together. The subject matter often suggests variable ordering or grouping. Such organizational themes should also be tried. At a higher organization level where sequences of alternative plots are to be viewed, following an experimental design can insure systematic coverage of the possibilities and call attention to particular contrasts.

2.6 Connectedness

When visual elements are connected by ink they become easier to lift off the page as a visual unit. The small dots in dot charts (Cleveland and McGill 1984a) provide a visual connection between the label and the data point. The solid line joining the bounds in the confidence interval symbol provides a stronger connection. The vertical lines in the box of box-and-whiskers plots provides connectedness which helps us process the symbol as a single visual unit of information. In profile plots, flooding the region below the profile with ink has much the same effect. Identification of visual units implies the control of visual processing. A remarkable example is to look at a 2-D or 3-D scatterplot and then to connect the points with a minimal spanning tree (Friedman and Rafsky 1979, 1981). The connecting lines seem to remove many eye traversal paths and make the display look much simpler.

2.7 Perceptual Flexibility

The spanning-tree example above simplifies interpretation by reducing the viewer's options. Many techniques simplify some perceptual tasks and complicate others. For example, the spanning tree makes it harder to assess point density. In general, we do not know how to produce plots that are optimal for any particular perceptual task. In any case, plots that accommodate many perceptual tasks are useful in exploratory analysis. After a question is raised by viewing some general-purpose display, more insight can be gained by producing a display more tailored to the task at hand.

3 ENHANCEMENT PROCEDURES FOR 2-D DISPLAYS

In this paper, the emphasis is on displaying data in dimensions higher than two. However, a brief discussion of the two-variable case is appropriate. When two variables are to be represented, the scatterplot is usually the most effective nontabular presentation method available (Tufte 1983). Enhancement methods can make scatterplots considerably more informative (Cleveland and McGill 1984b). This section briefly reviews selected enhancement techniques for single scatterplots.

Transformations are an extremely important tool for graphical data analysis. Frequently the information that can be extracted from a scatterplot is quite limited unless the data are transformed. Consider the sodium ion concentration versus chloride ion concentration plot in Exhibit 1a. Inclusion of extreme points squashes most of the data into the lower left corner of the plot, which hides the structure present in the body of the data. The same data are shown in Exhibit 1b after transforming concentrations to logarithmic scales. Now the structure in the data is quite clear. Thus a simple reason for transformations is to make better use of the available visual space. More sophisticated reasons for transformations relate to statistical modeling and curve fitting. Perhaps the most common reason for transformation is to derive the data analysis advantages that result from having an approximately symmetric univariate distribution. Other reasons include variance stabilization and, in the bivariate situation, transformation to linearity (Daniel and Wood 1980). Univariate transformations are discussed in more detail by Box and Cox (1964), Tukey (1977), and Chambers et al. (1983). Multivariate transformations are discussed by Gnanadesikan (1977) and Tukey and Tukey (1981).

The graphical representation of missing data can be a problem. When one or both values of a point pair are missing, it can still be advantageous to represent the point. Such points can be positioned on the plot frame. That is, the missing value is effectively replaced by the coordinate for the frame posi-

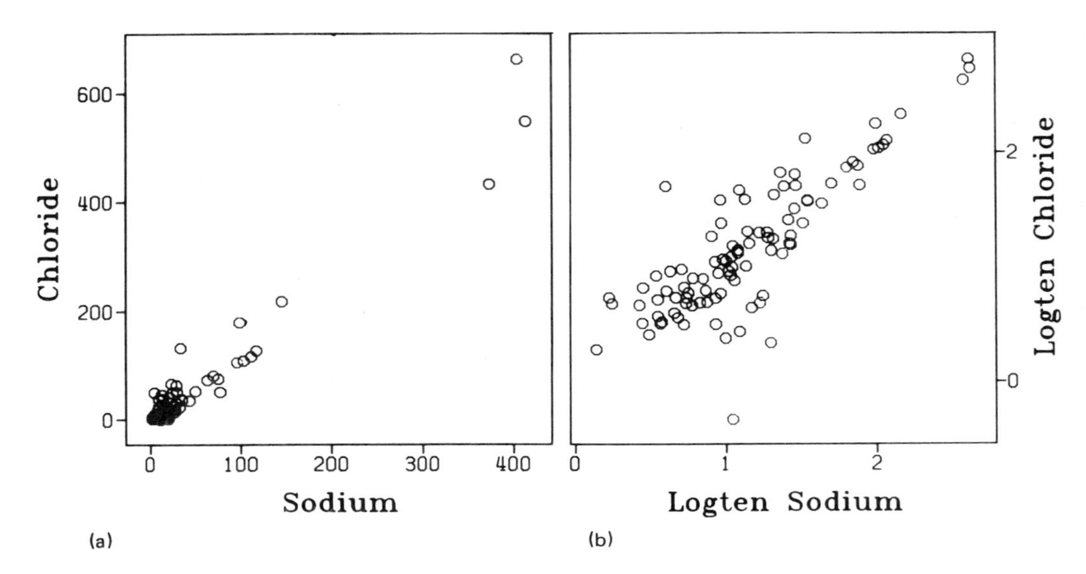

Exhibit 1a — Poorly displayed data. Data are paired sodium and chloride ion concentration measurements in micro-moles per liter from individual rain samples collected at Acid Deposition Site 152A — Indian River, Delaware (Watson and Olsen 1984). Note that a few extreme points squash the body of data into a small region.

Exhibit 1b — Transformed data. Logarithmic transformations have been applied to the data in Exhibit 1a. Note the improved use of plotting space.

tion. (A similar approach plots off-scale data on the frame and missing data just outside the frame.) Since white space is used, missing values are unambiguously represented. Exhibit 2 shows a plot of pH versus logarithm of volume. Note that the missing values correspond to small volumes though not exactly to the smallest volumes. This raises questions about the small-volume data that are not missing.

Scatterplots are often used to judge whether the dependence of y on x is linear or nonlinear. This task is considerably simplified by plotting a set of points $[x(i), y(i)]$, called "smoothed values," where $y(i)$ summarizes the middle of the y distribution for the data in a neighborhood around $x(i)$. Such smoothed values are sometimes called a "middle smoothing." The positive residuals obtained from subtracting the middle smoothing from the data can also be smoothed. When this second set of smoothed values is added onto the middle smoothing the result is called an "upper smoothing." A "lower smoothing" can be obtained in a similar fashion. A variety of smoothing algorithms are available. Exhibits 3a and 3b shows the results of the three smoothings using locally weighted regression (Cleveland 1979). A more complete description of smoothing techniques for enhancing scatterplots is given by Cleveland and McGill (1984b).

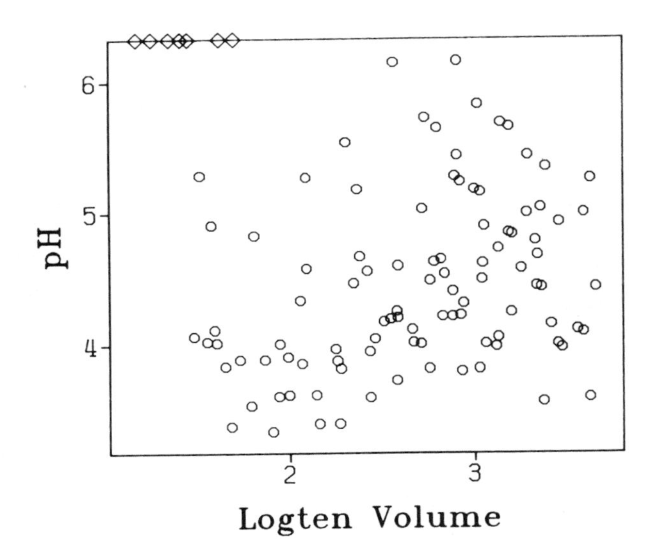

Exhibit 2 — Missing data. Data are paired pH and volume measurements from individual rain samples collected at Acid Deposition Site 152A — Indian River, Delaware (Watson and Olsen 1984). Volume measurements are in logarithms of milliliters. The diamonds on the top plot frame represent points with missing pH values. Note that corresponding volume measurements are small.

The recurrent problem of overplotting of points in a scatterplot can greatly increase the chance of interpretation errors. Two methods for dealing with overplotting are 1) jittering the data and 2) counting the data in local regions and representing the count. Jittering, as discussed by Chambers et al. (1983), involves the addition of noise to variables to increase the number of unique plot positions. This is particularly effective when one variable has been measured on a categorical or ordinal scale. The second approach introduces a third variable, the count, or local density and requires a display technique for the third variable. We recommend binning the data using hexagonal regions and displaying hexagonal symbols. Exhibit 4b shows the plot when an embedded hexagon is used to represent the count. The area of the embedded hexagon is chosen proportional to the count. In the example, points are plotted exactly if three or fewer occur in a region. Exhibit 4b can be compared against the original data in Exhibit 4a, in which little overplotting occurs. Close study reveals that some clusters in Exhibit 4a get split into different bins. Unless binning occurs at device resolution level, some distortion of fine structure is inevitable. Exhibit 4b does capture the general impression given in Exhibit 4a. The advantage of binning becomes obvious as overplotting increases. Since hexagons are similar to circles and the comparison hexagons are in view, the perceptual response should be roughly linear with area (Cleveland, Harris and McGill 1982), but further study may be warranted. The motivation for hexagonal bins, the binning algorithm and alternative symbol scaling procedures are discussed by Carr, Littlefield and Nicholson (1985).

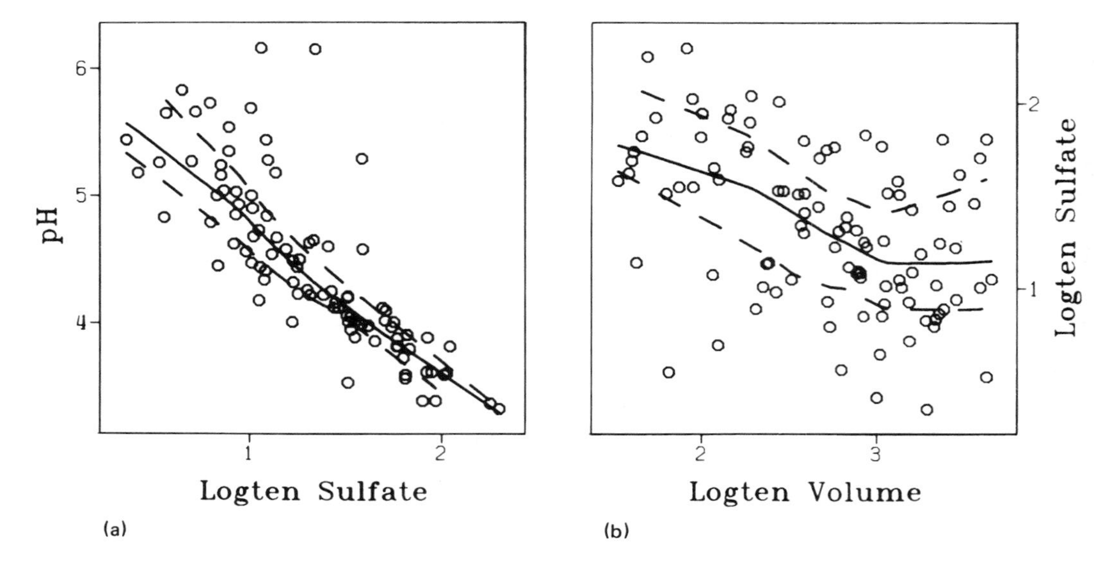

Exhibit 3a — Smooths. Data are paired pH and sulfate measurements from individual rain samples collected at Acid Deposition Site 152A — Indian River, Delaware (Watson and Olsen 1984). Sulfate ion concentration measurements are in logarithms of micro-moles per liter. Middle, lower, and upper smooths based on locally weighted regression are represented by lines. The middle smooth shows a candidate functional relationship between pH and sulfate ion concentration. Lower and upper smooths summarize how close the data are to the functional relationship.

Exhibit 3b — A volume effect. Data are paired sulfate and volume measurements on the same rain samples as in Exhibit 3a. The sulfate ion concentration measurements are in logarithms of micro-moles per liter. Volume measurements are in logarithms of milliliters. Clearly, sulfate ion concentration changes with volume.

4 TECHNIQUES FOR DISPLAYING ADDITIONAL VARIABLES

For displays of more than two variables, the scatterplot is typically the foundation. That is, two variables sometimes called "front" variables are represented by their point position. Methods for representing additional variables, sometimes called "back" variables, include color, static stereo, glyphs, sequential views and juxtaposed views. These are discussed in turn.

4.1 Color

For displaying one additional variable, color is a very tempting choice because no additional space is required. The arrival of color graphics devices and hard copy units has made color plots easy to produce and record. However, color has been slow to catch on in the statistical community because of the high cost of publication in traditionally black-and-white journals and because of eye catching examples of injudicious use (Wainer and Francolini 1980). This challenges those who want to use color. ALDS papers (Carr 1981; Nicholson and Littlefield 1983) identify four uses of color: aesthetics, categories, ordered scales, and anaglyphic stereo.

Aesthetics

Use of color as accent or in conjunction with a geometric cue can make a statistical graphics display easier to look at and possibly more informative. Tufte (1983) notes the harsh moire' effects induced when striping/cross hatching is used to distinguish adjacent regions. His positive examples using tinted brown and grey are gentle to view. Information concerning aesthetic color choice (in view of tremendous individual differences) for both direct- and reflected-light displays would be helpful. However, for our analysis purposes, aesthetic choices are not critical, just more pleasing.

Categories

Color is an excellent means to distinguish subsets of points in a scatterplot. Morse (1979), in an article on effective data display, references a number

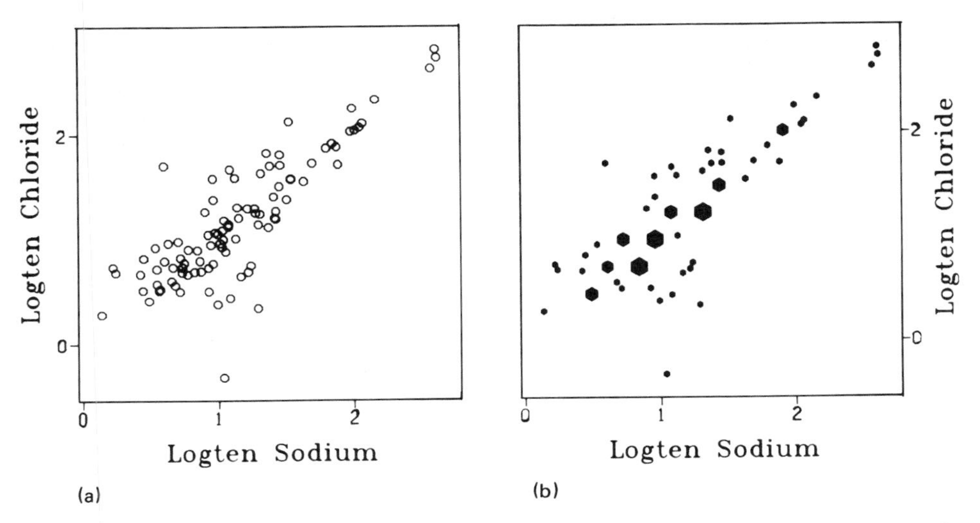

Exhibit 4a — Individual data points. Data are paired sodium and chloride ion concentration measurements in micro-moles per liter from individual rain samples collected at Acid Deposition Site 152A — Indian River, Delaware (Watson and Olsen 1984). Note the strong linear relationship.

Exhibit 4b — Hexagon binning regions and symbols. The data in Exhibit 4a have been binned and the count is represented by area of filled hexagons in Exhibit 4b. The strength of the linear relationship appears about the same in the two plots. Square binning regions are more prone to distortion.

of papers that substantiate this claim. Color is particularly useful because it is perceived independently of other graphical variables such as symbol shape (with minor exceptions; see Cleveland and McGill 1983). For scatterplots with many points, say in the thousands, there does not seem to be any effective alternative to color for distinguishing categories.

The number of categories that are well represented by color is limited to around five or six. More colors than this bog down mental processing. As for color choice, there is some physiological basis for using blue as a background and light-addition mixtures of red and green for foreground points (Murch 1984). For two categories we often choose yellow and cyan. These colors have full-intensity green, which shows up well both on the CRT and in prints. In our experience with a dark background, blue is inadequate as a foreground color and the visibility of red much poorer than expected. Another use of color is to identify overlapping points in different categories. On many display devices, colors for different categories can be chosen to form distinct combinations. For example, overlapping portions of red and green points can be displayed as yellow. If the number of categories is small (two or three), all of the distinct overlap combinations can be identified uniquely.

Ordered Scales

The use of hue and/or saturation to represent an ordered variable is a poor choice when viable alternatives are available. However, space requirements often eliminate alternatives. Furthermore, color can be used as a redundant or distributional reference coding. Thus, an ordered color scale has some uses. We have experimented with paths through the color cube as illustrated in Exhibit 5. Before discussing our preferred path it is helpful to describe the red, green and blue color cube coordinate system. Transformations to other color systems are available (Smith 1978; Joblove and Greenberg 1978).

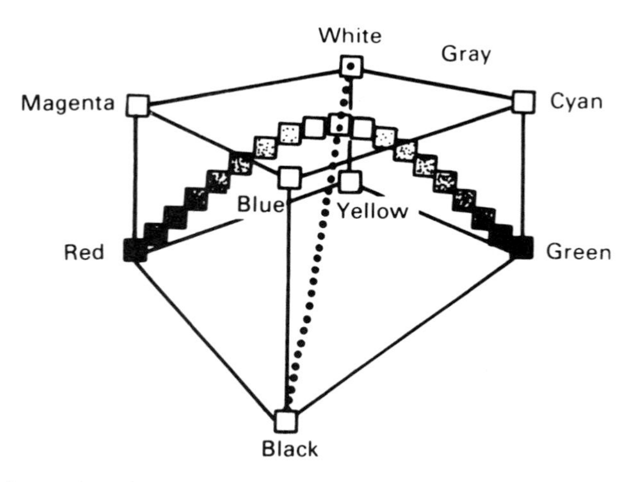

Exhibit 5 — An ordered color scale. The color cube depicts the red/green/blue color coordinate system for a CRT display. Full intensity colors using one or more color guns produce the cube vertices. Ordered color scales can be defined as ordered sets of positions on smooth curves through the cube.

The eight vertices of the cube are the full-saturation mixtures of all possible subsets of the primary colors red, green and blue. These primaries correspond to the three guns in a color cathode ray tube (CRT) display. Thus, to the resolution of the intensity scale in the display device, all colors in the interior and on the boundary of the color cube are possible. The major diagonal connecting the black and the white vertices is the same lightness axis as in the ISCC-NBS standard color solid. The remaining two dimensions of hue and saturation in the standard color solid are nonlinearly transformed into the color cube. Saturation increases with distance from the black/white axis and hue is the remaining dimension. Since saturation does not remain constant on the surface of the cube, color progression on the surface represents more than change in hue.

A path that conveys order for us is a red-gray-green arch-like curve (Nicholson and Littlefield 1983). Using percents of full intensity as the red, green and blue coordinates, the path endpoints are $E_1 = (100, 0, 0)^T$, red, and $E_2 = (0, 100, 0)^T$, green. The vertex is $V = (70, 70, 70)^T$, gray. A parametric representation of the path is

$$C(t) = (-t + |t|^\alpha)E_1/2 + (t + |t|^\alpha)E_2/2 + (1-|t|^\alpha)V \tag{1}$$

for

$$-1 \leqslant t \leqslant 1.$$

The shape parameter α determines the angular acceleration of the curvature at V. Our experience is that the curve in Exhibit 5 with $\alpha = 1.5$ provides more discernible levels near the vertex than does the simple parabola with $\alpha = 2$. The path from each endpoint to the vertex can be thought of as a saturation path although the lightness is also varying. Resolution is increased by joining two such paths. This type of color representation is particularly advantageous when there is a natural center to the quantitative variable being displayed. An example would be in plotting residuals, where gray corresponds to zero and full-saturation red and green correspond to the most extreme positive and negative residuals. As always, judicious transformation of the data before setting up the correspondence to the display variable can make the display more effective.

The color path (1) is restricted to two hues. Ordering hues appears to be much more complicated. Gnanadesikan (1977) describes several experiments to obtain ordered hues using multidimensional scaling. While there was some success, the fact that a complex algorithm was needed for the ordering suggests a certain futility to the process. There are partial successes in the sense that some hue scales are well recognized within particular professions. Examples include the visible spectrum for physicists and Vauthier's mountain-to-the-sea colors for cartographers. The degree of success is dependent on learning. In our use of color to display a quantitative back variable, we want an almost immediate understanding of the scale and only a minimal amount of effort in extracting the information conveyed by its use. We have tried other curves, such as the red/green diagonal and a red-yellow-green parabolic curve. These choices do not give the uniformity of interpretation obtained from the curve in Exhibit 5.

Overlapping data points must be handled differently from categorical data displayed with color. With ordered color scales, combinations of colors are not meaningful. With raster-scan display devices, a crude solution is to display only the last color plotted at each position. This method has some versatility because the plotting sequence can be ordered according to an arbitrary back variable.

4.2 3-D Scatterplots

Several investigators (Fisherkeller, Tukey and Friedman 1974; Donoho, Huber and Thoma 1981; Friedman, McDonald and Stuetzle 1982) have used motion-cued, 3-D scatterplots to represent three variables. For people with eye-dominance problems the preferred production methodology is true 3-D hardware (Stover 1981, 1982). Common production alternatives include side-by-side presentation, alternating images viewed through opto-electrical shuttering glasses (Schmandt 1982), polarized light stereo, and color anaglyph stereo (Carr and Littlefield 1983). Books describing stereo production and perception include those of Lipton (1982), Ferwerda (1982), and Julesz (1971). While perhaps the poorest for viewing, color anaglyph stereo is easy to produce on many color CRTs and is a useful diagnostic tool for those who can fuse the two images.

To produce anaglyph stereo plots all that has to be done is to plot each data point (x, y) twice, once in red as $(x - dx, y)$ for the left-eye image and once in cyan as $(x + dx, y)$ for the right-eye image. Half the separation between the two images of the point is given by

$$dx = k * \text{range}(x) * [\max(z) - z]/\text{range}(z). \quad (2)$$

Thus the point separation is primarily a function of the depth variable z. The value for k depends on the physical size of the plot and the viewing distance. For scatterplots on a CRT, $k = 0.026$ works comfortably. Equation (2) causes the maximum z value to appear in the display surface and other z values to appear behind the surface. Modifications of Equation (2) can make the data appear in front of or on both sides of the display surface. Putting the points behind the display has some technical advantages. Many people like to touch data points floating in space, so putting the points in front is common. More people seem to have difficulty fusing the images when the points should appear on both sides of the display surface, so we generally avoid that alternative even though it adds resolution and a natural center.

The major alternative to Equation (2) is the classical perspective projection as described by Newman and Sproull (1979). We note that the resultant image is technically correct only when looking at the focal point for the projection. In scatterplots, the eye is drawn to individual data points. This can result in difficult/incorrect fusion of the images when points are far from the projection focal point. Transformation (2) holds up as the eye wanders the screen and the nonlinear stretching in depth does not cause serious distortion.

The difficult part of producing color anaglyph stereo is in selecting left and right eye display colors to match the viewing glass filters. Without the

EXHIBIT 6. Stereo Anaglyph Color Intensities for Redundant Depth Cueing (Red, Green, and Blue in percent of full intensity)

Intensity Level	Red Surface R	Red Surface G	Red Surface B	Cyan Surface R	Cyan Surface G	Cyan Surface B
1	67	20	0	0	53	67
2	73	20	0	0	53	73
3	80	20	0	0	60	80
4	87	13	0	0	60	87
5	93	13	0	0	67	93

NOTE: Background is $R = 13$, $G = 27$, $B = 0$.

proper colors crosstalk occurs. That is, both views are seen with the same eye. Systematic experimentation is required to minimize the problem. First, red and cyan intensities are chosen so that when viewed through the opposite-colored filter against a black background the light leakage is not too intense. Next, a background is selected to match the intensity of the leakage. To hide the cyan leakage through the red filter, a little red is added to the background. The red leakage is treated similarly. When the background intensity matches the leakage, small hue differences are easily ignored.

We sometimes vary the color intensity with depth to strengthen the stereo depth cue. Red/green/blue percentages for five such intensity levels are given in Exhibit 6. The numbers can be converted to other color coordinate systems, but they will undoubtedly have to be adjusted for the specific filter and CRT combination. Fortunately, the problem need be solved only once for each combination. When using multiple intensities, the data points are plotted in depth order. For overplotted points, close red points overwrite distant points in the red surface. The cyan image is handled similarly. The two surfaces are then merged using light-addition so that overplotted red and cyan points are displayed as white.[3]

Not everyone can fuse left and right images to produce a stereo image. People with eye-dominance problems usually fail. Other reasons for failure affect only a small percentage of people (Julesz 1971). For those who potentially can see stereo, learning is sometimes required. Exhibit 7 is a good training plot, since lines are easier to fuse than points. Viewing through the glasses with the green filter over the left eye should cause the fix depth contour lines to appear in front of the page. The exhibit shows that stereo can be used to represent fixed density contours for 3-D data because such contours form shells in space (Scott and Thompson 1983; Scott 1984). Another useful training plot is a random point surface as shown in Exhibit 8. This plot appears behind the page. Helpful variations of the exhibit for beginners also include redundant intensity and point size cues. For viewing plots of arbitrary three-variable data sets, plotting a minimal spanning tree that connects near-neighbor points can be helpful even for experienced viewers.

---

[3]The text describes CRT stereo using direct red/cyan light with the red lens on the left. The exhibits illustrate printed stereo using reflected orange/green light with the green lens on the left. See Plate 4 for Exhibits 7 and 8.

Stereo scatterplots are a natural way to view 3-D data. Exhibit 9 is an interesting example.[4] The raw data was a set of 21,862 short time series. The problem was to determine the number and shape of distinct temporal patterns that were present. From the physics of the problem, we knew several things about the data. The first few points in each series were irrelevant. Several competing phenomena could be present simultaneously in the early part of the series. The late part of each series was due to a different physical phenomenon from earlier portions and this phenomenon varied in intensity. Finally, consecutive points were highly correlated so that the patterns were smooth.

To reduce the dimensionality of the data while preserving the physics, we constructed three time averages for each series over an early, a middle, and a late time interval. Thus, each of the 21,862 series was reduced to variables representing averages across three disjoint time intervals.

Our initial stereo plot used late, middle, and early as the abscissa, ordinate, and depth dimensions, respectively. Excessive overplotting masked the structure in the lower-left corner. Moving the viewing position up and to the right produced the plot of Exhibit 9 which does a much better job of showing the upward-hooking cloud in the lower left. Due to saturation in the lower left, much of the structure is still hidden and possibly distorted because of erroneous left eye/right eye image linkage in the brain. However, this simple stereo plot does show an extraordinary amount of structure and can be used to subset the time series into distinct groups of similar temporal shape.

Some imposing structures in the display contain large numbers of points. An important consideration is that other distinctive structures, for example, the linear cloud lying in front of the pipe-like configuration and the thin needle passing horizontally through the upper part of the pipe bowl, are made up of a small fraction of the total number of points. Algorithmic decompositions, such as projection pursuit, focus on dominant structure and would not necessarily bring out such minor, though distinctive, structure early in the sequence of projections. In this particular case direct viewing allows quick subsetting into major structure groups. With stereo this viewing is done with three variables as opposed to two, which means that each group is more precisely defined.

For 3-D point clouds the visual tasks that are performed include looking for: 1) unusual clusters as in Exhibit 9, 2) a candidate modeling surface, 3) multivariate outliers, and 4) conditional relationships. An example of a conditional relationship question is, given that x and y are small, how does the z distribution compare to the rest of the z data? While alternative display techniques may better handle specific instances of such tasks, the multiple-purpose aspects of stereo views are advantageous. Furthermore, quite reasonable data sets can be constructed in which a stereo view shows multivariate outliers that would remain hidden in any linear-projected view.

Stereo plots are also useful for showing surfaces. A simple procedure for showing $z = f(x, y)$ is to generate uniform random points over the domain of x and y and to compute z. As an additional cue, short line segments can be plotted orthogonal to steepest ascent as in Exhibit 8. Of course, contour lines

---

[4]See Plate 5 for Exhibit 9.

can be plotted in stereo, and color side-by-side displays are effective (Phillips and Odel 1984). With smooth contour lines, dashed lines reduce the possibility of incorrect fusion.

Color anaglyph stereo plots are not without problems. A difficulty specific to color anaglyphs is the cost and color control required to produce acceptable printed versions. A second problem for 3-D views in general is that no fully acceptable solution to the overplotting problem has been found. Sampling, binning, slicing up the display, and looking from different directions can help, but deficiencies can be cited for each of these approaches. The final problem is more of a challenge. Much is likely to escape or fool our eyes because in three dimensions more is to be seen and we have little experience in looking at 3-D point clouds compared to 2-D point clouds.

4.3 Glyphs

Our definition of "glyph" is rather broad. We consider a glyph to be any plotting symbol that displays data by changing the appearance of the symbol. This concept encompasses a wide variety of graphical forms, including the original Anderson (1957) glyphs and metroglyphs and Chernoff (1973) faces, plus weathervanes, *k*-sided polygons, and castles and trees developed by other investigators. Fienberg (1979) gives an overview and references many of the specific forms. In this paper, we will discuss general aspects of the glyph technique.

As with many other graphical techniques, glyphs have two main uses: exploration (EDA) and presentation. The motivations and corresponding glyph designs are different. For presentation use, the display is intended to convey a specific impression. Some glyph designs, particularly Chernoff faces, seem well suited for this task. One can carefully assign data dimensions to glyph features, and there is no need to maintain neutrality or independence among dimensions. In addition, one can select the number of data points and dimensions as necessary to best display the important aspects.

For exploratory use, the purpose of the glyph is to display as many dimensions and points as possible, in a neutral fashion. An ideal EDA glyph would enable the viewer to consider any combination of dimensions simultaneously, from singly to all combined. It is important that the glyph display the data dimensions independently, and that no dimensions be allowed to overwhelm others.

Another important consideration is the tradeoff between glyph complexity and number of data points displayed. A glyph which works acceptably with 50 data points can become useless with 500 data points. There are two problems. The first is that the human visual system can handle only a few features (at a glance). If the glyph is so complicated that study is required to understand it, then it is too complicated for use with large numbers of points. For interactive exploratory analysis we are dubious about glyphs that represent data in dimensions much above five. The second problem is that glyphs overlap as the number of data points increases, and the overlapping glyphs can be difficult to interpret.

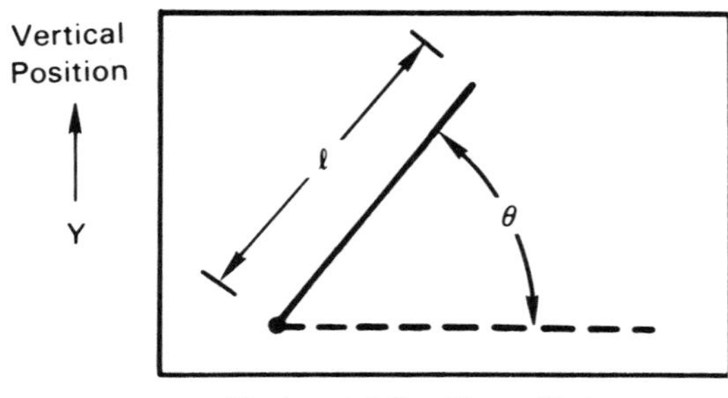

Exhibit 10 — Ray glyph definition. The ray glyph displays upto six variables: x — horizontal position of dot, y — vertical position of dot, θ — orientation of ray, l — length of ray, and color of symbol. Our order of preference for display of continuous variables is position, orientation, length and color. For a categorical variable color is our first choice.

Besides restricting the dimensions, ALDS investigations have considered selecting the glyph and assigning variables to glyph characteristics. With respect to the glyph, elements should be easy to compare against a scale. For example, the "ray" glyph in Exhibit 10 can display six dimensions: horizontal position, vertical position, ray orientation, ray length, dot size, and color. When many data points are viewed in the aggregate, global patterns are seen. By concentrating on the glyph horizontal position, vertical position, orientation, length, dot size and/or color one can consider the six dimensions either independently or together.

When viewing several variables, it is common to restrict the range of some variables and focus attention on the relationships among two or three other variables. Mentally restricting the range of variables is difficult except for those represented by position or distinct colors. Thus with respect to assignment of variables to glyph characteristics it should not be arbitrary. Our preferences for glyph characteristics are point position, stereo depth (if stereo is used), orientation, linear size, areal size, and color (if desperate). Stereo provides point position and avoids a shifting of visual/mental gears even though relative depth may be assessed less accurately than relative orientation. For some people stereo is not an option.

As an example of the ray glyph Exhibit 11 shows the Anderson (1935) iris data, consisting of four measurements on each of 50 flowers from each of three varieties. The scatterplot (point position) shows the relationship between sepal length and petal length for these flowers. The remaining two measurements, sepal width and petal width, are indicated by the orientation and length of the ray, respectively. The dot size represents the three varieties. A better display uses distinct colors to code for variety since different dot sizes make ray lengths harder to compare. Extreme orientations of 0° and 135° were selected so that there would be no confusion between 0° and 180°. The maximum length for the ray is determined by systematic experimentation so that the display best conveys four dimensions. Here a maximum length of 0.7 in. on

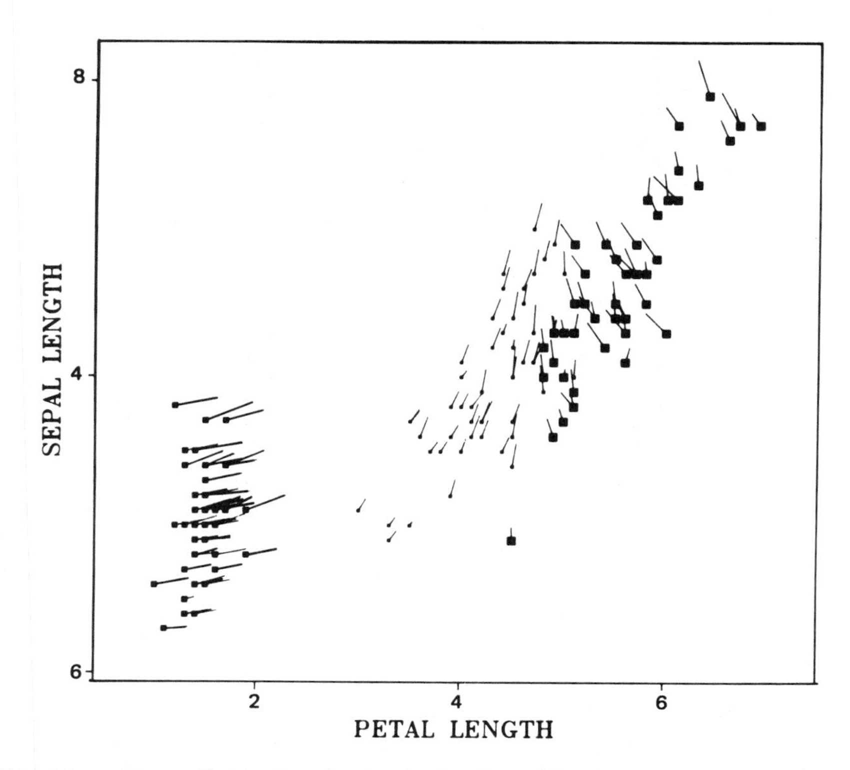

Exhibit 11 — Four dimensional ray glyph plot. The basic scatterplot (position of ray dot) displays sepal length versus petal length for three varieties of iris from the Gaspe Peninsula (Anderson 1937). Ray orientation displays petal width and ray length displays sepal width. Ray dot size identifies the variety — small is Virginica, medium is Setosa and large is Versicolor.

the CRT display screen gave enough resolution that the relative sepal widths for the three varieties could be discerned and yet the rays did not overlap excessively.

From scatterplots of all possible pairs, it is very easy to see the positive correlations among petal length, petal width, and sepal length and the negative correlations of these three with sepal width. These pairwise correlations are also evident in Exhibit 11. The ray orientation increases from left to right and from bottom to top, while ray length decreases from left to right and from bottom to top. While Exhibit 11 may not convey new information, it is a concise way of plotting the four measurements and of identifying the three iris varieties. Even 2-D plots make it clear that setosa irises are a separate cloud. However, Exhibit 11 clearly shows that virginica and versicolor irises are not separate even in four dimensions. Also, there is some question whether a casual view of a number of 2-D plots would pick out some of the higher-dimensional characteristics. For example, the four virginica runts and the single versicolor runt are perhaps more obvious in the 4-D glyph plot.

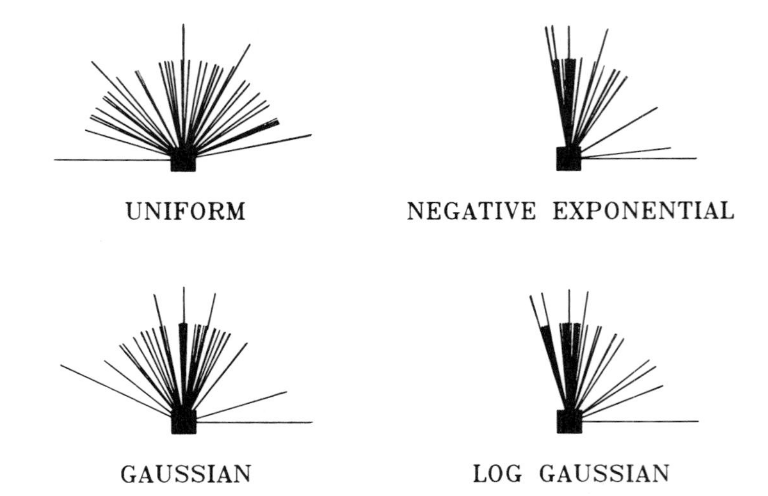

Exhibit 12 — Ray orientation representation of samples. Samples of size 41 from four distributions are displayed using orientation of a ray glyph. Longer rays distinguish extremes, quartiles, and median. Note that orientation can convey both peakedness and skewness.

The ray glyph can be carried a bit further. Easily discernible horizontal and vertical orientations can convey the distributional aspects, central tendency and extremes, respectively. Consider the collections of ray symbols illustrated in Exhibit 12. Each ray orientation represents the value of an observation in a sample. Think of the ray as a pointer on a measuring instrument, such as a speedometer, increasing from left to right. A linear transformation of the sample codes the median as vertical (90°) and the minimum as left (0°) for a skewed left distribution or the maximum as right (180°) for a skewed right distribution. Thus, all the values fall between 0° and 180°. Values near the median are discernible as near vertical. Skewness in the distribution is also discernible as a single orientation for near-horizontal rays. Preventing the rays from dropping below horizontal avoids ambiguity.

Exhibit 12 illustrates the use of ray orientation to represent samples of 41 observations taken from uniform, negative exponential, Gaussian and log-Gaussian distributions. To help see gross features of the distributions, quartiles and median (with 41 observations ordered positions 1, 11, 21, 31, 41) are plotted with longer rays. The central peakedness of Gaussian relative to uniform is evident. The skewness of negative exponential and log-Gaussian is also evident, as is the fact that negative exponential and log-Gaussian have very different left-tail characteristics.

Ray length is less quantitative than orientation but can be used for nonspatial representation of an additional dimension. Length and color together do better. We have tried two approaches with varying degrees of success. Both use a linear ray length scale. For the *first* approach, length and color are used redundantly. Since a positive length is necessary for the ray to have an orientation, the minimum length for easy visualization of orientation corresponds to the minimum of the data. If rays are too long then the scatterplot becomes cluttered. Trial and error selects minimum and maximum lengths to correspond to the minimum and maximum of the data. Color can separate the ray length distribution into quantiles. For example, red and green can split lengths at the median with light and dark hues splitting at the quartiles. Variations using color ordering are reasonably successful at reinforcing a finer length discrimination. The *second* approach is a variation on the thermometer. Now all the rays are the same length. With color, say yellow, the "mercury" portion of the ray represents the quantity, with the rest of the thermometer closer to the background color. The thermometer approach has the advantage of a common reference scale for cross comparison of distinct rays (Cleveland and McGill 1984a). Without color, the mercury portion can be drawn thicker. However, the additional thickness consumes space and accentuates saturation problems.

Stereo methods can also apply to the display of glyphs. Exhibit 13[5] shows three views of five variables selected from seven variables which define each of 500 replications of a high-energy particle physics experiment at the Stanford Linear Accelerator. The four-body experiment and data reduction are described by Friedman and Tukey (1974) and by Tukey and Tukey (1983). Negative variables are replaced by absolute values, then all variables are transformed to natural logarithms. These transformed variables are designated V1, V3, V4, V5, and V7. Exhibit 13 presents three different spatial views of these 200 5-D data points. Exhibit 13 plots V1 as abscissa, V3 as ordinate, and V5 as stereo depth. Exhibit 13 plots -V5 as abscissa, V3 as ordinate, and V1 as stereo depth; thus from a spatial standpoint, Exhibit 13 is the view looking into Exhibit 13 from exterior right. Exhibit 13 plots V1 as abscissa, −V5 as ordinate, and V3 as stereo depth; thus Exhibit 13 is the view looking into Exhibit 13 from exterior top. In each of the views ray orientation is V7 and ray length is V4. Thus in all three views the ray symbols for a particular point look the same. To help the viewer spatially orient between views two benchmark points (the point with the largest V1 value and the point with the largest V5 value) are plotted with larger ray dots. Exhibit 13 clearly shows two spatial clusters—top right rear and bottom right front, respectively. The same two spatial clusters appear in Exhibit 13 and in Exhibit 13. The similarity of the ray symbols in the first-mentioned cluster means that it is a 5-D cluster while the wide variety of ray symbols in the second means that it is only a cluster in the three spatial variables. Ray orientation codes the median of V7 as vertical and largest V7 value as horizontal right; ray length increases with V4. Thus the first cluster (from looking at Exhibit 13) has very large V1 values (abscissa), V3 values (ordinate), V4 values (length), and V7 values (orientation), and very small V5 values (stereo depth). Correlation structure involving both spatial and glyph element variables is quite apparent in Exhibit 13. For example, in Exhibit 13 there is a smooth progression of increasing ray orientation as the points come forward out of the page (that is, as the stereo depth variable increases), suggesting a positive correlation between V5 (stereo depth) and V7 (ray orientation). The shape

[5]See Plate 5 for Exhibit 13.

of the cloud in Exhibit 13 is the result of a logarithmically transformed constraint in that the untransformed points tend to lie close to coordinate planes with all points constrained to a 5-D simplex.

Exhibit 13 is an example of a very small subset of all possible coordinate views of a 5-D data set. An interesting question is the selection of a representative set of such coordinate views. For example, one wonders if the views in Exhibit 13 might be more informative if some position and shape variables were switched. A simple approach to selecting a systematic subset of all possible views is provided by the latin square experimental design. Let the rows represent distinct views, the columns distinct variables, and the treatments plotting symbol characteristics. When plotting n-variable data, any such latin square assignment defines n distinct views where each variable is assigned once to each plotting symbol characteristic. This would appear to be a minimal-sized subset of coordinate views with any degree of symbol assignment symmetry.

Aggregation is one approach to overplotting and glyphs can be used to represent properties of the aggregate. For example, a histogram is an aggregation across one variable, with one summary statistic (count) displayed by a particularly simple glyph (box with variable height). Similarly, the box-and-whisker plot is an aggregation across one variable, with five summary statistics (minimum, maximum, median, and two hinges) displayed with a somewhat more complicated glyph. For saturated scatterplots the data are aggregated or grouped using two variables. Bachi (1978) and Tukey and Tukey (1981) illustrate the technique using a regular grid to group the data. In each cell formed by the regular grid a distributional summary of a third variable is plotted using a glyph. The aggregation approach is not entirely satisfactory. Sequential displays as discussed below provide an alternative.

4.4 Sequential Views

Sometimes it is helpful for the analyst to see different displays close together in time. There are several distinct uses of this technique, which can be loosely classified based on the relationship between the various displays:

1. Alternation, in which the various displays show data selected from different classes.
2. Slicing, in which the data are selected based on an ordinal parameter. The different displays, or slices, correspond to different values of the selection parameter.
3. Rigid motion, such as 3-D rotation, in which the displays differ in values of the projection parameters.
4. Other parametric transformations, such as changing glyph appearance based on rotation of the back variables.

In alternation and slicing the various displays show different subsets of the data. In alternation, points are selected based on some natural division, typically a categorical variable such as sex or experiment number. In this case, alternating the displays allows the analyst to quickly compare the corresponding distributions. The technique seems to be effective for detecting situations where the distributions have features, such as holes or edges, that are similar in nature but occur at slightly different positions. Areas of the display that have similar density appear to flicker as the displays switch back and forth, while areas that have different density are perceived as pulsing.

Although slicing is superficially similar to alternation, it is used for a different purpose. Whereas alternation is used to compare two or more different distributions, slicing is used to understand a single distribution having one more dimension than the basic display format. To take the simplest example, a 3-D distribution can be viewed as a sequence of 2-D scatterplots, each composed of just the points contained in a slice of the third (nondisplayed) dimension. As the slices are displayed in sequence from one extreme to the other of the additional depth variable, the analyst develops some understanding of the overall 3-D distribution. This example is obviously an alternative to other 3-D display techniques such as rotation and stereo. The slicing approach seems to have some advantage if the data set is so large that displaying all the points at once causes unacceptable overlap (Nicholson and Littlefield 1982). Otherwise either rotation or stereo would probably be preferred.

The other two uses of sequential views, rigid motion and parametric transformations, are discussed in more detail in Section 5. In general, these two techniques require complete recomputation of new displays as the transformation parameters change. However, if the motion or transformation is restricted to a few discrete values, then one can use the same display techniques (described below) that are appropriate to alternation and slicing. In this case, we cycle repeatedly back and forth through the range of parameter values, a procedure which we call "rocking."

In all the sequential display techniques, it is important that the transitions between images be clean. That is, the drawing process must be hidden so that the user sees complete images back to back, with no blackouts or partial displays. This requires at least double buffering. On some display devices, multiple buffering provides a particularly efficient implementation. The idea is to partition the bit-mapped display memory into several sections, each of which can be displayed individually. Each display is then generated once, into its own section of display memory. After all displays have been generated, they can be cycled indefinitely in a "film loop" at essentially no cost.

Most display devices have the hardware features required for film loops. Either hardware zoom/pan or a color map can be used. Briefly, one can partition the memory either spatially (Exhibit 14a) or across bit planes (Exhibit 14b). Spatial partitioning uses zoom/pan to select individual displays, and provides the maximum number of colors. Bit-plane partitioning uses the color map to select displays, and provides the maximum resolution. Shoup (1979) and Baecker (1979) discuss the techniques at length.

4.5 Multiple Juxtaposed Views—Scatterplot Matrices

An alternative to glyph plots and sequential plots is the simultaneous display of juxtaposed views. Juxtaposition plots and overlays are often effective for com-

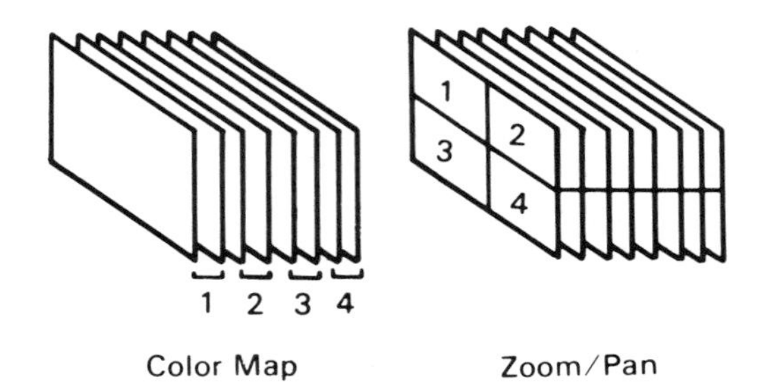

Exhibit 14a — Memory partitioning by bit planes. Partitioning the memory by bit planes reduces the number of colors that can be displayed. The plotting resolution is not affected.

Exhibit 14b — Memory partitioning by zoom/pan. Partitioning the memory by zoom/pan reduces the plotting resolution. The color options are not affected.

parison purposes. Several variations and generalizations of juxtaposition for scatterplots have been presented in the literature. These include regular polytope projection and multiwindow displays (Tukey and Tukey 1981), casement displays (Chambers et al. 1983) and the scatterplot matrix. The scatterplot matrix in conjunction with subset representation is a powerful display technique and is described below.

A scatterplot matrix for p variables is obtained by arranging the scatterplots for all $p(p - 1)$ ordered pairs of variables in a matrix so that each marginal scale applies to $p - 1$ plots. Exhibit 15 illustrates such a display for six variables where the variables are chemical analyses on rain samples collected at a particular acid deposition monitoring site. The results for 106 individual rain samples are displayed. For all the variables except pH, the original measurements are ion concentrations in micromoles per liter and log concentrations are plotted. Negative pH (already a log scale) is plotted to increase the number of positive correlations in the matrix. Despite substantial overplotting, several patterns are evident. The most obvious pattern is the high positive correlation between sodium, chloride, and magnesium in the lower right. For many analysis purposes one variable would suffice for all three. That is nice to know since other variables are not in the plot. The positive correlation between sodium and chloride raises confidence in the data since the site is reasonably close to an ocean salt source. The three variables together raise a question, "Does ocean water contain high magnesium ion concentration?" The positive correlation between -pH, sulfates and nitrates is also interesting since the correlation is less evident with many ion concentrations not being shown. The missing data shown in Exhibit 15 consistently fall in the top-right portion of the plot frames. Missing data are associated with high ion concentrations. By referring to Exhibits 2 and 3b a possible explanation emerges. Small sample volumes yield missing data and heavy rains yield low concentrations. Thus the scatterplot matrix proves useful even without tools that help bring out additional relationships.

The originator of the scatterplot matrix does not seem to be known. Tukey and Tukey (1981) described an organized collection of two-variable scat-

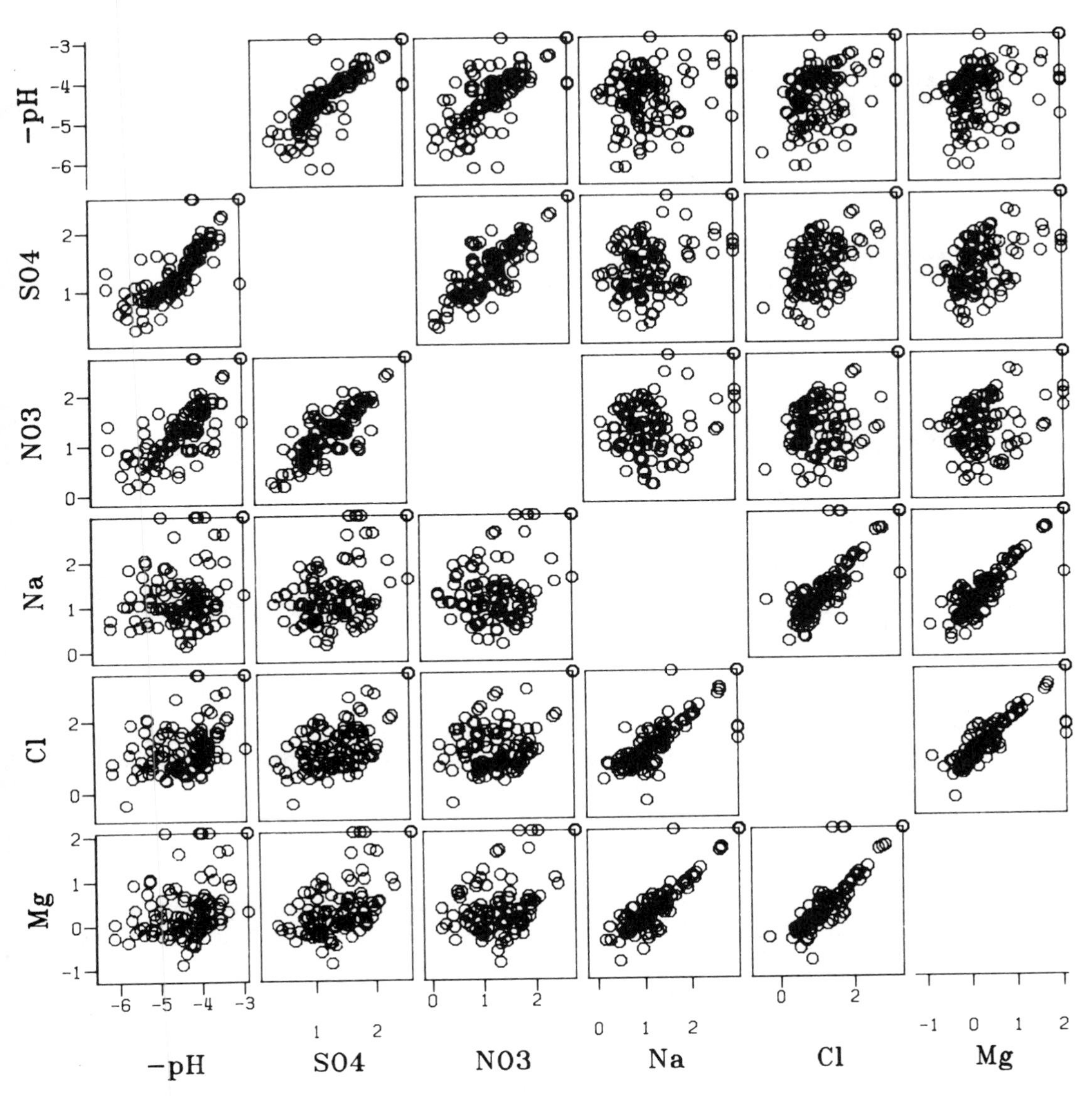

Exhibit 15 — Scatterplot matrix. Data are multiple measurements on individual rain samples collected at Acid Deposition Site 152A — Indian River, Delaware (Watson and Olsen 1984). Except for -pH all measurements are ion concentrations in logarithms of micro-moles per liter. Missing data are shown on the plot frames. Note the high correlations among -pH, sulfate and nitrate and among sodium, chloride and magnesium, with low correlation between the two sets.

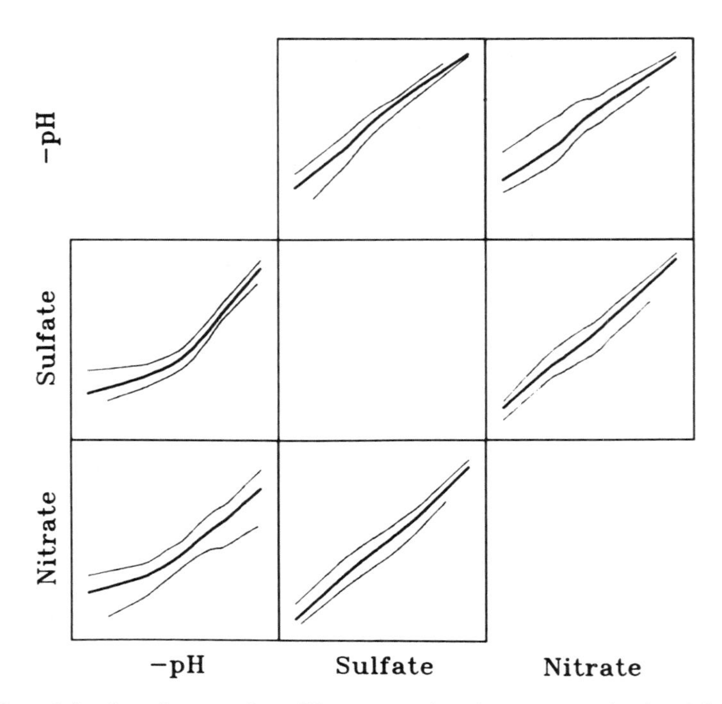

Exhibit 16 — Matrix of smooths. The smooths shown are obtained by applying locally weighted regression to pH, sulfate and nitrate measurements as shown in Exhibit 15. The plot of sulfate versus pH suggests a nonlinear relationship. The plot of pH versus sulfate suggests the expected linear relationship. This demonstrates the non-equivalence of smooths for the plot pairs.

terplots and called it a generalized draftsman's display. Further descriptions are found in Chambers et al. (1983), Carr and Nicholson (1984), and Becker and Cleveland (1984). The display of multiple microplots facilitates the rapid scanning of many dimensions. In some cases extreme points can be followed from plot to plot. However, the disconnected representation of multiple aspects of the same multivariate point complicates the discovery of higher-dimensional patterns. In an attempt to get at higher-dimensional aspects of the data, Diaconis and Friedman (1980) suggested connecting the several reproductions of selected points with line segments. More recently Friedman, McDonald and Stuetzle (1982) demonstrated the use of color as a linkage to identify selected points in a matrix of plots. The following summarizes various approaches to high-performance interaction, color representation of subsets and enhancement procedures as described by Carr and Nicholson (1984) and Becker and Cleveland (1984).

Enhancement procedures and additional variable representation methods can be applied to individual plots in the scatterplot matrix. As an example, Exhibit 16 shows a scatterplot matrix with lower, middle, and upper smooths. Such plots can show nonlinearities and are a valuable addition to correlation matrices which indicate linear structure. Most scatterplot matrices have pairs of equivalent plots with coordinate axes interchanged. However, in this case, the smoother treats independent and dependent variables differently so that the plot are not equivalent.

Because individual plots in the matrix are small and space is at a premium, techniques involving color are particularly advantageous. One of the most useful techniques is interactive subset selection and representation using color. Approaches to subset selection include algorithmic methods such as stratifying on the basis of percentiles and direct graphical methods such as drawing polygons around points (Littlefield 1984) and brushing over points. The brushing approach to subset selection and various corresponding display options including point identification have been well described by Becker and Cleveland (1984). For high-performance systems with small sample sizes, brushing is the preferred graphical method for subset selection since it encompasses polygon selection and provides for rapid treatment of rectangular regions. In the color context the brush can be multicolored and can simultaneously define several subsets.

Derived variables such as principal components can also be plotted in the scatterplot matrix. Fixed-count brushing is obtained by plotting a univariate ranking of points as a thin strip beside the matrix and sliding a colored brush over the strip. For large sample sizes and several variables, real-time displays are not possible, but sequences of subsets can still be viewed in rapid succession using film-loop technology (Section 4.4) in which the interaction controls the display rates.

The way color can be used to indicate subsets is of some interest. Exhibit 17 is the matrix scatterplot display of four variables of high-energy particle physics data that were illustrated in Exhibit 13.[6] Here we look at pairwise structure of variables V1, V2, V6 and V7. Note that in Exhibit 13 we are looking at 5-D structure of variables V1, V2, V4, V5 and V7; hence, we have not seen variable V6 before. In Exhibit 17 pencil-like clusters are outlined in the third and fourth columns of Row V1. The cluster of points in the scatterplot of V1 versus V6 is colored red and the one in the scatterplot of V1 versus V7 is colored green, with the remaining points colored blue. The same color coding is used in the other scatterplots to identify the subset points. Overplotting of red subset points and background points is shown in magenta and overplotting of green subset points with background points is shown in cyan. In all the scatterplots the two subsets are disjoint, thus there is no overplotting of red and green points. In the general case the two subsets might have points in common. Then the color coding is more complicated. First, there are four types of points—those in the first set, those in the second set, those in both and those in neither. Second, with four types of points, there are 11 ways of overplotting with at least two distinct types.

Clearly, in Exhibit 17 color does an excellent job of identifying the two subsets and of showing the interesting parallels in the relationships among V6, V1 and V3 and among V7, V1 and V3. In this instance it would be difficult to locate the correct subsets in the other scatterplots without the unique, indivi-

---

[6]See Plate 15 for Exhibit 17.

dual subset identifier. While the locations are logical once identified in all the scatterplots, they are not obvious from the geometry of the plots.

For the scatterplot matrix, display tools in addition to subset selection and subset display variations are helpful. Some of the more obvious tools are:

1. Quick access to graphically defined subsets so that histories can be maintained and logical operations can be performed in the redisplay process.
2. Easy variable reordering so that highly related variables appear together.
3. Pan and zoom with variable resolution so that detail can be observed.

Many minor variations on the scatterplot matrix structure have been suggested. Obviously space can be saved if only the lower triangle is plotted. Unfortunately, this complicates looking at all plots involving the same variable. Exhibit 17 omits labels on the diagonal and scales. The variable ranges can also be added to the diagonal regions (Becker and Cleveland 1984). Other uses of the diagonal region such as for histograms tend to make the display look busy. A final option is to run the diagonal from lower left to upper right. This makes the matrix and equivalent plot pairs symmetric about 45° lines. The symmetry in our exhibits corresponds to the familiar correlation matrix.

Scatterplot matrices can be compared by juxtaposition or by using overlays. In Exhibit 18, the lower-left and upper-right triangles are used to obtain a comparison of two acid deposition monitoring sites. The scaling for the plots is based on the combined data sets. Note that the order of the variables is switched in the lower-left and upper-right triangles so that the plot pairs have the same orientation and the plot pairs are placed symmetrically with respect to the matrix center. Multivariate measurements on duplicate samples can also be displayed using the two triangles. The replicates for the same variable appear in the diagonal.

For comparison purposes scatterplot matrices can be overlayed. This requires scaling for the combined data sets. If binning is applied to control overplotting, the same bin center lattice should be used for both data sets. Distinct points (or area symbols representing counts) from the two matrices are plotted in two different colors. Overplotted regions are represented with a third color.

Selected variables can be removed from the p variables in the scatterplot matrix and added as stereo or symbols to the remaining plots. This adds resolution to the scatterplot variables and can enhance perception of higher-dimensional relationships. For example, a simple stereo plot can show a 3-D relationship that is impossible to deduce with certainty from the marginal views. However, with little plotting space, overplotting is a serious problem. In addition, many simple higher-dimensional patterns can also be found with subset representations. Thus high-dimension representations in scatterplot matrices have difficulties and competitors.

PLATE 13 *(Gabriel and Odoroff)*

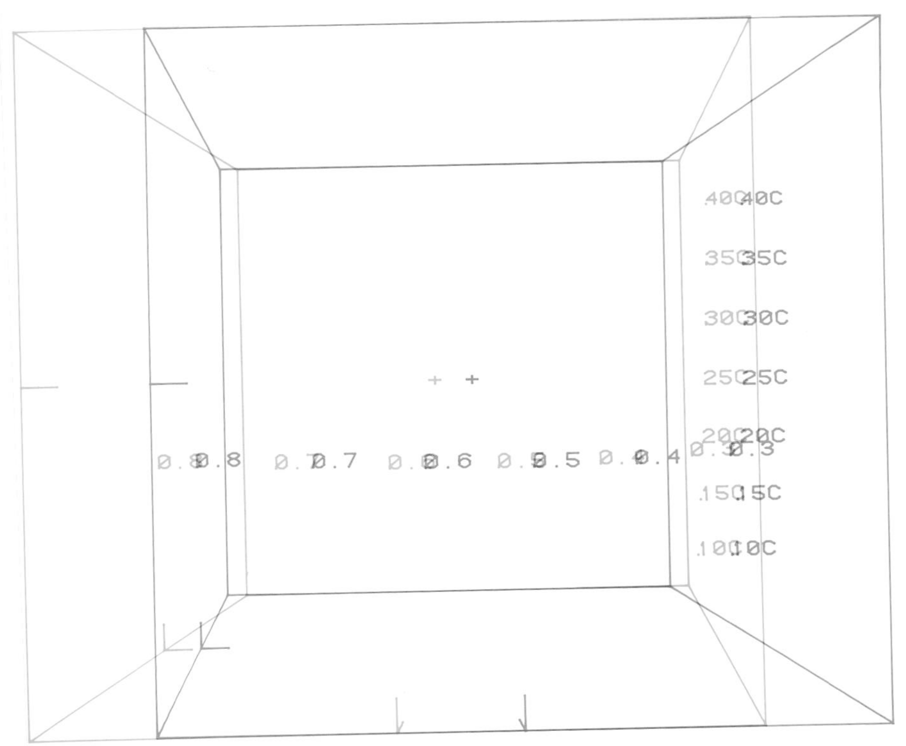

Figure 18 Three-dimensional SQRT biplot of the alcohol density data.

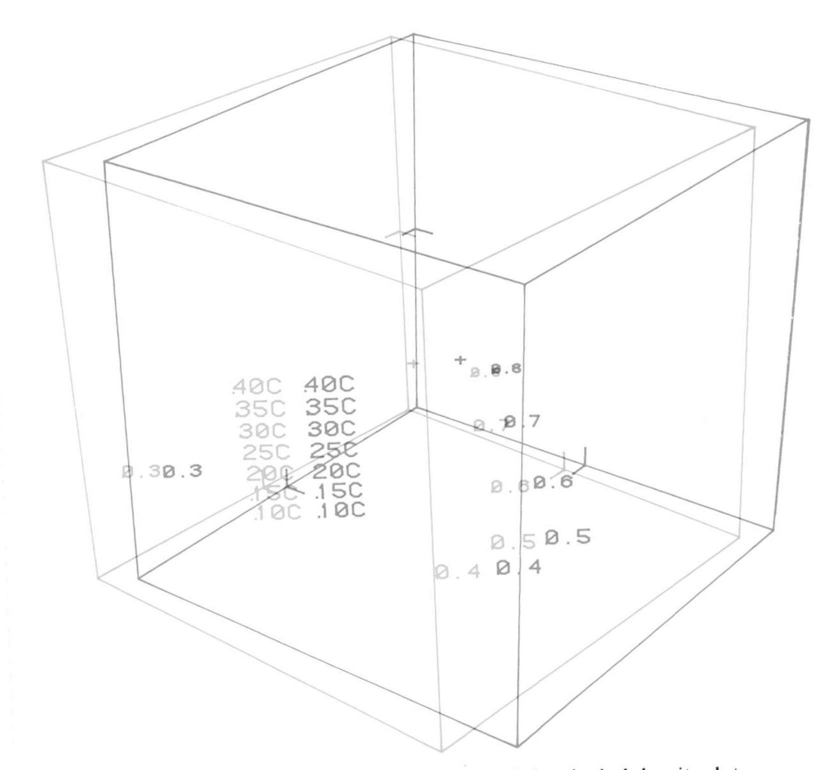

Figure 19 A view of the three-dimensional GH-biplot of the alcohol density data.

PLATE 14 *(Gabriel and Odoroff)*

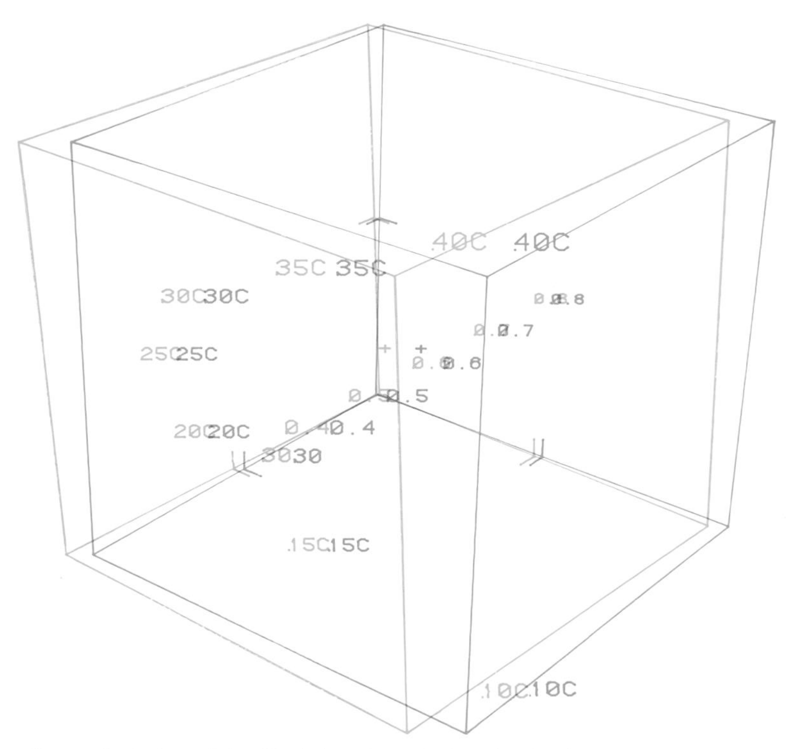

Figure 20 A view of the JK-biplot of the alcohol density data.

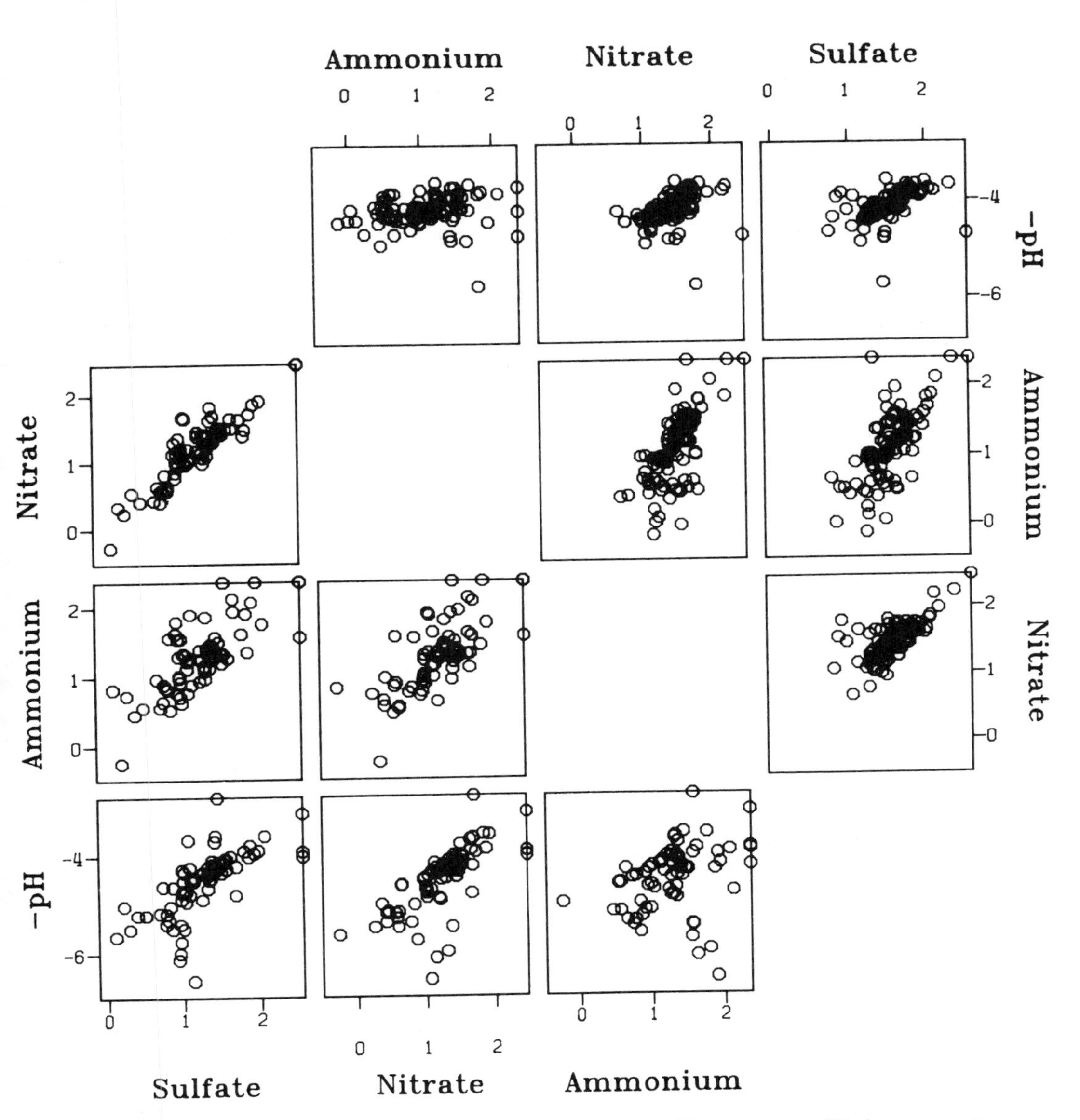

Exhibit 18 — Comparison of scatterplot matrices. Data are multiple measurements on individual rain samples collected at Acid Deposition Sites 053C and 153A—Raleigh North Carolina and Zanesville, Ohio, respectively (Watson and Olsen 1984). Except for -pH all measurements are ion concentrations in logarithms of micro-moles per liter. Missing data are shown on the plot frames. The lower left plot (Raleigh data) has variables in reverse order from those in the upper right plot (Zanesville data). Corresponding plots are symmetrically placed about the matrix center and have the same variables as abscissa and ordinate. This allows direct comparison. Note for example the difference in range in the nitrate versus sulfate plots at the two sites.

Because few people can easily visualize in more than three dimensions, what can be learned about high dimensions is limited. The scatterplot matrix breaks down the problem. Given two variables in a restricted domain, the matrix shows what happens to other variables. By varying the domain an attempt is made to build up the global picture. This can lead to finding simple relationships. The beauty of the scatterplot matrices of basic variables is that when relationships are found, interpretation/explanation is relatively easy. Although explanation in terms of linear combinations of variables is not simple, projection methods such as principal components and projection searches (Friedman and Tukey 1974, Asimov 1983) can find patterns that are readily missed by scatterplot matrices. Thus while scatterplot matrices are often useful, they are limited windows into higher dimensions.

5 LINEAR COMBINATIONS OF DATA

5.1 Projection into Two or Three Dimensions

Projection has long been a classic approach for the display of higher-dimensional data. Coupled with an optimization algorithm, projection provides structure insight in many applications. However, even the most familiar of such techniques, principal components, demands understanding of theoretical aspects or the results may be misleading. A classical example concerning 10 characteristic measurements on 375 grasshoppers is described by Gnanadesikan (1977). Here several arbitrary decisions on how the algorithm is to be applied produce a hodgepodge of dissimilar "optimum" 2-D and 3-D projection views. Possibly, if this analysis were being done today, insight from multidimensional graphics would allow a more meaningful and unequivocal application of principal components. A particularly rewarding use of projection is described by Friedman and Tukey (1974). They apply a projection pursuit algorithm to the 7-D particle physics data that we transformed and illustrated here in Exhibits 13 and 17. The resulting 2-D projection view consists primarily of several fuzzy curves, which suggests that the outcomes of most of the scattering events are constrained to lower dimensional manifolds.

Difficulty in understanding the physical significance of the particular linear combination of variables selected as viewing axes by projection algorithms often precludes use of the results beyond the exploratory analysis stage. Also, particularly with many data points, projection algorithms tend to focus on dominant structure in the data. Patterns of outliers or even well defined, low-dimensional structure of a small percentage of the data are only found by these algorithms through iteration based on residuals. On the other hand, a simple graphics presentation of all the data as in Exhibit 9 may allow quick identification of important minor structure. A challenge of statistical graphics is to combine the judgment of the analyst, based on his limited visualization of higher dimensions, with the power of computational algorithms to obtain better understanding of high-dimensional structure.

5.2 Rigid Motion

To date, rigid motion has been used almost exclusively to create a depth cue. In principle, there is no reason why the motion cannot include nonspatially represented variables. For example, in a quadruple scatterplot with two spatial variables and two nonspatially represented variables, a simple rotation is defined by a fixed 2-D subspace, "the axis," and an orthogonal rotation 2-D subspace. If the fixed subspace is defined by the nonspatially represented dimensions, then the motion appears as a revolving of the display about a fixed point. On the other hand, if the fixed subspace is defined by the two spatial dimensions then the motion appears as a smooth change in the characteristics of each plotting symbol with the location of symbols remaining fixed. Neither of these motions is very informative in the sense of providing additional information over that in the static display. Now, suppose that the fixed subspace involves all four dimensions. For example, it might be the subspace spanned by the two vectors (1,1,1,1) and (1,-1,1,-1). The motion appears as a combination of plotting-symbol spatial movement and plotting-symbol change of characteristics. Experience suggests that rigid motion is a useful tool for discerning high-dimensional clusters and low-dimensional manifold structure. For example, distinct clusters appear as similar spatial movement/characteristic changes in the plotting symbols. Low-dimensional manifolds appear as a smooth gradation of spatial position/characteristic change of the plotting symbols.

For a glyph-enhanced scatterplot, whether it is spatially 2-D or 3-D, the simplest rigid motion is defined by an origin 0 and an $N - 2$ dimensional fixed subspace spanned by linearly independent vectors $D_1, \ldots, D_{N-2}$. Two additional orthonormal vectors, V_1 and V_2, span the rotation subspace. Define the 2-D polar coordinate rotation matrix as

$$R(\theta) = \begin{bmatrix} \cos(\theta) & -\sin(\theta) \\ \sin(\theta) & \cos(\theta) \end{bmatrix}. \tag{3}$$

The position $P(\theta)$ of the point P after rotation through angle θ is expressable as

$$P(\theta) = 0 + VR(\theta)V^T(P - 0) + D(D^TD)^{-1}D^T(P - 0) \tag{4}$$

where $V = (V_1, V_2)$ and $D = (D_1, D_2, \ldots, D_{N-2})$. Since $VV^T + D(D^TD)^{-1}D^T = I$, the rotated position $P(\theta)$ can be completely defined in terms of the rotation subspace.

Let $\theta(t) = \theta_0 \, sin(\pi t/2)$ be the relationship between the rotation angle and time t. The display of a sequence of rotated, N-dimensional points as glyph-enhanced scatterplots for a systematic, uniformly spaced sample of t values that includes the integers creates a rocking motion through angle $2\theta_0$ centered on the original static view. The appearance is a repetitive to-and-fro motion of glyph symbols coupled with a repetitive smooth change of glyph symbol characteristics. The motion is quite easy to follow even with a crude sampling grid of two intermediate times between the extremes.

A more complex rigid motion is created by replacing simple 2-D polar coordinate rotation with 3-D spherical coordinate motion. The fixed subspace is $N - 3$ dimensional. Now motions like a figure-8 trace on the unit sphere are possible. Clearly, rigid motions of arbitrary complexity can be defined in a similar manner. Their utility is an open question. As described here, the rigid motion is a redundant cue, helping to define interrelationships that involve nonspatially represented dimensions. In a static N-dimensional display, subtle structure may be missed with the gross quantification of nonspatially represented dimensions. Adding rigid motion can accentuate such relationships. A more ambitious use of rigid motion is to bring in additional dimensions in a fashion analogous to rotation-cued 3-D. That is, the glyph-enhanced scatterplot is an N-dimensional projection of an M-dimensional space. With the rotation defined in the full space, interrelationships among all M dimensions will begin to appear as the rocking angle increases.

The choice of the fixed subspace clearly influences the utility of rigid-motion displays. Our early investigations suggest that an algorithm such as principal components shows promise for this selection. For example, choosing the fixed subspace as the smallest principal components maximizes the movement in the display. On the other hand, if the interest is on the behavior of points that do not fit the general pattern, then the general pattern should define the fixed subspace. For example, choosing the fixed subspace as directions with the largest principal components helps classify outliers by the pattern of their discrepant motion.

6 CHALLENGES

This chapter presents a range of display techniques that help to reveal patterns in multivariate data. With most display techniques there are technical details that must be surmounted before the techniques are useful. For techniques such as the anaglyph stereo, where the analyst might tackle the implementation, we have provided substantial detail. For others, such as 5-D rigid motion, programming assistance is likely to be required and the details have been omitted. We hope that other authors will make technical details available so that software developers can transform the display techniques to generally available tools.

Most techniques presented here have repeatedly proved helpful in looking at data. The static display techniques are generally minor adaptations of the better techniques described in the statistical graphics literature. The dynamic display techniques, sequential views, high-performance interaction with scatterplot matrices, and higher-dimensional rigid motion have not yet stood the test of time. They represent a blending of ideas from statistics, image processing, animation, human factors, etc. and can expect to undergo rapid evolution. We challenge others to investigate this fertile area.

Since Fienberg's (1979) statement that there is no theory of graphics, progress has been made. Many papers have appeared which discuss graphics principles. Principles like the maximizing data-ink ratio (within reason) and using display variables that have high perceptual accuracy of extraction provide general guidelines for improving any plot. However, a list of specific techniques for particular objectives, such as looking for multivariate outliers, has yet to be established. As such guidance is collected in an integrated knowledge base, the development of expert systems for statistical graphics becomes possible. Attempts to construct such systems should help clarify what we know and what we do not know. The current theory of statistical graphics notably lacks improvement principles for dynamic displays. We would like to know more about eye movements, competing muscular movements, memory, and human perception over time. We would like to know more about what specific visual aspects of plots analysts respond to (appropriately or not). We challenge all disciplines with the requisite expertise to join in the development of a theory of statistical graphics.

The importance of particular graphical techniques is hard to establish. The difficulties are multifaceted. Once a pattern is found, it is typically easy to develop other displays that show the pattern. The probability of detecting significant patterns and false error rates are dependent on ease of use as well as optimal display design. The interplay between graphics and modeling is not well understood. Graphics does not always come first. Our mathematical perception can exceed our visual perception. Decisions as to what graphics to display are often driven by statistics. A lot of hard work is ahead to establish the importance and limitations of particular graphical techniques.

Historically, the major developments in graphics have come from specific applications. A current problem with graphical display is the difficulty in showing data while conveying the constraints of particular physical/mathematical realities. There is much to be learned from maps of DNA, geological resources, the stars, and correlated residuals. Having task-specific graphics displayed in the statistics literature, whether or not statistical considerations are evident, would benefit both statistical graphics and the application area.

REFERENCES

1. Anderson, E. 1935. "The Irises of the Gaspe Penninsula." *Bulletin of the American Iris Society* 59:2-5.

2. Anderson, E. 1957. "A Semigraphical Method for Analysis of Complex Problems." *Proceedings of the National Academy of Sciences* 13:923-927. Reprinted in *Technometrics* 2(1950):387-391.

3. Asimov, D. 1983. *The Grand Tour.* PUB-3211, Stanford Linear Accelerator Center, Stanford University, Stanford, California.

4. Bachi, R. 1978. "Graphical Statistical Methodology in the Automation Era." Graphical Presentation of Statistical Information: Presented at 136th Annual Meeting of American Statistical Association, Soc. Stat. Sect. Sess. *Graphical Methods of Statistical Data*, Boston, Massachusetts, 1976. Technical Publication 43:13-68.

5. Baecker, R. 1979. "Digital Video Display Systems and Dynamic Graphics." *Computer Graphics* 13(2):48-56.

6. *Becker, R. A., and W. S. Cleveland. Brushing a Scatterplot Matrix: High-Interaction Methods for Analyzing Multidimensional Data.* Statistics Technical Report No. 7, AT&T Bell Laboratories, Murray Hill, New Jersey. 1984.

7. Box, G. E. P., and D. R. Cox. 1964. "An Analysis of Transformations." *J. Royal Statistical Society* B(26):211-252.

8. Carr, D. B. 1981. "Raster Color Displays—Examples, Ideas, and Principles." *Proceedings of the 1980 DOE Statistical Symposium*, CONF-801045, October 29-31, 1980, Berkeley, California. pp. 116-126.

9. Carr, D. B., and R. J. Littlefield. 1983. "Color Anaglyph Stereo Scatterplots—Construction Details." *Computer Science and Statistics: Proceedings of the 15th Symposium on the Interface*, March 1983, Houston, Texas, North Holland Publishing Company, New York, pp. 295-299.

10. Carr, D. B., and W. L. Nicholson. 1984. "Graphical Interaction Tools for Multiple 2- and 3-Dimensional Scatterplots." *Computer Graphics '84: Proceedings of the 5th Annual Conference and Exposition of the National Computer Graphics Association, Inc.* ISBN 0-941514-05-6, Vol. 2, May 1984, Anaheim, California, pp. 748-752.

11. Carr, D. B., R. J. Littlefield and W. L. Nicholson. 1985. "Scatterplot Matrix Techniques for Large N." PNL-SA-1311, Pacific Northwest Laboratory, Richland, Washington. To appear in *Computer Science and Statistics: Proceedings of the 17th Symposium on the Interface.* March 1985, Lexington, Kentucky.

12. Chambers, J. M., W. S. Cleveland, B. Kleiner, and P. A. Tukey. 1983. *Graphical Methods For Data Analysis.* Duxbury Press, Boston, Massachusetts.

13. Cleveland, W. S. 1979. "Robust Locally Weighted Regression and Scatterplots." *JASA* 74(368):829-836.

14. Cleveland, W. S., C. S. Harris and R. McGill. 1982. "Judgments of Circle Sizes on Statistical Maps." *JASA* 77(379):541-547.

15. Cleveland, W. S., and R. McGill. 1983. "A Color-Caused Optical Illusion on a Statistical Graph." *The American Statistician* 37(2):101-105.

16. Cleveland, W. S., and R. McGill. 1984a. "Graphical Perception: Theory, Experimentation, and Application to the Development of Graphical Methods." *JASA* 79(387):531-554.

17. Cleveland, W. S., and R. McGill. 1984b. "The Many Faces of a Scatterplot." *JASA* 79(388):807-822.

18. Daniel, C., and F. S. Wood. 1980. *Fitting Equations to Data.* 2nd ed. John Wiley & Sons, New York.

19. Diaconis, P., and J. H. Friedman. 1980. *M and N Plots.* PUB-2495, Stanford Linear Accelerator Center, Stanford University, Stanford, California.

20. Donoho, D., P. J. Huber and H. M. Thoma. 1981. "The Use of Kinematic Displays to Represent High-Dimensional Data." *Computer Science and Statistics: Proceedings of the 13th Symposium on the Interface*, Pittsburgh, Pennsylvania, March 1981, Springer-Verlag, New York. pp. 274-278.

21. Ferwerda, J. G. 1982. *The World of 3-D--A Practical Guide to Stereo Photography.* Netherlands Society for Stereo Photography, Haren, The Netherlands.

22. Fienberg, S. E. 1979. "Graphical Methods in Statistics." *The American Statistician* 33:165-178.

23. Fisher, S. 1982. "Viewpoint Dependent Imaging: an Interactive Stereoscopic Display, Processing and Display of Three-Dimensional Data." *Proceedings SPIE* 367:41-45.

24. Fisherkeller, M. A., J. H. Friedman and J. W. Tukey. 1974. *PRIM-9: An Inter-Active Multidimensional Data Display and Analysis System.* PUB-1408, Stanford Linear Accelerator Center, Stanford University, Stanford, California.

25. Friedman, J. H., and L. C. Rafsky. 1979. "Multivariate Generalizations of the Wald-Wolfowitz and Smirnov Two-Sample Tests." *Annals of Statistics* 7:697-717.

26. Friedman, J. H., and L. C. Rafsky. 1981. "Graphics for the Multivariate Two-Sample Problem." *JASA* 76(374):277-295.

27. Friedman, J. H., and J. W. Tukey. 1974. "A Projection Pursuit Algorithm for Exploratory Data Analysis." *IEEE Transactions on Computer* C-23:811-890.

28. Friedman, J. H., J. A. McDonald and W. Stuetzle. 1982. "An Introduction to Real-Time Graphical Techniques for Analyzing Multivariate Data." In *Proceedings of the 3rd Annual Conference and Exposition of the National Computer Graphics Association, Inc.* Vol. 1, June 1982, Anaheim, California, pp. 421-427.

29. Gnanadesikan, R. 1977. *Methods for Statistical Data Analysis of Multivariate Observations.* John Wiley & Sons, New York.

30. Hall, D. L. 1983. "ALDS 1982 Panel Review." PNL-SA-11178, Pacific Northwest Laboratory, Richland, Washington.

31. Joblove, G. H., and D. Greenberg. 1978. "Color Spaces for Computer Graphics." *Computer Graphics* 12(3):20-25.

32. Julesz, B. 1971. *Foundations of Cyclopean Perception.* University of Chicago Press, Chicago, Illinois.

33. Kleiner, B., and J. A. Hartigan. 1981. "Representing Points in Many Dimnensions by Trees and Castles." *JASA* 76(374):260-269.

34. Kruskal, W. H. 1982. Criteria for Judging Statistical Graphics." *Utilities Mathematica* B(21B):283-310.

35. Lipton, L. 1982. *Foundations of the Stereo-Scopic Cinema, A Study in Depth.* Van Nostrand Reinhold, New York.

36. Littlefield, R. J. 1984. "Basic Geometric Algorithms for Graphic Input." In *Computer Graphics '84: Proceedings of the 5th Annual Conference and Exposition of the National Computer Graphics Association, Inc.* ISBN 0-941514-05-6, Vol. 2, May 1984, Anaheim, California, pp. 767-776.

37. Morse, A. 1979. "Some Principles for the Effective Display of Data." *Computer Graphics* 13(2):94-101.

38. Murch, G. M. 1984. "Physiological Principles for the Effective Use of Color." *IEEE Computer Graphics and Applications* 4(11):49-54 (ISSN 0272-1716).

39. Newman, W. M., and R. F. Sproull. 1979. *Principles of Interactive Computer Graphics.* 2nd ed. McGraw-Hill, New York.

40. Nicholson, W. L., and R. J. Littlefield. 1982. "The Use of Color and Motion to Display Higher-Dimensional Data." In *Proceedings of the 3rd Annual Conference and Exposition of the National Computer Graphics Association, Inc.* Vol. 1, June 1982, Anaheim, California, pp. 476-485.

41. Nicholson, W. L., and R. J. Littlefield. 1983. "Interactive Color Graphics for Multivariate Data." *Computer Science and Statistics: Proceedings of the 14th Symposium on the Interface*, July 1982, Troy, New York, Springer-Verlag, New York, pp. 211-219.

42. Phillips, M. B., and G. M. Odell. 1984. "An Algorithm for Locating and Displaying the Intersection of Two Arbitrary Surfaces." In *IEEE Computer Graphics and Applications* 4(9):48-58 (ISSN 0272-1716).

43. Schmandt, C. 1982. "Interactive Three-Dimensional Computer Space, Processing and Display of Three-Dimensional Data." *Proceedings SPIE* 367:155-159.

44. Scott, D. W. 1984. "Multivariate Density Function Representation." *Computer Graphics '84: Proceedings of the 5th Annual Conference and Exposition of the National Computer Graphics Association, Inc.* ISBN 0-941-514-05-6, Vol. 2, May 1984, Anaheim, California, pp. 794-801.

45. Scott, D. W., and J. R. Thompson. 1983. "Probability Density Estimation in Higher Dimensions." *Computer Science and Statistics: Proceedings of the 15th Symposium on the Interface*, March 1983, Houston, Texas, North Holland Publishing Co., New York, pp. 173-197.

46. Shoup, R. 1979. "Color Map Animation." *Computer Graphics* 13(2):8-13.

47. Smith, A. R. 1978. "Color Gamut Transform Pairs." *Computer Graphics* 12(3):12-19.

48. Stover, H. S. 1981. "Terminal Puts Three-Dimensional Graphics on Solid Ground." *Electronics* 6:150-155.

49. Stover, H. S. 1982. "True Three-Dimensional Display of Computer Data, Processing and Display of Three dimensional Data." *Proceedings SPIE* 367:141-144.

50. Tufte, E. 1983. *The Visual Display of Quantitative Information.* Graphics Press, Cheshire, Connecticut.

51. Tukey, J. W. 1977. *Exploratory Data Analysis.* Addison-Wesley, Reading, Massachusetts.

52. Tukey, J. W., and P. A. Tukey. 1983. "Some Graphics for Studying Four- Dimensional Data." *Computer Science and Statistics: Proceedings of the 14th Symposium on the Interface.* July 1982, Troy, New York, Springer-Verlag, New York. pp. 60-66.

53. Tukey, P. A., and J. W. Tukey. 1981. "Graphical Display of Data Sets in Three or More Dimensions." Chapters 10, 11, and 12 in *Interpreting Multivariate Data*, ed. V. Barnett. Wiley, Chichester, United Kingdom.

54. Wainer, H. 1984. "How to Display Data Badly." *The American Statistician* 38(2):137-147.

55. Wainer, H., and C. M. Francolini. 1980. "An Empirical Inquiry Concerning Human Understanding of Two-Variable Color Maps." *The American Statistician* 34(2):81-93.

56. Watson, C. R., and A. R. Olsen. 1984. *Acid Deposition System (ADS) for Statistical Reporting: System Design and Users Code Manual.* EPA/600-8-84-023, U.S. Experimental Protection Agency, Research Triangle Park, North Carolina.

Orion I: Interactive Computer Graphics in Statistics[1]

John Alan McDonald[2]

Department of Statistics
Stanford University
Stanford, CA

ABSTRACT

The Orion I workstation is an experimental computer graphics system designed and built by Jerry Friedman and Werner Stuetzle at the Stanford Linear Accelerator Center (SLIC) in 1980-81[1] with the support of the Office of Naval Research, the U.S. Army Research Office, and the Department of Energy. It is being used to explore applications of computer graphics—especially interactive motion graphics—for statistical data analysis.[2,3]

The methods being developed on Orion I are among the first in the rapidly growing field of computer intensive statistical methods. These new methods, stimulated by the evolution of computing technology, are creating a new paradigm for statistical research; graphical and other computer intensive methods have a strong engineering flavor, in contrast to the more rigorous mathematical nature of traditional statistics.

1 TRADITIONAL STATISTICAL GRAPHICS

Graphical methods are among the oldest in statistics. Examples of graphical representation of quantitative data are known from 3000 B.C.[4] Any elementary text in statistics will discuss in detail the standard statistical graphics, such as the histogram and the scatterplot. However, until recently, the development of new graphical methods has been outside the mainstream of statistical research.

This is, in part, due to limitations of technology. The potential in pencil and paper graphics was thoroughly exploited early in the nineteenth century;

---

[1] This research was supported by ONR contract N00014-83-K-0472.

[2] *Current affiliation*: Department of Statistics, University of Washington, Seattle, Washington.

most of the standard techniques (histogram, scatterplot, etc.) were in use by the 1830's. Statistics acquired a real identity separate from mathematics in the twentieth century, especially in the period after World War II. Thus the successful graphical methods were well entrenched by the time statistics emerged as a field; these established methods left little room for improvement. Only unusually creative researchers were able to suggest useful additions to a statistician's graphical tool kit that could be used with pencil and paper.

2 EXPLORATORY DATA ANALYSIS

One of these unusually creative statisticians was John W. Tukey, who, in the 1960's, developed a family of both graphical and non-graphical statistical methods referred to collectively as exploratory data analysis.[5,6,7] The underlying philosophy of exploratory data analysis is very different from that of classical statistics.

Classical statistics emphasizes mathematical analysis. Problems that arise directly from real data are usually impossible to analyze mathematically, so only simplified, idealized problems are considered. Simplifying assumptions are made so that it is possible to determine rigorously the "best" solution. The implicit hope is that the "best" solution to an idealized problem will be a reasonable solution in more realistic circumstances.

The collection of methods characterized as "exploratory data analysis" approaches statistical problems from the opposite direction. These methods are not optimal in any precise sense. Their use is justified by the fact that they work in practice with real data.

3 DESCRIPTION AND INFERENCE

One reason that graphical methods are so different from classical statistics is that they address different problems.

Using graphical methods in statistics means looking at pictures of data; the purpose of graphical methods is primarily descriptive. Statistical description is concerned with discovering and summarizing features of a particular data set.

Classical statistics concentrates on mathematical analysis of formal methods for inference. An inference is a generalization from a given data set to some larger population (real or hypothetical) from which the data set is presumed to be a sample.

Formal inference is necessarily based on mathematical models. Description is necessary, first, to help construct a model, and later, to check to see if a chosen model is really appropriate.

4 PRIM-9

The introduction of modern computer graphics to statistics began at SLAC, in 1972, when Mary Ann Fisherkeller, Jerome H. Friedman, and John W. Tukey developed a system called Prim-9.[8] Prim-9 pioneered the use of real-time motion in statistical graphics, most notably to display three-dimensional scatterplots.

The traditional statistical graphics, the histogram and the scatterplot, are excellent pictures of data in one and two dimensions. However, it is very difficult to see relationships among three variables by looking at histograms and scatterplots.

With Prim-9, it was possible to see the three-dimensional shape of a data set using real-time rotation. If a point cloud is subjected to small rotations, and the rotated point cloud is projected on the display screen quickly enough for the motion to seem smooth; a viewer gets a convincing and accurate perception of the shape of the point cloud as a three-dimensional object.

The three-dimensional scatterplot is a simple idea. It is an obvious and natural extension of the standard, two-dimensional scatterplot. However, this simple first step in new statistical graphics gives statisticians a powerful new tool. It is trivial to see and understand structure in data that is very difficult or impossible to see in any number of two-dimensional pictures.

On Prim-9 in 1972, the display of three-dimensional scatterplots stretched the limits of the available computing power. The computation was done on SLAC's large, central, mainframe computer, and IBM 360/91. The total cost for hardware in Prim-9—the 360/91, the graphics device, and a high speed channel to transfer data from the 360/91 to the graphics device—was millions of dollars.

Prim-9 made a strong positive impression when it first appeared, but it did not have much immediate impact on the practice of statistics. The expense of the hardware made it impractical for such systems to be developed except at special institutions such as SLAC.

Successor to Prim-9 were developed by Peter Huber and his students in Switzerland (Prim-ETH) in 1978 and at Harvard (Prim-H) in 1979-1980. They were (and are —Prim-H is in use and still under development) similar to Prim-9 in basic computing technology and graphical capabilities, though perhaps one tenth as expensive.

5 ORION I

The latest descendant of Prim-9, Orion I, represents a substantial advance over the Prim systems, at least in hardware. Orion I is both more powerful and much less expensive that the Prim systems because it takes advantage of advances in microprocessor and computer graphics technology in the last decade.

Orion I cost $60,000 in 1981. It has much more computing power than the Prim systems. Orion I uses a color graphics device, as opposed to the strictly black and white displays in the Prim systems. Color is essential for several new methods that are used to look at pictures of more than three-dimensional data.

An important implication of Orion I is that such graphics systems will soon be widely available. The price of systems like Orion I is dropping rapidly; a better system could be built today for $25,000, less than half of what is cost eighteen months ago. It is likely that in a few years such graphics systems will be in the price range of personal computers.

6 APPLICATIONS OF ORION I

The methods we are developing on Orion I emphasize real-time motion and interaction. So it is difficult to do them justice in a verbal description. To compensate for this, we have made two movies, which are available for viewing.[9,10]

Almost any statistical problem will benefit from the addition of interactive graphics. An interactive graphics system makes it possible to bring human intelligence to bear directly on the problem. Humans have powerful abilities for pattern recognition that are so far not equaled by computers. Further, a human can adapt the analysis to unforeseen features of a data set, using judgement and a knowledge of the context of the problem from which the data arose. These points are clearly illustrated in our two films.

Orion I is being used to develop interactive graphical methods for many statistical problems, including: general multivariate analysis, regression, classification, clustering, time series, signal processing, seismic data (multivariate point processes), and analysis of digitized x-rays (interactive image processing).

The first problem, multivariate analysis, refers to the analysis of data sets in which there are many variables. A typical data set will have 100-1000 observations, and for each observation we will have values of perhaps 4-20 variables.

The basic graphical problem posed is this: how can we draw pictures that show us relationships involving 4 or more variables at once? If the data set had only two variables we would just look at an ordinary scatterplot. With three variables we can use a three-dimensional scatterplot. Looking at three-dimensional pictures is a very natural activity for human beings because we live in a three-dimensional world. However, drawing pictures that show four or more variables and are also easily understood is very difficult.

We are pursuing two basic approaches to more than three-dimensional data. The first approach is to find ways of effectively representing as many variables as possible at once in a single picture. For example, we take a three-dimensional scatterplot and color each point according to the value of an additional variable. This gives us a picture that shows some relationships among the four variables. We can add other features to the points in the scatterplot, such as size, shape, or orientation, to code values of more variables. Each additional feature will, unfortunately, add less information to the picture and will make the picture harder to interpret.

Another method puts several different scatterplots on the screen, and uses interactive coloring schemes to show relationships between the contents of the scatterplots.

The second approach to many-dimensional data is called Projection Pursuit.[11,12,13] The basic idea of Projection Pursuit is to compress (or project) the information in the many-dimensional data into a two or three-dimensional picture, which is easy to look at. We choose the projection to try to capture any interesting structure in the many-dimensional data.

REFERENCES

1. Friedman, J.H. and Stuetzle, W., "Hardware for Kinematic Statistical Graphics," Tech. Rep. Orion 003, Dept. of Statistics, Stanford University, (1982b).

2. Friedman, J.H., McDonald, J.A., and Stuetzle, W., "An Introduction Real Time Graphical Techniques for Analyzing Multivariate Data," *Proceeding 3rd Annual Conference of N.C.G.A.* (1982a).

3. McDonald, J.A., "Interactive Graphics for Data Analysis," Ph.D. thesis, Department of Statistics, Stanford University. Available as Technical Report Orion 011, Department of Statistics, Stanford, (1982).

4. Beniger, James R. and Robyn, Dorothy L., "Quantitative Graphics in Statistics: A Brief History," *The American Statistician,* 32 (1978).

5. Tukey, John W., "The Future of Data Analysis," *Annals of Mathematical Statistics,* 32, (1962).

6. Tukey, J.W., *Exploratory Data Analysis,* Addison-Wesley (1978).

7. Mosteller, F. and Tukey, J.W., *Data Analysis and Regression,* Addison-Wesley (1977).

8. Fisherkeller, M.A., Friedman, J.H., and Tukey, J.W., "Prim-9, An Interactive Multidimensional Data Display and Analysis System," *Proceeding 4th International Congress for Stereol* (1975).

9. Friedman, J.H., McDonald, J.A., and Stuetzle, W., (1982b) "Exploring Data with the Orion I Workstation," a 25 minute 16mm sound film, may be borrowed from: J.H. Friedman, Bin-88, Computation Research Group, Stanford Linear Accelerator Center, P.O. Box 4349, Stanford, CA 94305.

10. Friedman, J.H., McDonald, J.A., and Stuetzle, W., (1982c) "Projection Pursuit Regression with the Orion I Workstation," a 20 minute 16mm sound film, may be borrowed from J.H. Friedman, Bin-88, Computation Research Group, Stanford Linear Acclerator Center, P.O. Box 4349, Stanford, CA 94305.

11. Friedman, J.H. and Stuetzle, W., "Projection Pursuit Regression," *JASA, 76* (1981).

12. Friedman, J.H. and Stuetzle, W., (1982a) "Projection Pursuit Methods for Data Analysis," in *Modern Data Analysis,* Launer, Robert L. and Siegel, Andrew F., eds., Academic Press, 1982.

13. Friedman, J.H. and Tukey, J.W., "A Projection Pursuit Algorithm for Exploratory Data Analysis," *IEEE Trans. Comput. C-23* (1974).

Illustrations of Model Diagnosis by Means of Three-Dimensional Biplots[1]

K. Ruben Gabriel and Charles L. Odoroff

Department of Statistics and Division of Biostatistics
University of Rochester
Rochester, NY

ABSTRACT

The biplot is a graphical device that permits inspection of approximations to data matrices and is helpful in indicating models to fit the data. Its 3-D analogue, or bimodel, provides a closer approximation to the data and may thus be more revealing. In some examples it has led to the diagnosis of non-linear 3-D models that could not be directly inferred from planar biplots. To be useful to the researcher, representation of this 3-D bimodel must be both simple and revealing. The researcher must get a "feel" of the bimodel and must be able to use it to "feel" the data's structure. A variety of representations have been tried out, including projections on selected planes, projections on the faces of a dodecahedron, interactive pursuit of projection planes, stereo views, two-color anaglyphs and polarized anaglyphs. Imaginative use of color and real-time rotation are yet to be tried out. The choice of a suitable technique of representation must be such as to provide the researcher with an easy way to view his/her data and get a "feel" for its structure. This requires the representation to be both simple and revealing. Experimentation with such representation continues.

1 INTRODUCTION

This paper illustrates the use of three-dimensional (3-D) biplots for the diagnosis of models to fit data matrices. Diagnosis of models is an important, though often neglected, part of statistical data analysis. Biplots are especially suitable for such diagnostics because they display the elements of the data matrix and not merely properties of the rows or columns separately.

[1]This research was supported by ONR contract N00014-80-C-0387.

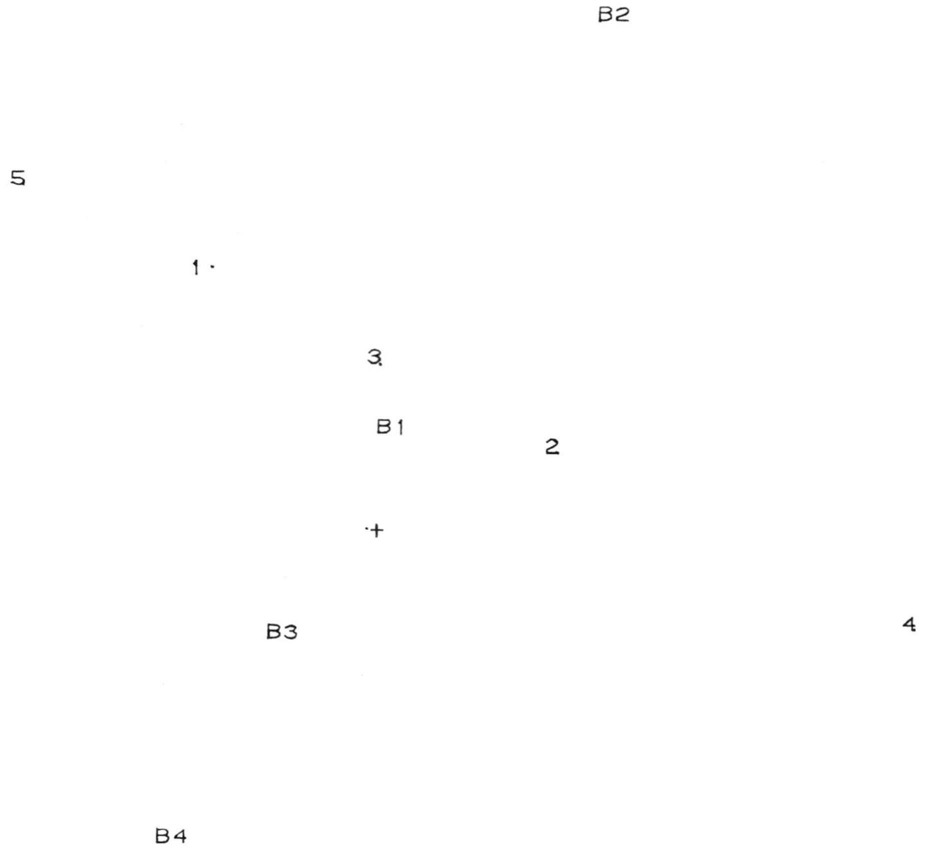

Figure 1 Biplot of the data of Table 1.

Biplots seem to be the only graphics display which simultaneously show both the scatter of the units (rows) and the configuration of the variables (columns), and do so in a way that also recovers the observations (matrix elements) themselves (Gabriel, 1981b). By contrast, most other techniques display properties of EITHER the rows OR the columns separately, but not of both together. For example, multidimensional scaling plots show inter-row differences as distances but do not display the columns. Correspondingly, factor analytic plots show the correlations between columns but ignore the rows. Another joint display is provided by correspondence analysis which superimposes row markers and column markers in a manner that does not allow recovery of the elements (see, for example, Greenacre, 1984).

The 3-D biplot represents a data matrix $Y(n \times m)$ through a factorization into the product of two matrices AB^T which approximates Y, where A is $n \times 3$ and B is $m \times 3$. The rows $a[i]$, $i = 1, \ldots, n$ and $b[j]$, $j = 1, \ldots, m$ of the A and B matrices form the 3-D markers of the biplot. This representation is only as good as the rank 3 approximation that is used to obtain AB^T (see Gabriel,

Table 1 A Very Simple Example

Y				A		B	
3	11	-1	-5	-2	3	0	1
1	9	-3	-7	2	1	2	5
2	10	-2	-6	0	2	-1	-1
-1	-7	-5	9	6	-1	-2	-3
4	12	-4	-4	-4	4		

1978, Gabriel and Zamir, 1979, and Gabriel and Odoroff, 1984, for methods of approximation). The explicit mathematical form of representing the elements $y[i,j]$ of Y as inner products $y[i,j] = a[i]^T b[j]$ of the markers allows the diagnosis of models fitting the $y[i,j]$'s from regularities observed in 3-D displays of the $a[i]$'s and/or $b[j]$'s.

As a first example, consider the 2-D biplot of Figure 1, which displays the data of the (5×4) matrix Y of Table 1 by means of the factor matrices A(5×2) and B(4×2). The marker $a[1]$ has biplot coordinates -2 and 3, the marker $a[2]$ coordinates 2 and 1, and so on until the marker $b[4]$ which has coordinates -2 and -3. To see how this displays the elements of Y, take, for an example, the inner product of $a[1] = (-2,3)$ with $b[1] = (0,1)$; this is clearly $a[1]'b[1] = 3$, which is indeed equal to $y[1,1]$. Other elements can be similarly obtained as inner products and can therefore be visualized on the biplot. (To visualize inner products recall that $a^T b$ is the product of the length of b times the length of the projection of a on b and it takes a negative sign if the projection is in the opposite direction from that of the vector projected upon.)

This graphical display makes it easy to see the main features of the matrix. Thus, $a[1]$, $a[3]$ and $a[5]$ are on the same side of the origin on which all the b's are; so all the elements in the first, third and fifth rows are positive. Since $a[5]$ is farthest out in that direction, the elements in the fifth row are the largest. By the same reasoning, the position of $a[4]$ indicates that the fourth row has the largest negative elements. Analogously, the position of $b[2]$ indicates large elements in the second column, and so forth.

Another obvious feature is that the row markers $a[1], \ldots, a[5]$ are on one line and the column markers are on another line and the two lines are at right angles. That is characteristic of matrices with additive structure, i.e., matrices whose elements can be decomposed as $y[i, j] = r[i] + v[j]$, for some $r[i]$, $i=1,\ldots,n$ and $v[j]$, $j=1,\ldots,m$. This was shown by Bradu and Gabriel (1978), who also discussed features of 2-D biplots which are characteristic of matrices with other structures. They proposed the inspection of (approximate) biplots of data matrices for the presence of such features as colinearity and orthogonality with the aim of diagnosing a model that may fit the matrix. (See Section 3, below.)

The above example was two dimensional: the matrix Y was of rank 2 and could be biplotted exactly in the plane. That is not always possible: most data matrices are of higher rank and the plane biplot of their rank 2 approximations will not always reveal their salient features. 3-D biplots may do better.

The second simple example is that of the 8x10 matrix Y of Table 2. Each

Table 2 Another Simple Example

Y									
4.	-1.	4.	9.	4.	9.	6.	9.	14.	13.
0.	-5.	0.	5.	0.	5.	2.	5.	10.	9.
-2.	7.	-2.	3.	-2.	3.	0.	3.	8.	7.
-6.	-11.	-6.	-1.	-6.	-1.	-4.	-1.	4.	3.
-4.	9.	2.	-5.	8.	1.	10.	7.	0.	5.
-6.	7.	0.	-7.	6.	-1.	8.	5.	-2.	3.
-10.	3.	-4.	-11.	2.	-5.	4.	1.	-6.	-1.
-14.	-1.	-8.	-15.	-2.	-9.	0.	-3.	-10.	-5.

A			B		
1.	3.	1.	-4.	2.	2.
1.	1.	1.	3.	2.	-4.
1.	0.	1.	-2.	2.	0.
1.	-2.	1.	-1.	2.	4.
1.	2.	-2.	0.	2.	-2.
1.	1.	-2.	1.	2.	2.
1.	-1.	-2.	2.	2.	-2.
1.	-3.	-2.	3.	2.	0.
			4.	2.	4.
			5.	2.	2.

of its factor matrices A and B has three columns and their biplot is displayed in 3-D - Figure 2.[2] This biplot also shows very systematic features: The column markers $b[1]$, $b[2]$, ..., $b[10]$ are on a plane and the row markers are on two parallel lines, $a[1]$, $a[2]$, $a[3]$ and $a[4]$ are on one line and $a[5]$, $a[6]$, $a[7]$ and $a[8]$ are on another line. Moreover, these two lines are at right angles to the plane of the $b[j]$'s. These particular biplot features were shown by Tsianco (1980) to be characteristic of a matrix with two additive structures, i.e., for which there exist $r[1], \ldots, r[8]$ and $v[1,1], \ldots, v[1,10]$ as well as $v[2,1], \ldots, v[2,10]$, such that $y[i,j] = r[i] + v[k,j]$ where $k=1$ if $i=1,2,3,4$ and $k=2$ if $i=5,6,7,8$.

It is instructive to compare the model diagnosed from Figure 2 with what would have been diagnosed from an approximate 2-D biplot of the same data - Figure 3 - in which the third coordinates of the a's and the b's were ignored. In this biplot one can see only two colinear sets of markers, the row markers' single line appearing at right angles to the line of the column markers. One would therefore have wrongly concluded that, as in the example of Table 1 and Figure 1, a single additive structure fitted the entire matrix.

In the study of real data, we would not expect to meet many matrices that are exactly of rank 2 or 3, but in many instances a low rank approximation is quite useful and, indeed, often very close to the data. Planar biplots, using the rank 2 approximation, have found many applications (See references in Gabriel, 1981b); however, there are instances where an additional dimension can provide not only a closer fit, but additional insight as well.

[2]See Plate 6 for Figure 2.

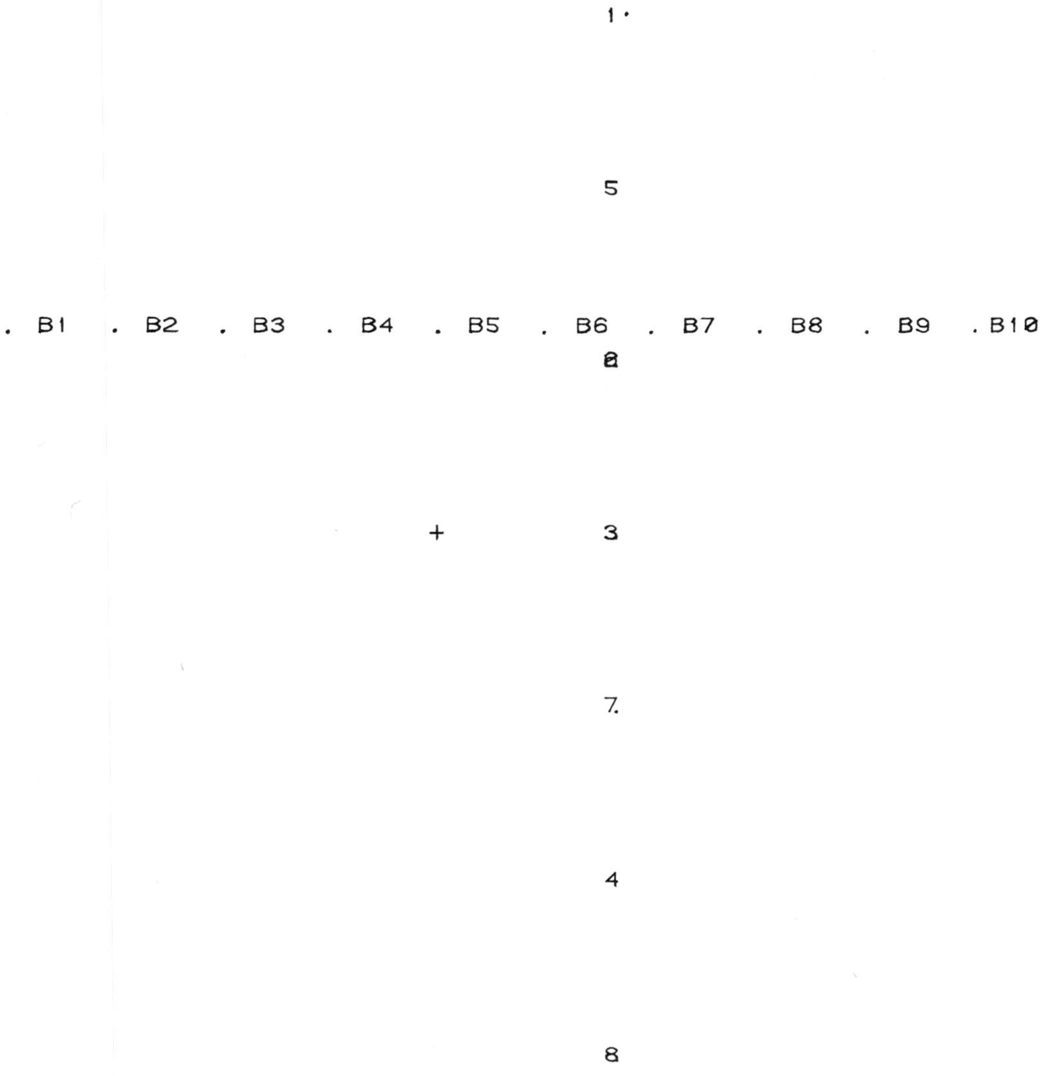

1 ·

5

. B1 . B2 . B3 . B4 . B5 . B6 . B7 . B8 . B9 . B10

2

\+ 3

7.

4

8

Figure 3 Two-dimensional biplot of Table 2.

As an example, consider 1951-52 monthly average temperatures at 50 stations in North, Central and South America which were studied by Tsianco and Gabriel (1984). Figure 4 shows their 2-D biplot, with stations represented by asterisks and months by their number. (For that, and the following, analyses, the data have been centered on the means of all 50 stations). They found the scatter of the station(row)-markers to reveal geographical groupings which are associated with similarity of temperatures. Our present interest focuses on the configuration of the month(column)-markers which is very regular, indeed almost colinear. The sequence of months along that line is indicative of a seasonality effect, except that the proximity of April and October, and of May

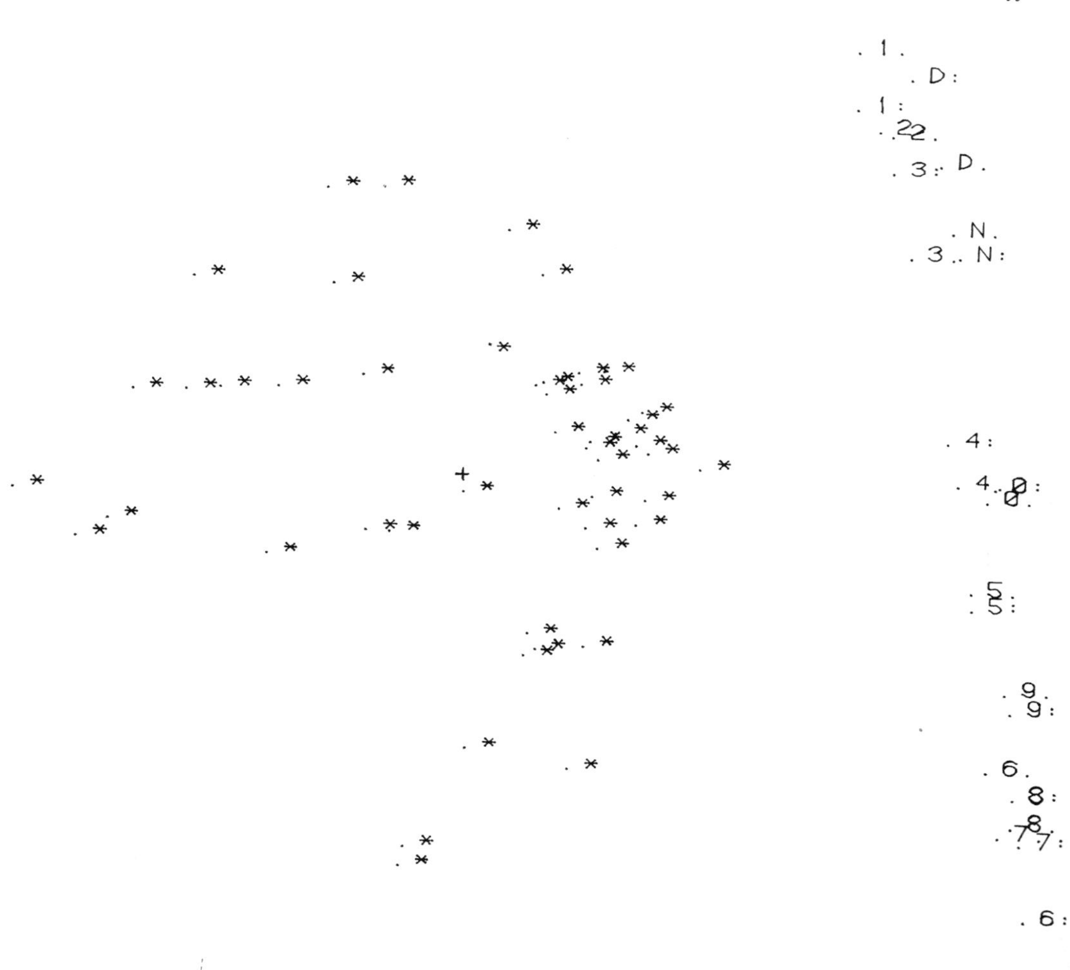

Figure 4 Two-dimensional biplot of American temperature data.

and September, does not agree with that. This suggests that another dimension is needed for a meaningful picture. We therefore present the 3-D biplot of these temperatures in Figure 5.[3]

Now the apparently linear configuration of the column markers is revealed to be the side view of an elliptical pattern, and the markers for the 12 months of each year are situated at fairly regular angular intervals on its circumference. That is a much more appealing picture and it suggests parametrization of the column markers as

$$b[j] = c + d\sin(\theta[j]) + e\cos(\theta[j]),$$

[3]See Plate 6 for Figure 5.

where c is a marker for the center of the ellipse, d for its major axis and e for its minor axis. If this is rewritten $b[j] = Qp[j]$, where $Q = (c, d, e)$ and $p[j] = (1,\sin(\theta[j]), \cos(\theta[j]))^T$, then the factorization of Y becomes $(A^*)P^T$, where $A^* = AQ$ and $p[j]$ is the j-th row of P. Thus, the elements of Y are represented by

$$y[i,j] = a^*[i,1] + a^*[i,2] \sin(\theta[j]) + a^*[i,3] \cos(\theta[j]),$$

which, after some trigonometry, can be parametrized as

$$y[i,j] = k[i] + q[i] \cos(\phi[i] + \theta[j]).$$

That is surely a very reasonable harmonic model for the temperature of month j at station i.

Note how the observation of an elliptic pattern on the 3-D biplot has led to the diagnosis of a harmonic model for the data themselves. Because the biplot displays a factorization of the data, the elliptical parametrization of the column markers could be transformed into a harmonic model for the data. Note also how important the third dimension was in revealing the essential features of this data set.

2 METHODS OF GRAPHICAL DISPLAY OF 3-D BIPLOTS

The diagnostics discussed in this paper are useful only to the extent that effective methods of displaying 3-D biplots are readily available. Color analglyphs are used for this purpose in the present volume, but alternative methods are also available. (Polarized analglyphs were used at the initial presentation of this paper at Luray, May 1983). This Section describes various 3-D display techniques (See also Gabriel, 1984) and the resources required to implement them.

The natural way to display any 3-D object would be to construct a physical model. That is obviously too cumbersome, slow and expensive to be used in standard applications. Instead one must use two-dimensional simulations of the three-dimensional objects as either static or dynamic displays. (A dynamic display is simply a static one updated very frequently to simulate motion.) For data analysis, the interactive use of dynamic displays is most effective, but static displays have a role for printed presentation and when dynamic devices are not accessible. Dynamic display requires much more expensive devices than static simulations. At the time of this writing a static device costs about \$1,000 (pen plotter or micro computer with bit-mapped graphics display) and a dynamic device costs \$10,000–50,000 (color raster display devices with many frame buffers and sufficient local computing power to refresh a display rapidly).

Three-dimensional displays are not difficult to implement. Their requirements are 1) a set of graphics routines to perform basic functions (points, lines, axes, labelling, etc.) 2) higher level routines to create displays, 3) display devices and the software device drivers with which to display them. Dynamic display requires that the computations be performed rapidly. If a good high resolution display has 1,000 by 1,000 picture elements and must be refreshed every 1/30 of a second, that may require about 30,000,000 transformations per second. This is not beyond the power of inexpensive display devices.

The mathematics of simulation of 3-D configurations in two dimensions has been reduced to a simple calculus (Newman and Sproull, 1983). It involves the reduction of the coordinates of three-dimensional objects to two dimensions in a way that preserves the impression of three dimensions. Simple matrix computations involving the circular functions suffice.

Many packages of low level graphics primitives are available now. There has recently been a great deal of work to standardize the primitives in hardware and software (GKS and Core). We have used a set of portable FORTRAN routines available in the public domain from the National Center for Atmospheric Research (McArthur, 1983). For biplot display we have used these graphics primitives in a set of higher level routines tied together by a simple command language called BGRAPH (Tsianco et al, 1981). BGRAPH incorporates the following options for both static and dynamic displays.

A Dynamic Displays

A powerful means to achieve the perception of depth of a display is to create motion and view the object from a rapidly changing sequence of vantage points. (Alternatively, one could move the object that is being viewed from a fixed vantage point.) This kind of motion can be simulated by presenting a rapid succession of views, with small angular separation between them. We have found that display times of 1/10 of a second per view, and angular separation of up to 20 degrees gives an adequate simulation of motion and impression of three-dimensionality.

Implementation of such simulation of motion may require the recomputation of all the coordinates of a display and refreshing of the image within less than 1/30 of a second. That can be done only by a dedicated mainframe or by the bit-mapped color raster graphics devices which are incorporated in some single user workstations. A much cheaper alternative is to store the coordinates of a succession of prechosen views in separate frame buffers of the display device and then flip through these in rapid succession. This can be used for "rocking" a view up and down, and right and left, of a prechosen vantage point. That gives the viewer a sense of depth without moving far from the viewing direction that has been chosen. We have implemented this technique on a color raster graphics device with twenty four frame buffers and have added color and perspective for further depth cues. We are finding this to be quite an effective tool for inspection of 3-D biplots and uncovering models to fit data matices.

B Static Displays

(i) PERSPECTIVE PLOTS. BGRAPH has an option to display 3-D data in a viewing cube (drawn in perspective on the viewing surface) into which the data is projected using the viewing transformations mentioned above. Another BGRAPH option displays the biplot on the faces of a viewing cube or dodecahedron. Dimensionality is simulated by cues from the perspectively drawn viewing body and by making character size a function of distance.

(ii) STEREO USING GLASSES. BGRAPH allows the selection of a

variety of options to produce stereo pairs. Side-by-side stereo pairs can be viewed by means of stereo glasses, either directly on the screen or as reproduced by means of a printer or a simple plotter. Better resolution can be obtained by electing to draw stereo pairs on separate pages and using a copier to reduce them to a size suitable for viewing. (Stereographic glasses are available under the name of Stereoscope from Hubbard Scientific Company, Northbrook, IL.)

(iii) STEREO USING POLARIZING LENSES. A stereo display that can be viewed simultaneously by several people can be created by using two slide projectors with polarizing filters in slip rings fitted over the lenses. The filters must be oriented so that the planes of polarization of the two projectors are perpendicular to each other. The projected images are then decoded by polarizing filters worn by the observers. An aluminized screen must be used since ordinary projector screens diffuse and depolarize the light beams. (Suitable polarizing filters of Polariod Type HN22 can be obtained from the Polaroid Corporation, Polarizer Division, One Upland Road, Norwood, MA 02062 and stereo glasses from Marks Polarized Corporation, Deer Park, NY 11729.)

(iv) STEREO USING COLOR ANALGLYPHS. The displays reproduced in this paper are produced by printing right and left views in different colors and are to be viewed through colored glasses which decode the stereo effect. (For a description of the construction and use of color analglyphs see Pearce, 1977.)

3 BIPLOT DIAGNOSTICS

The use of biplots for diagnosing mathematical models that fit, or approximate, data matrices was discussed first by Bradu and Gabriel (1978), who considered only 2-D biplots. They noted that colinearity of row markers characterizes Columns-Regression models,

$$y[i,j] = u[j] + s[i]v[j],$$

and similarly, that colinearity of column markers characterizes Rows-Regression models,

$$y[i,j] = r[i] + s[i]v[j].$$

(These models are due to Mandel, 1961, and to Finlay and Wilkinson, 1963). They further noted that if row markers and column markers are separately colinear, the angle between the two lines characterizes the model: a right angle is characteristic of the Additive model

$$y[i,j] = r[i] + v[j],$$

and any other angle is characteristic of the Concurrent model (A reparametrized form of the one-degree-of-freedom-for-non-additivity model due to Tukey, 1969),

$$y[i,j] = m + s[i]v[j].$$

They also made some attempts at modelling more complex patterns on 2-D biplots, as did Kester (1979) in a more systematic fashion. Tsianco (1980) expressed some reservations about the possibility of determining graphically whether an angle was of exactly ninety degrees and consequently doubted whether the biplot would be useful in distinguishing the Additive model from the Concurrent one.

These biplot diagnostics are useful for data matrices that can be closely approximated by a rank 2 matrix. More general diagnostics can be obtained by considering three-dimensional biplots, as these are also suitable for data which need a rank 3 matrix for adequate approximation. The extra dimension also greatly enriches the collection of models that may fit.

Tsianco (1980) studied a variety of such models and the biplot features which diagnose them. Some of these are listed in Table 3 where the following notation is used. The most general 3-D model

$$y[i,j] = r[i]u[j] + s[i]v[j] + t[i]w[j],$$

is abbreviated as $ru + sv + tw$. A general 2-D model is correspondingly denoted by $ru+st$. If any one factor becomes a constant it is simply omitted, so that, for example, the Columns-Regression model would be written $u+sv$ and the Rows-Regression model $r+sv$. If two factors are constant their product is denoted by m, so that the Concurrent model is written $m+sv$, etc. When two subsets of row vectors induce separate models, as in the example of Table 2 and Figures 2 and 3, above, their separate sets of parameters are indicated by an asterisk. For the above example, the $y[i,j] = r[i]+v[k,j]$ model would be written $r+v^*$.

Other workers in this field (Gower, 1983; Bradu, 1984) have arrived at similar results. In addition, Kester (1979) produced diagnostics for biplots of three-way tables. Bradu (1984) also suggested more general functional diagnostics and gave an example which would show up on the biplot as two parabolas; he diagnosed its model as

$$y[i,j] = m + r[i] + v[j] + kr[i]v[j] + pr[i]^2 + qv[j]^2.$$

Work is under way to systematize these and other rules and to make general statements about the diagnostic meaning of various general characteristics of biplot marker configurations. (For a preliminary report, see Chuang, Gabriel and Therneau, 1984.)

4 SOME EXAMPLES OF 3-D BIPLOTS AND THE MODELS THAT CAN BE DIAGNOSED [4]

This section presents a number of examples of 3-D biplots of matrices of rank 3, each of which is constructed to fit some linear or quadratic model. Inspection of each biplot should reveal patterns of row and/or column markers which can be used to reconstruct the model underlying the matrix elements. Thus, the matrix Y of Table 2 is constructed by fitting two additive models, one to the first 4 rows and another to the last 4 rows. Inspection of the 3-D biplot of Figure 3 reveal a structure that is known to be characteristic of precisely that model (Last entry of bottom row of Table 3).

---

[4]See Plates 7 to 12 for Figures 6 to 17.

Table 3 Some 3-D Biplot Linear Diagnostics

	Configuration of column markers			
	ANY	COPLANAR		
Row Marker Configuration		angular relation to row configuration		
		NOT APPLIC.	NON-PERP.	PERP.
Any	$ru+sv+tw$	$r+sv+tw$		
Plane	$u+sv+tw$		$m+sv+tw$	$r+v+tw$
Parallel planes	$u^*+sv+tw$		$m^*+sv+tw$	$r+v^*+tw$
Line	$u+sv$		$u+sv$ or $m+sv$	$r+v$
Parallel lines	u^*+sv		u^*+sv or m^*+sv	$r+v^*$

The object of these examples is to show that the 3-D biplot can reveal models underlying the elements of a matrix. Of course these examples are limited in size and somewhat artificial in that they are generated exactly according to certain 3-D models. Real data matrices can only be approximated by rank 3 models since their elements do involve errors. So the diagnosis of models from biplots of real data matrices is likely to be more difficult.

Each of the examples of this section is a biplot of a 20x15 matrix of rank 3. On the analglyph displays of the 3-D biplots, the 20 row markers are labelled by X's and the 15 column markers by C's. The notation for the models we diagnose follows that of Section 3, above; the names assigned to the models are similar to those given there for rank 2 models, except that the added dimension is indicated by the words plus-One. Thus, we refer to a model obtained by adding a third dimension to a Rows-Regression model as a Rows-Regression-plus-One model.

The first example, Figure 6, has the row markers spread all over the three-dimensional display, but the column markers are clearly on a plane. Checking Table 3 in the first row (row markers have no special pattern) and the second column (column markers are on a plane) we learn that a Rows-Regression-plus-One model is appropriate, i.e.,

$$y[i,j] = r[i] + s[i]v[j] + t[i]w[j].$$

The next two examples, the biplots of Figures 7 and 8, have the same planar column marker configuration as Figure 6, but both have the row markers lying on another plane. In Figure 7 the two planes appear to be perpendicular, whereas in Figure 8 they do not. According to Table 3 the former is characteristic of an Additive-plus-One, or FANOVA, model (Gollob, 1968),

$$y[i, j] = r[i] + v[j] + t[i]w[j],$$

and the latter of a Concurrent-plus-One, or Vacuum Cleaner, model (Tukey, 1962),

$$y[i,j] = m + s[i]v[j] + t[i]w[j].$$

Actually, it is not easy to ascertain from a plot whether two planes are at right angles or not. To try and make sure, we rotated the plots until our viewpoint allowed us to look straight down the line on which the two planes intersect. This could of course not be the viewpoint of both eyes, so we chose that of the right eye. (The viewer of the analglyph can see this by closing the left eye as she/he views the display.) Such rotations are not easy to do unless one has devices for real-time rotation, especially since biplot displays do not really show the planes but only a small number of markers that lie on them. However, the rotations of the biplots of Figures 7 and 8 into those of Figures 9 and 10, respectively, seem to meet the requirement reasonably well. And indeed we do see the right angles in the former but not in the latter, which confirms our diagnosis.

The biplot of Figure 11 has the same column markers configuration as the three preceeding examples, but the row markers are very clearly on two separate lines, which are parallel to each other but not perpendicular to the column markers' plane. As we can see from the last line of Table 3, this indicates a model of two Columns-Regressions, i.e.,

$$'y[i,j] = u[k,j] + s[i]v[j],$$

where $k=1$ or 2 according to whether i belongs to one or the other of the groups of markers represented by the two lines observed on the biplot. Note that this diagnosis is equivocal on whether the $u[k,j]$ terms might be independent of j, in which case this would be a model of two Concurrents, i.e.,

$$y[i,j] = m[k] + s[i]v[j].$$

(Actually, only the first model fits the data underlying this biplot.) Note also how this diagnosis differs from the Additive model example of Table 2 in which the two row marker lines were at right angles to the plane of the column markers.

The next few examples illustrate non-linear configurations. For simplicity we have chosen parabolas. Both the row markers and the column markers appear on parabolas in Figures 12 through 16. Each set of markers therefore must be coplanar. Inspection of the 3-D biplots suggests that these planes are not at right angles in any one of them (though that is never easy to determine graphically, as already mentioned). The models must therefore be specializations of the Concurrent-plus-One model, above, with quadratic relationships between the $t[i]$ and the $s[i]$, and between the $v[j]$ and the $w[j]$. In Figure 12 the directrices of the two parabolas appear to be parallel, which is characteristic of the model

$$y[i,j] = m + s[i]v[j] + s[i]^2 v[j]^2.$$

In Figure 13, on the other hand, the directrices of the two parabolas appear to be at right angles, and that is characteristic of model

$$y[i,j] = m + s[i]w[j]^2 + s[i]^2 w[j].$$

The differential diagnosis of these two examples depends on identification of a particular angle, and that is not easy to do by the present graphical techniques. Indeed, we display these biplots in a rotation that makes it relatively easy to diagnose the model.

Another data set generated by the generalized parabolic model

$$y[i,j] = \sum\sum c[a,e]s[i]^a v[j]^e, \quad a = 0, 1, 2 \; e = 0, 1, 2$$

is biplotted in Figure 14 and again in Figures 15 and 16 after certain rotations. The presence of the two non-perpendicular planes and the parabolas on them is already apparent from the original biplot in Figure 14. The rotations were used in attempts to find out more about the angular relationships. Thus, Figures 15 and 16 resulted from trying to look at the biplot from the direction of the line common to both planes (the row markers' plane and the column markers' plane) and from the direction perpendicular to it. Figure 15 strongly suggests non-perpendicularity of the planes. Figure 16 indicates that the two parabolas' directrices are neither parallel nor perpendicular. The appropriate model is therefore found to be

$$y[i,j] = m + \sum\sum c[a,e]s[i]^a v[j]^e. \quad a = 1,2 \; e = 1,2$$

Evidently, there still is more to parabolas than meets the eye.

A different pattern is evident in the 3-D biplot of Figure 17. Row markers are on one plane and column markers are on another and the two are clearly not at right angles. That indicates a Concurrent-plus-One model. However, the column markers are noticed to be on a lattice, on five parallel lines (labelled 1, 2, ..., 5) intersecting with three other parallel lines (labelled +,*,-). Kester (1979) has shown this to be characteristic of a three-way layout in which the columns categorize two factors (at five and three levels, respectively) and there is no interaction between those two factors. The Concurrent-plus-One model, with double column subscripts g,j for those two factors, would be

$$y[i,j] = m + s[i]v[g,j] + t[i]w[g,j],$$

but lack of interaction between the g and j classifications reduces this to

$$y[i,j] = m + s[i]v[g,*] + t[i]w[*,j].$$

5 THE EXAMPLE OF ALCOHOL DENSITY [5]

For an illustration of the insights biplot display can provide for modelling, we turn to the well known data on the density of solutions of alcohol in water due to Osborne, McKelvey and Bearce (1913). As in previous analyses of these data (Mandel, 1971; Bradu and Gabriel, 1978; Hoaglin, Wong and Emerson, 1983; Mandel, 1984; Gabriel and Odoroff, 1985), we concentrate on the subset of the data reproduced in Table 4.

Figure 18 is the 3-D biplot of these data centered on their overall mean of 0.896765. (Such centering makes inspection of a data set's features easier, much as windowing does.) We are immediately struck by the coplanarity both

[5] See Plates 13 and 14 for Figures 18 to 20.

Table 4 Density of Solutions of Alcohol in Water

(Mean observed density D_4^t)

	Columns are averages of pairs of determinations at temperatures in Centigrade						
% alcohol by weight	10	15	20	25	30	35	40
30.086	.959652	.956724	.953692	.950528	.947259	.943874	.940390
39.988	.942415	.938851	.935219	.93150	.927727	.923876	.919946
49.961	.921704	.917847	.913922	.909938	.905880	.901784	.897588
59.976	.899323	.895289	.891202	.887049	.882842	.878570	.874233
70.012	.875989	.871848	.867640	.863378	.859060	.854683	.850240
80.036	.851882	.847642	.843363	.839030	.834646	.830202	.825694

From Tables XXIV-XXXV of Osborne, McKelvey and Bearce (1913).

of the row markers (for concentrations) and of the column markers (for temperatures) and the almost perfect perpendicularity of these two planes. This strongly suggests that we should start by fitting an additive model. Indeed, such a model reduces the sum of squares of residuals to 26,111E-9 (as compared to 64,260E-6 for deviations from the overall mean) and obviously provides an excellent fit. Addition of one multiplicative term results in the Additive-plus-One model with a sum of squared residuals of only 3,484E-12; evidently the third dimension can play a substantial role in representing these data.

Since the perpendicularity of the two planes may be somewhat in doubt, one may prefer to fit the Concurrent-plus-One model,

$$y[i,j] = m + s[i]v[j] + t[i]w[j].$$

At the cost of introducing a single extra parameter, this further reduces the sum of squared residuals to 1,494E-12. (Note that least squares fitting of this model involves iteration between fitting the constant m - which is NOT the overall mean - and fitting the two multiplicative terms $sv+tw$ - no closed form is available for this fit. See similar remarks about the ordinary Tukey model in Gabriel, 1978).

A closer look at the 3-D biplot suggests there may be some curvature in each set of markers. This is mainly in the third dimension which fits only a very small percentage of the overall variability. In order to inspect this in more detail, we present two more biplots of these data, using factorizations which give equal scaling to each dimension of the row markers - Figure 19 - or of the column markers - Figure 20.

The column (temperature) markers are now seen to be situated at regular temperature intervals along a parabolic curve. The row (concentration) markers, on the other hand, are not as close to a plane; rather, there is a linear, or very slightly curvilinear trend from concentration 80% to 40%, but the marker for the 30% concentration is well away from this sequence.

The parabolic character of the column markers' configuration can be modelled by introducing a quadratic

$$w[j] = k + pv[j] + qv[j]^2$$

into the preceding model. It is less clear how to model the row markers' configuration. If they were approximated by another parabolic relationship

$$t[i] = f + gs[i] + hs[i]^2,$$

one would arrive precisely at the model that Mandel (1971) fitted so closely: the sum of squared residuals is reduced to 43E-9 for his 16-parameter model. If, instead, one used a cubic relationship, as Gabriel and Odoroff (1985) have done, one might improve the fit slightly: a sum of squared residuals of 40E-9 with a 10-parameter model.

This example illustrates how much can be seen from inspection of the 3-D bimodel, and how appropriate it may be for the choice of a suitable model to fit to the data. The fitting has to be done by numerical methods, be they least squares or resistant analogues, but the diagnosis of the mathematical form of the model to be fitted can be guided by the regularities observed on the biplot.

6 CONCLUDING REMARKS

This paper deals with the diagnosis of models to fit data matrices. Much statistical data analysis consists of the search and fitting of models to summarize and simplify complex data sets. Statistical texts devote much attention to methods of estimating the parameters of models and of testing hypotheses about them. They devote little or no space to the process of searching for models. It is to this neglected process of searching for models that this paper is addressed.

The methods of diagnosis described here take advantage of a unique feature of biplot display. The data on these displays may be recovered as accurately as the rank 2 or rank 3 approximation permits. Other methods do not usually permit recovery of the data, because they do not have an explicit connection between the display and the data matrix. The biplot recovers the matrix elements as inner products of the markers displayed. Only on a biplot can observed patterns of row and column markers be translated into models for the data.

ACKNOWLEDGMENTS

The authors gratefully acknowledge Michael Tsianco's permission to quote some of the biplot diagnostics which were presented in his doctoral thesis. They are also appreciative of the critical discussions of the methods of this paper with their colleagues Christy Chuang and Terry Therneau and the excellent computational work of Sandy Plumb-Scanlon.

REFERENCES

Bradu, Dan (1984). Response surface model diagnosis in two-way tables. *Communications in Statistics,* **13,** No. 24., 3059-3106.

Chuang, J.C., Gabriel, K.R., and Therneau, T. (1984). Geometrical diagnosis of linear/bilinear models with the biplot. Paper presented at the October 1984 ASA-IASC-SIAM Conference on Frontiers in Computational Statistics at Boston, Massachusetts.

Bradu, D. and Gabriel, K.R. (1978). The biplot as a diagnostic tool for models of two-way tables. *Technometrics,* **20,** 47-68.

Corsten, L.C.A. and Gabriel, K.R. (1976). Graphical exploration in comparing variance matrices. *Biometrics,* **32,** 851-863.

Cox, C. and Gabriel, K.R. (1982). Some comparisons of biplot display with pencil-and-paper E.D.A. methods. In *Modern Data Analysis* (R. L. Launer and A. Siegel, eds.). New York, Academic Press, pp 45-82.

Finlay, K.W. and Wilkinson, G.N. (1963). The analysis of adaptation in a plant breeding programme. *Australian Journal of Agricultural Research,* **14**, 742-754.

Gabriel, K.R. (1971). The biplot - graphic display of matrices with application to principal component analysis. *Biometrika,* **58,** 453-467.

Gabriel, K.R. (1978). Least squares approximation of matrices by additive and multiplicative models. *Journal of the Royal Statistical Society, B,* **40,** 186-196.

Gabriel, K.R. (1981a). Biplot. In *Encyclopedia of Statistical Sciences,* **Vol. I,** (S. Kotz, N.L.Johnson and C.Read, eds.). New York, John Wiley, pp. 262-265.

Gabriel, K.R. (1981b). Biplot display of multivariate matrices for inspection of data and diagnosis. In *Interpreting Multivariate Data,* (V. Barnett, ed.). London, John Wiley, pp. 147-173.

Gabriel, K.R., Multivariate Graphics. *In Encyclopedia of Statistical Sciences,* **Vol. V**, (S. Kotz, N.L.Johnson and C.Read, eds.). New York, John Wiley.

Gabriel, K.R. and Odoroff, C.L. (1984). Resistant Lower Rank Approximation of Matrices. In *Data Analysis and Informatics, III,* (E.Diday, M.Jambu, L.Lebart, J.Pages and R.Tomassone, eds). Amsterdam, North Holland, pp. 23-30.

Gabriel, K.R. and Odoroff, C.L. (1985). Some reflections on strategies of modelling: How, when and whether to use principal components. In *Biostatistics: Statistics in Biomedical, Public Health and Environmental Sciences* (P.K.Sen, ed.). Amsterdam, North Holland, 315-331.

Gabriel, K.R. and Zamir, S. (1979). Lower rank approximation of matrices by least squares with any choice of weights. *Technometrics,* **21**, 489-498.

Gollob, H.F. (1968). A statistical model which combines features of factor analytic and analysis of variance techniques. *Psychometrika,* **33,** 73-115.

Gower, J.C. (1983). Three-dimensional biplots. (Unpublished manuscript).

Greenacre, M.J. (1984). *Theory and Applications of Correspondence Analysis.* London, Academic Press

Hoaglin, D.C., Wong, G.Y., and Emerson, J.D. (1983). Resistant diagnosis of interaction in two-way tables. Annual Meeting of the American Statistical Association, Section of Physical and Engineering Sciences, Philadelphia, August 15, 1984.

Kester, N.K. (1979) . Diagnosing and fitting concurrent and related models for two-way and higher-way layouts. Ph.D. Dissertation at the University of Rochester, NY.

Mandel, J. (1961). Non-additivity in two-way analysis of variance. *Journal of the American Statistical Association,* **56**, 878-888.

Mandel, J. (1971). A new analysis of variance model for non-additive data. *Technometrics,* **11,** 411-429.

Mandel, J. (1984). Evolution of a model for physical data appearing in two-way tables. Annual Meeting of the American Statistical Association, Section of Physical and Engineering Sciences, Philadelphia, August 15, 1984.

McArthur, G.R. (1983). *The System Plot Package, NCAR Technical Note-TN/162.* Boulder, CO, National Center for Atmospheric Research, Scientific Computing Division.

Newman, W.M. and Sproull, R.F. (1980). *Principles of Interactive Computer Graphics (Second Edition).* New York, McGraw-Hill.

Pearce, G.F. (1977). *Engineering Graphics and Descriptive Geometry in 3-D.* Toronto, Macmillan.

Osborne, N.S., McKelvey, E.C., and Bearce, H.W. (1913). Density and thermal expansion of ethyl alcohol and of its mixtures with water. *Bulletin of the Bureau of Standards,* **9**, 327-474.

Tsianco, M.C. (1980). Use of biplots and 3D-bimodels in diagnosing models for two-way tables. Ph.D. Dissertation at the University of Rochester, NY.

Tsianco, M.C. and Gabriel, K.R. (1984). Modeling Temperature Data: An Illustration of the Use of Biplots in Nonlinear Modeling. *Journal of Climate and Applied Meteorology,* **23**, No. 5, pp. 787-799.

Tsianco, M.C., Odoroff, C.L., Plumb-Scanlon, S., and Gabriel, K.R. (1981). BGRAPH--A Program for Biplot Multivariate Graphics. Technical Report 81/20. Department of Statistics, The University of Rochester.

Tukey, J.W. (1949). One degree of freedom for non-additivity. *Biometrics,* **5**, 232-242.

Tukey, J.W. (1962). The future of data analysis. *Annals of Mathematical Statistics,* **33**, 1-67.

Multivariate Thin Plate Spline Smoothing with Positivity and Other Linear Inequality Constraints[1]

Grace Wahba

Department of Statistics
University of Wisconsin
Madison, WI

ABSTRACT

The relationship between smoothing splines and Butterworth filters, both on the line and on the plane, is reviewed. Splines may be thought of as the extension of low pass filters to noisy, unequally spaced data from nonperiodic functions. Furthermore, linear inequality constraints such as nonnegativity and monotonicity may be imposed on the spline. Numerical results in the smoothing of scattered noisy discrete non-Gaussian data on the plane subject to the constraint that the smoothed surface represent a probability (i.e., be between 0 and 1) are given. These results provide a method of estimating posterior probabilities in the classification problem, for bivariate data.

1 INTRODUCTION

It has been suggested (e.g. Wahba (1982), Wegman (1984), Wegman and Wright (1983)) that spline smoothing and the imposition of linear inequality constraints such as nonnegativity, monotonicity, etc. could be combined to fit smooth (nonparametric) curves and surfaces to discrete, noisy data.

In this paper we will first describe relationship between splines and low pass (Butterworth) filters, both in one and two dimensions. This relationship allows one to extend the notion of "low pass filter" in both one and two dimensions, to irregularly spaced data and nonperiodic source, via spline theory. Then we will discuss the (smooth) imposition of families of linear equality constraints (for example, nonnegativity) on the filter output, which is the spline.

[1] This research was sponsored by ONR under Contract No. N00014-77-C-0675.

We will discuss the (adaptive) choice of the bandwidth parameter(s) from the data by the method of generalized cross validation for constrained problems (CGCV). Finally, we will present an example of the use of a two dimensional thin plate (TP) smoothing spline, subject to the linear inequalities $0 \leqslant f(x,y) \leqslant 1$, $(x,y) \in \Omega$, for the estimation of posterior probabilities in the classification problem.

2 SMOOTHING POLYNOMIAL SPLINES (ONE DIMENSIONAL)

The model is

$$z_i = f(x_i) + \epsilon_i, \quad 1 = 1,2,\ldots, n$$

where $0 \leqslant x_1 \leqslant x_2, \ldots, \leqslant x_n \leqslant 1$, the ϵ_i are independent, normally distributed random variables with common, unknown variance σ^2, and f is supposed to be in the (Sobolev) space W_2^m, where

$$W_2^m = \{f: f, f', \ldots, f^{(m-1)} \text{abs. cont.}, f^{(m)} \in L_2[0,1]\}.$$

Given the data $z = (z_1, \ldots, z_n)'$, the smoothing spline estimate f_λ of f is the minimizer, in W_2^m, of $I_\lambda(f)$ defined by

$$I_\lambda(f) = \frac{1}{n}\sum_{i=1}^{n}(f(x_i) - z_i)^2 + \lambda\int_0^1 (f^{(m)}(u))^2 du.$$

It is known that if $\lambda > 0$ and there are at least m distinct x_i's among the $x_1, \ldots, x_n$, then $I_\lambda(f)$ is strongly convex over W_2^m and hence has a unique minimizer, call if f_λ. The properties of this minimizer were described by Schoenberg (1964), and are as follows:

$$f_\lambda \in \pi^{2m-1} \text{ for } x_i \leqslant x \leqslant x_{i+1}, \quad i = 1,2,\ldots,n-1$$
$$f_\lambda \in \pi^{m-1} \text{ for } 0 \leqslant x \leqslant x_1 \text{ and } x_n \leqslant x \leqslant 1$$
$$f_\lambda \in C^{2m-2}$$

where π^k are the polynomials of degree k or less, and C^k are the functions with k continuous derivatives. Thus f_λ is piecewise polynomial, with the pieces joined so that all derivatives up to and including the 2m-th are continuous. The parameter λ may be chosen adaptively from the data by the method of generalized cross validation (GCV). The GCV estimate $\hat{\lambda}$ is an an estimate of λ^*, where λ^* is that λ which minimizes predictive mean square error $R(\lambda)$ defined by

$$R(\lambda) = \frac{1}{n}\sum_{i=1}^{n}(f_\lambda(x_i) - f(x_i))^2.$$

Properties of $\hat{\lambda}$ may be found in Craven and Wahba (1979), Li (1983a,b), Speckman (1982), Wahba (1983b). A computer program for computing $f_{\hat{\lambda}}$ is available in IMSL. See also Kimeldorf and Wahba (1971), Wahba (1978, 1983a).

3 SMOOTHING POLYNOMIAL SPLINES AS THE OUTPUT OF A BUTTERWORTH FILTER

Suppose that $x_i = i/n$, $i = 1,2,\ldots,n$, and suppose it is known that f satisfies the periodicity conditions $f^{(\nu)}(0) = f^{(\nu)}(1), \nu = 0,1,\ldots,m-1$. Then one can minimize $I_\lambda(f)$ subject to these constraints. Then, to a good approximation, f_λ can be shown to be the output of a Butterworth filter, with filter function $\psi(\nu) = 1/(1 + (2\pi\nu)^{2m}), \nu = 1,2,\ldots,$. More precisely, let us find f_λ of the form

$$f_\lambda(x) = \hat{\alpha}_o + 2\sum_{\nu=1}^{n/2-1} \hat{\alpha}_\nu \cos 2\pi\nu x + 2\sum_{\nu=1}^{n/2-1} \hat{\beta}_\nu \sin 2\pi\nu x + \hat{\alpha}_{n/2} \cos \pi\ nx \tag{3.1}$$

to minimize $I_\lambda(f)$. It can be shown that the coefficients (α_ν,β_ν) of the minimizer satisfy

$$\alpha_\nu = a_\nu/(1 + \lambda(2\pi\nu)^{2m}) = a_\nu\,\psi(\nu), \nu = 0,1,\ldots,n/2$$

$$\beta_\nu = b_\nu/(1 + \lambda(2\pi\nu)^{2m}) = b_\nu\psi(\nu), \nu = 1,2,\ldots,n/2-1$$

where

$$a_\nu = \frac{1}{n}\sum_{j=1}^{n} \cos 2\pi\ \nu\frac{j}{n} z_j \quad \nu = 0,1,\ldots,n/2$$

$$b_\nu = \frac{1}{n}\sum_{j=1}^{n} \sin 2\pi\ \nu\frac{j}{n} z_j \quad \nu = 1,2,\ldots,n/2-1.$$

(For more details, see Craven and Wahba (1979), Wahba (1982). Thus, f_λ may be thought of as the output of a low pass filter with filter function $\psi(\nu) = 1/(1 + \lambda(2\pi\ \nu)^{2m})$, and, more to the point, smoothing splines can be thought of as a generalization of the output of a low pass filter to non periodic problems with unequally spaced data.

4 SMOOTHING POLYNOMIAL SPLINES WITH NONNEGATIVITY CONSTRAINTS

Suppose $f \in W_2^m$ and it is known that $0 \leqslant f(x)$, $0 \leqslant x \leqslant 1$. Since W_2^m is a reproducing kernel Hilbert space, (a space where the evaluation functionals are bounded), the set C of nonnegative functions

$$C = \{f: f \in W_2^m, f(x) \geqslant 0, 0 \leqslant x \leqslant 1\}$$

is closed convex. Thus (if there are m distinct values among the x_i's) I_λ has a unique minimizer (call it f_λ), in C. Two problems remain: (1) how to choose λ and (2) given λ, how to compute (a good approximation to) f_λ.

Let $C_L = \{f: f \in W_2^m, f(\frac{j}{L}) \geqslant 0, j = 0,1,\ldots,L$ where L is fairly large. It can be argued that for L large and λ given, the minimizer of I_λ over C_L is a good approximation to the minimizer of I_λ over C. (See for example Wahba (1973).

The problem of minimizing I_λ over C_L (with λ fixed) can be reduced to a standard quadratic programming (QP) problem in $n + m + L$ unknowns with

L linear inequalities. (See Kimeldorf and Wahba (1971)). We have found a computer program developed by Gill, Gould, Murray, Saunders and Wright (1982) to be highly suitable for solving the QP's arising in constrained spline problems.

The method of generalized cross validation for constrained problems (CGCV) proposed in Wahba (1980b) is suitable for choosing λ from the data when the estimate f_λ is the minimizer of $I_\lambda(f)$ in C_L. See also Wahba (1982) and Villalobos (1983) for further details concerning CGCV and numerical results.

5 MONOTONE SMOOTHING SPLINES

Let $C = \{f: f'(x) \geqslant 0, 0 \leqslant x \leqslant 1$. If $m = 2,3,\ldots$, then it can be shown that C is a closed convex set in W_2^m, and hence (if there are m distinct x_i's), $I_\lambda(\cdot)$ has a unique minimizer in C. The set C can be approximated either by

$$C_L = \{f: f'(\frac{j}{L}) \geqslant 0, j = 0,1,\ldots,L\}$$

or

$$C_L' = \{f: f\frac{(j+1)}{L} - f(\frac{j}{L}) \leqslant 0, j = 0,1,\ldots,L-1\}$$

and the problem reduced to a finite dimensional QP via the theory in Kimeldorf and Wahba (1971). The minimizer of I_λ in C_L or C_L' can be found by following the algorithmic approach in Villalobos (1983) and the CGCV estimate of λ also obtained following Villalobos. However this program has not, to our knowledge, been carried out.

6 MULTIVARIATE THIN PLATE SMOOTHING SPLINES

There are several generalizations of univariate splines to two and higher dimensions. In this paper we will only discuss thin plate (TP) splines. Thin plates splines can in principle be computed in two, three and higher dimensions. However, as the number of dimensions increases, the number of data points required to obtain good nonparametric surface estimates will become prohibitive. In this paper we will only consider 2 dimensions. (See Wahba and Wendelberger (1980), Wendelberger (1981), Hutchinson et. al. (1983) concerning 3 dimensions). Partially splined TP models suitable for more than three dimensions are introduced in (1983c), see also Engle, Granger, Rice and Weiss (1983). The basic foundation for the TP splines was laid by Duchon (1976) and Meinguet (1979). See also Wahba and Wendelberger (1980), Wahba (1983d).

The model behind the TPS for 2 dimensions is

$$z_i = f(x_i, y_i) + \epsilon_i, \quad i = 1,2,\ldots,n, \tag{6.1}$$

where (x_i, y_i), $i = 1,2,\ldots,n$ are n points on the plane, the ϵ_i are as before and f is in an appropriate function space X (see Meinguet (1979)) of real valued functions on the plane satisfying

$$J_m(f) = \int_{-\infty}^{\infty}\int_{-\infty}^{\infty}\sum_{\nu=0}^{m}\binom{m}{\nu}\left(\frac{\partial^m f}{\partial x^\nu \partial y^{m-\nu}}\right)^2 dxdy < \infty.$$

For $m = 2$, this reduces to

$$J_2(f) = \int_{-\infty}^{\infty}\int_{-\infty}^{\infty}(f_{xx}^2 + 2f_{xy} + f_{yy}^2)\,dxdy < \infty.$$

(If $m = 1$ the evaluation functionals in X are not bounded and we will not consider this case.) The TPS is the minimizer in X of

$$I_\lambda(f) = \frac{1}{n}\sum_{i=1}^{n}(f(x_i,y_i) - z_i)^2 + \lambda J_m(f) \qquad m \geqslant 2. \tag{6.2}$$

Let $\phi_1(x,y),\phi_2(x,y) \ldots \phi_M(x,y)$ be the $M = \binom{m+1}{2}$ polynomials of degree less than m. (For example, if $m = 2$, $M = 3$ and $\phi_1(x,y) = 1$, $\phi_2(x,y) = x$, $\phi_3(x,y) = y$). Let T be the M x n matrix with $i\nu th$ entry $\phi_\nu(x_i,y_i)$. It can be shown that if T is of full rank, then $I_\lambda(f)$ is strongly convex over f and will have a unique minimizer, call it f_λ.

$J_m(f)$ is isotropic, that is, it is invariant under rotations of the (x,y) plane. An anisotropic version of $J_m(f)$ is given by

$$J_m^k(f) = \int_{-\infty}^{\infty}\int_{-\infty}^{\infty}\sum_{\nu=0}^{m}\binom{m}{\nu}\left(k^{m-\nu}\frac{\partial^m f}{\partial x^\nu \partial y^{m-\nu}}\right)^2 dxdy$$

where k is some positive constant. Changing k is equivalent to changing the units in which y is measured.

Letting $f_{\lambda,k}$ be the minimizer of

$$I_{\lambda,k}(f) = \frac{1}{n}\sum_{i=1}^{n}(f(x_i,y_i) - z_i)^2 + \lambda J_m^k(f),$$

a comparison between $f_{\lambda,k}$ and the output of a low pass filter can be made, just as in Section 3, by letting $(x_i,y_i) = (i/n^2,j/n^2)$, $i,j = 1,2,\ldots,n$, and specializing to the case that f is a doubly periodic function analogous to (3.1). Some calculations will show that the resulting filter function $\psi(\mu,\nu)$ is

$$\psi(\mu,\nu) = 1/(1 + \lambda[(2\pi\mu)^2 + (2\pi k\nu)^2]^m).$$

(See Wahba (1981).) In the remainder of this paper we will fix $k = 1$, although we note that k can also be chosen by GCV (see Hutchinson et al. (1983), Wendelberger (1982)).

An explicit representation for the minimizer f_λ of (6.2) is available, and is (letting $t = (x,y)),t_i = (x_i,y_i)$

$$f_\lambda(t) = \sum_{i=1}^{n}c_i E_m(t - t_i) + \sum_{\nu=1}^{M}d_\nu\phi_\nu(t)$$

where

$$E_m(\tau) = \theta_m|\tau|^{2m-1}\log|\tau|,$$
$$|\tau| = |x^2 + y^2|^{1/2}$$

$$\theta_m = \{2^{2m-1}\pi[(m-1)!]^2\}^{-1}$$

and $c = (c_1, \ldots, c_n)'$ and $d = (d_1, \ldots, d_m)'$ satisfy

$$(K + n\lambda I)c + Td = z \quad (6.3)$$

$$T'c = 0 \quad (6.4)$$

where K is the $n \times n$ matrix with ijth entry $E_m(t_i - t_j)$. See Wahba and Wendelberger (1980). Extensive algorithmic and numerical results are available on the GCV estimate of λ and the computation of f_λ. See Bates and Wahba (1982,1983), Wahba (1980a,1983d), Wendelberger (1981).

7 THIN PLATE SMOOTHING SPLINES WITH LINEAR INEQUALITY CONSTRAINTS

Suppose it is known that f in the model of (6.1) satisfies a family of linear inequality constraints of the form $f \in C$ where

$$C = \{f: f \in X, a(x,y) \leqslant f(x,y) \leqslant b(x,y), (x,y) \in \Omega,$$

where a and b are sufficiently nice functions defined over a "nice" bounded domain Ω. If $m \geqslant 2$, C will be a closed, convex set in X, and, provided the matrix T is of rank M, $I_\lambda(f)$ of (6.2) will have a unique minimizer in C. To compute an approximation to this minimizer, we approximate the set C by the set C_L:

$$C_L = \{f: f \in X, a(u_l, v_l) \leqslant f(u_l, v_l) \leqslant b(u_l, v_l), l = 1,2, \ldots, L$$

where (u_l, v_l) are a suitably large set of points in Ω.

Now C_L is also closed convex so I_λ has a unique minimizer over C_L, and it can be shown by the methods in Kimeldorf and Wahba (1971), that this minimizer has a representation

$$f_\lambda(t) = \sum_{i=1}^{n} c_i E_m(t - t_i) + \sum_{l=1}^{L} b_l E_m(t - s_l) + \sum_{\nu=1}^{m} d_\nu \phi_\nu(t)$$

where $s_l = (u_l, v_l)$, and the c_i's, b_l's and d_ν's are obtained by minimizing a certain quadratic form subject to a family of linear inequality constraints. See Villalobos (1983), Villalobos and Wahba (1983). Villalobos (1983) has found that the QP of Gill, Gould, Murray, Saunders and Wright (1982) quite satisfactory for solving this QP for n and L of the order of one or two hundred. The Gill et al. algorithm appears to converge rapidly when a good starting guess for the active constraint set is available. In the two dimensional smoothing spline examples tried by Villalobos, a good starting guess is obtained by solving the unconstrained problem and determining those constraints which were violated or nearly violated. To estimate λ by GCV for constrained problems, it is required to solve the QP for a sequence of different λ's therefore rapid convergence is important.

It is believed that larger data sets can be handled by combining the approaches in Bates and Wahba (1983) and Villalobos (1983) and using the Gill et al., QP.

8 SOME NUMERICAL RESULTS WITH INEQUALITY CONSTRAINED THIN PLATE SMOOTHING SPLINES

Miguel Villalobos and I have studied the use of constrained thin plate smoothing splines for the estimation of posterior probabilities in the classification problem. Let (X, Y) be a bivariate random variable with density $h_1(x,y)$ if the source of (X, Y) is population 1, and $h_2(x,y)$ if the source is population 2. Suppose that the prior probability of each population is 1/2, then it is desired to estimate $p(x,y) = h_1(x,y)/(h_1(x,y) + h_2(x,y))$. Here, $p(x,y)$ is the posterior probability that the source is population 1 given $(X, Y) = (x,y)$. (If the prior probability of population 1 is q then the posterior probability is $qp(t)/[qp(t) + (1-q)(1-p(t))]$). These posterior probabilities are frequently of direct interest, for example, in estimating the probability of a heart attack, given (for example) $X =$ blood pressure and $Y =$ cholesterol level. Given "training" samples $t_i = (x_i, y_i)$, $i = 1,2,\ldots,n/2$ from population 1 and $t_i = (x_i, y_i)$, $i = n/2+1,\ldots,n$ from population 2, let $z_i = z(t_i) = 1$ if t_i is from population 1 and 0 if t_i is from population 2. The estimate p_λ of p under study is the minimizer of

$$\frac{1}{n}\sum_{i=1}^{n}(p(t_i) - z_i)^2 + \lambda J(p)$$

subject to

$$0 \leqslant p(t) \leqslant 1 \ , t \in \Omega_L$$

where $t = (x,y)$ and Ω_L is a 15×15 point array of points uniformly spaced over a rectangle of interest. Although the motivation for this estimate is heuristic, the numerical results obtained were excellent and demonstrate the feasibility and effectiveness of bivariate, inequality-constrained thin plate smoothing splines, even with non-Gaussian data. Other spline and related estimates for p are discussed in Silverman (1978,1982), O'Sullivan (1983), Raynor and Bates (1983).

Figure 1 gives a plot of a "true" posterior probability $p(x,y)$, where

$$f_1 \sim N\begin{pmatrix}0 & 1 \ 0\\ 0\ , & 0 \ 1\end{pmatrix}$$

$$f_2 \sim \frac{1}{2} N\begin{pmatrix}1 & 1 \ 0\\ -2.5, & 0 \ 1\end{pmatrix} + \frac{1}{2} N\begin{pmatrix}1 & 1 \ 0\\ 2.5, & 0 \ 1\end{pmatrix}.$$

A pseudorandom sample of 70 observations from each population was drawn, and a plot of this simulated data appears in Fig. 2. Fig. 3 presents the constrained, cross validated thin plate spline estimate of $p(x,y)$, and Figure 4 presents a plot of the level curves of the estimated p superimposed on the data. Figure 5 presents a plot of the level curves of the estimated p along with those of the true p. It can be seen that if the level curves of the estimated p are used in a classification algorithm, the misclassification rate will be quite close to the rate achieveable with the true p. Furthermore the surface is readily interpreted visually.

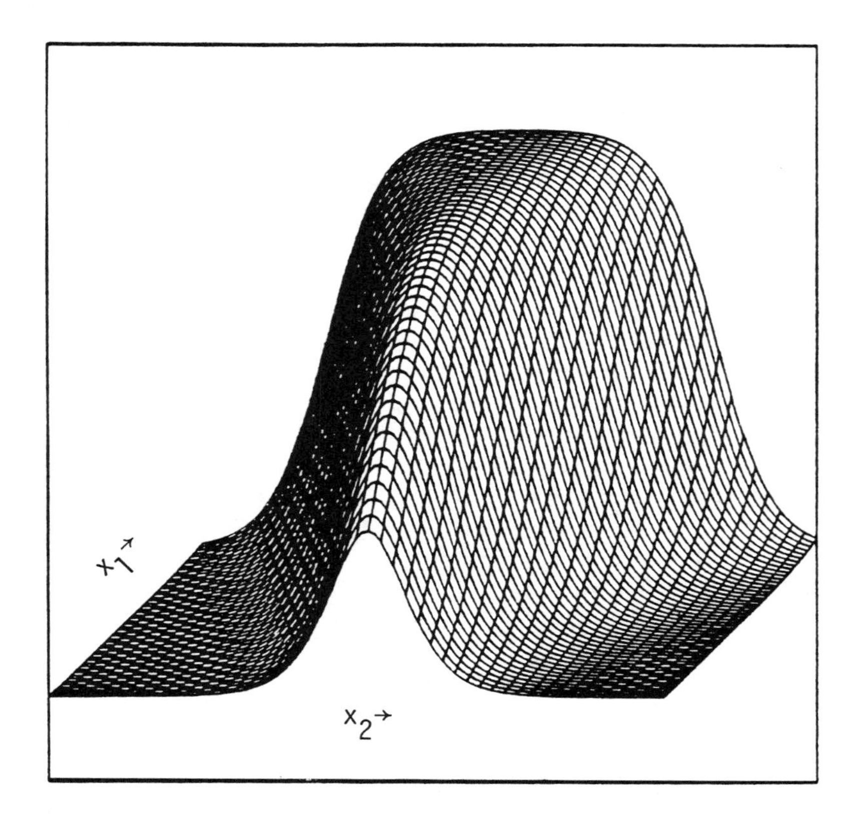

Figure 1 The true $p(x_1, x_2)$.

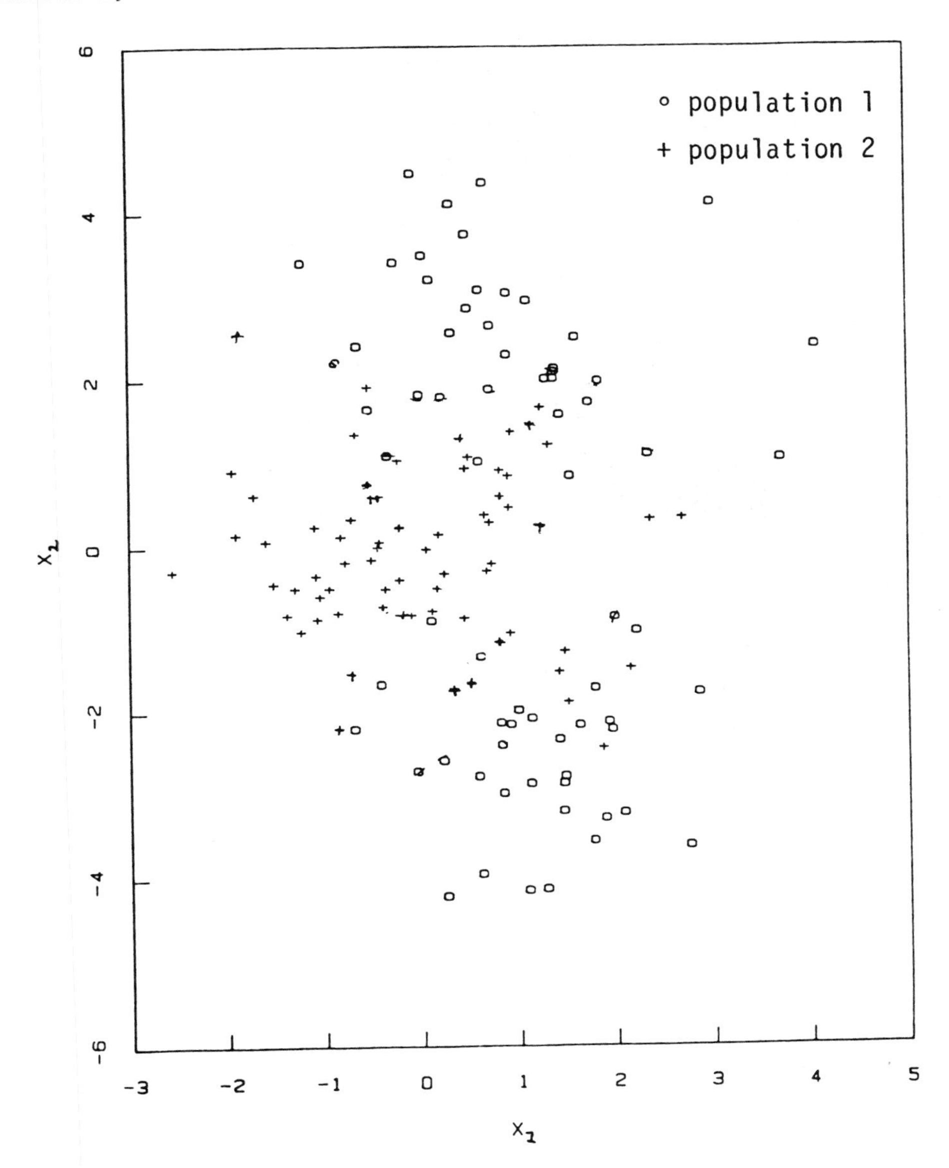

Figure 2 The pseudo data.

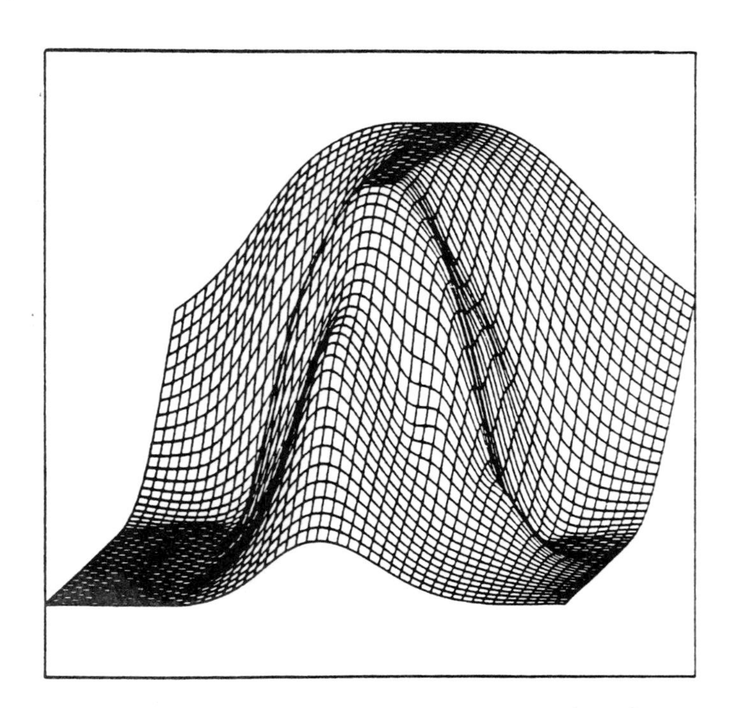

Figure 3 The constrained TP spline estimate $p_{\hat{\lambda}}(x_1, x_2)$.

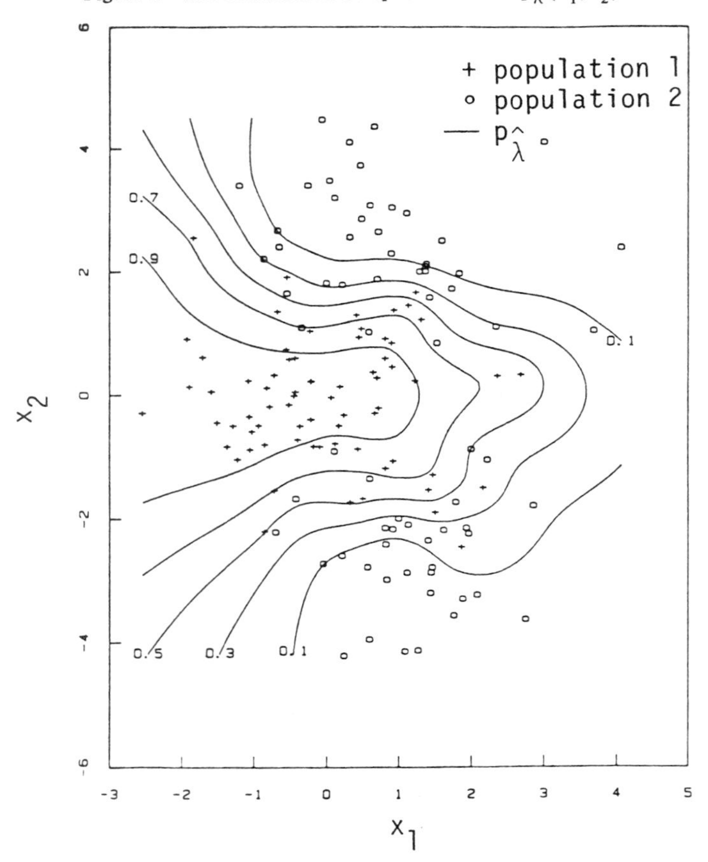

Figure 4 Level curves of $p_{\hat{\lambda}}$ and the pseudo data.

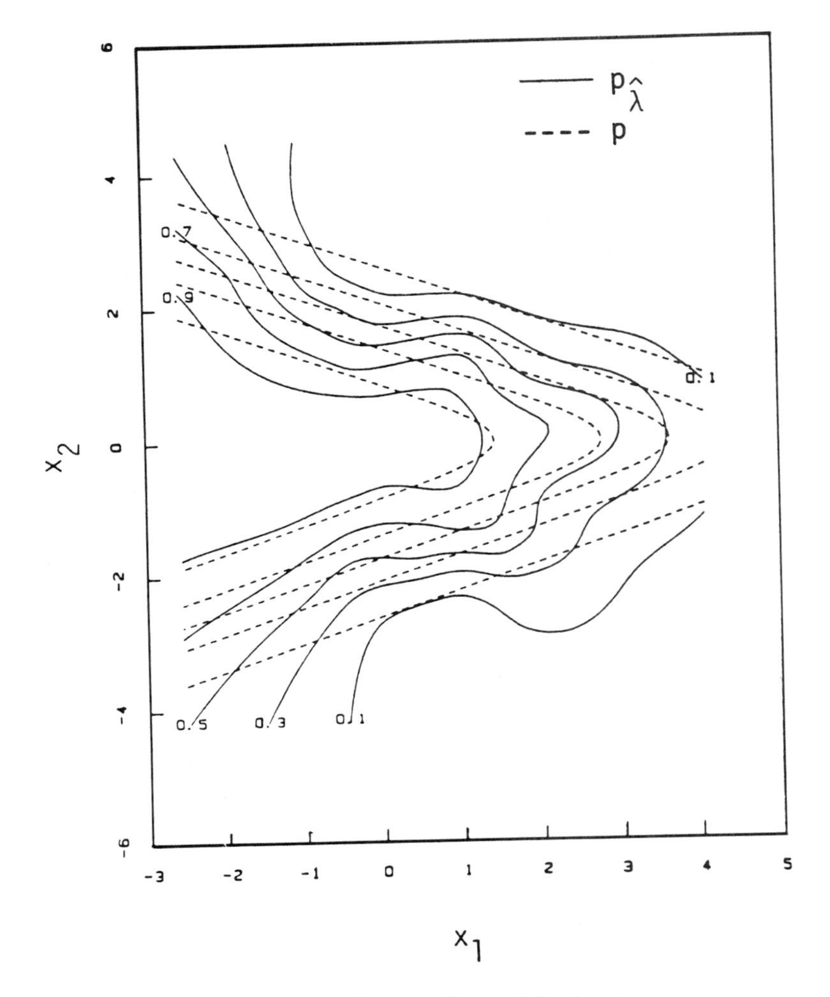

Figure 5 Level curves of $p_{\hat{\lambda}}$ and the true p.

Figure 6 plots some data from Smith (1947). The circles represent pairs of scores of 25 psychotics on a psychological test, and the crosses represent scores from a group of 25 normal persons. The solid curves represent the level curves of the constrained, cross validated thin plate spline estimate of p (assuming equal prior probabilities; the curves can also be obtained for other prior probabilities) and the dashed curves represent the level curves for a normal theory estimate of p. Figure 7 gives a picture of the spline estimate of p as seen from the south east corner of Fig. 6.

This method is relatively computationally intensive and would have been unthinkable a few years ago. It has not been particularly difficult or expensive to implement on our VAX-11/740 System with UNIX, however, and we believe these methods will see wide applicability in the future.

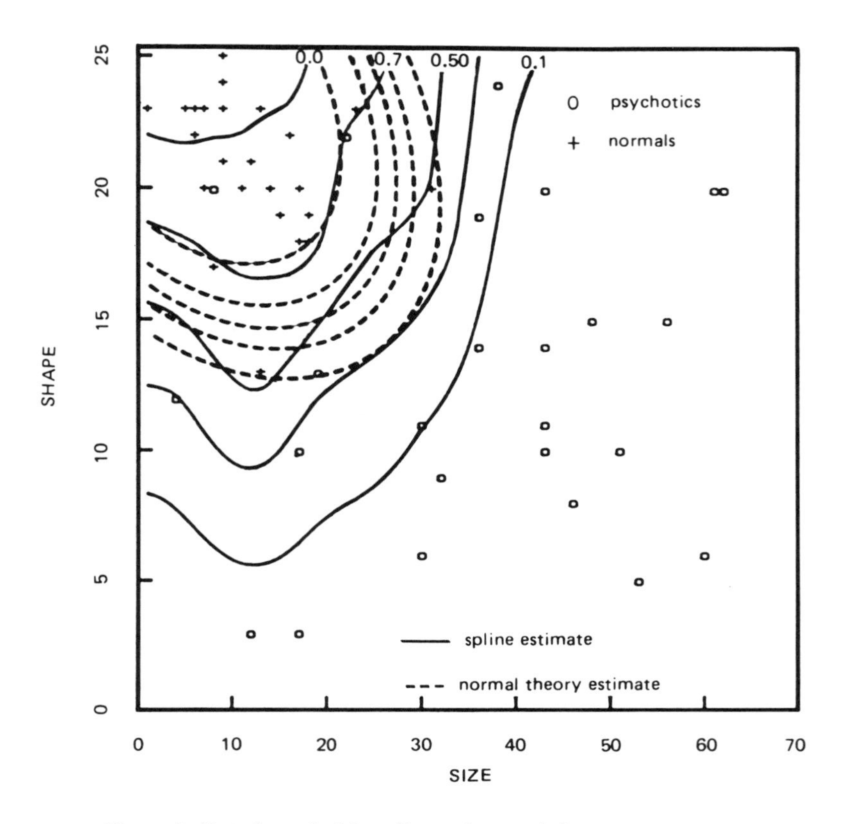

Figure 6 Data from Smith; spline and normal theory estimates of p.

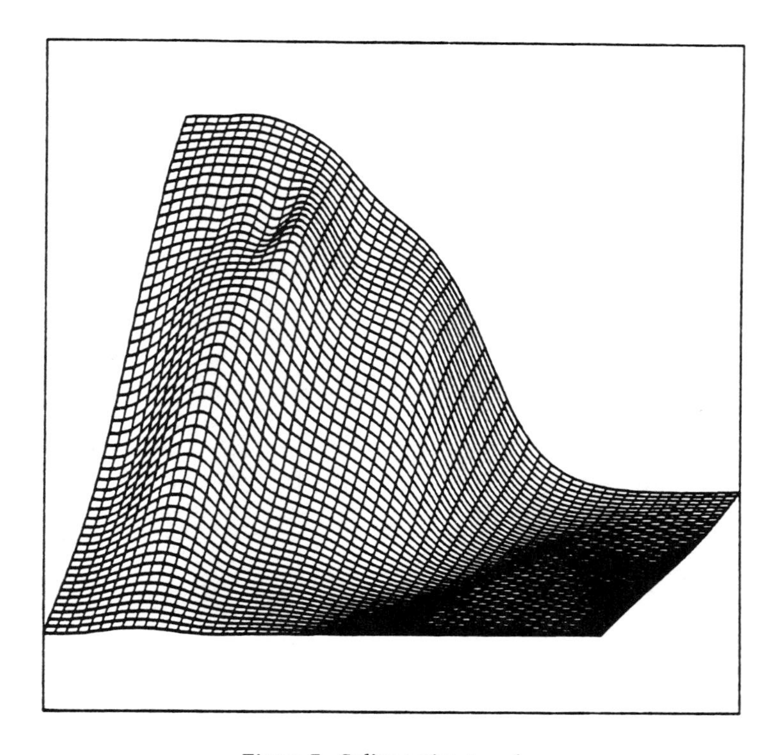

Figure 7 Spline estimates of p.

REFERENCES

Bates, D. and Wahba, G. (1983). *A truncated singular value decomposition and other methods for generalized cross-validation.* University of Wisconsin-Madison Statistics Dept. TR 715.

Bates, D.M. and Wahba, G. (1982). Computational methods for generalized cross-validation with large data sets. pp. 283-296. In *Treatment of Integral Equations by Numerical Methods*, C.T. H. Baker and G.F. Miller (ed.), London: Academic Press.

Craven, P. and Wahba, G. (1979). Smoothing noisy data with spline functions: estimating the correct degree of smoothing by the method of generalized cross-validation. *Numer. Math., 31*, 377-403.

Duchon, J. (1976). Splines minimizing rotation-invariant semi-norms in Sobolev Spaces. pp. 85-100. In *Constructive Theory of Functions of Several Variables*, K. Zeller (ed.),.

Engle, R., Granger, C., Rice, J., and Weiss, A. (1983). "Non-parametric estimates of the relation between weather and electricity demand." Discussion paper 83-17, Dept. of Economics, University of California, San Diego.

Gill, P., Gould, N., Murray, W., Saunders, M., and Wright, M., (1982). "Range space methods for convex quadratic programming." Systems Optimization Laboratory, Dept. of Operations Research Stanford University TR SOL 82-14.

Hutchinson, M., T. Booth, J. McMahon, and Nix, M., (1983). Estimating monthly mean values of daily total solar radiation for Australia, to appear, Solar Energy.

Kimeldorf, G. and Wahba, G. (1971). Some results on Tchebycheffian spline functions. *J. Math Anal. Applic., 33*, 82-95.

Li, K. (1983). Cross-validation in non-parametric regression. *Bull. I. M. S., 12*, 148.

Li, K. (1983). "From Stein's unbiased risk estimates to the method of generalized cross-validation." Purdue University Department of Statistics TR 83-34.

Meinguet, J. (1979). Multivariate interpolation at arbitrary points made simple. *J. App. Math. Phys. (ZAMP) 30*, 292-304.

O'Sullivan, F. (1983). *The analysis of some penalized likelihood schemes.* University of Wisonsin-Madison (thesis), TR 726.

Raynor, W. and Bates, D., (1983). Smoothing generalized linear models with applications to logistic regression, University of Wisconsin-Madison Statistics Department Technical Report No. 724.

Schoenberg, I. (1964). Splines functions and the problem of graduation. *Proc. Nat. Acad. Sci. U. S. A., 52*, 947-950.

Silverman, B. (1978). Density ratios, empirical likelihood and cot death. *J. Roy. Stat. Soc. C, 27*, 26-33.

Silverman, B. (1982). On the estimation of a probability density function by the maximum penalized likelihood method. *Ann. Statist., 10*, 795-810.

Smith, C. (1947). Some examples of discrimination. *Ann. Eugenics, 13*, 272-282.

Speckman, P. (1982). "Efficient non-parametric regression with cross-validated smoothing splines." Dept. of Statistics, University of Missouri-Columbia.

Villalobos, M. (1983). *Estimation of posterior probabilities using multivariate smoothing splines and generalized cross-validation.* Department of Statistics, University of Wisconsin-Madison Technical Report No. 725.

Villalobos, M. and Wahba, G. (1983). Multivariate thin plate spline estimates for the posterior probabilities in the classification problem. *Commun. Statist., 12*, 1449-1480.

Wahba, G. (1973). On the minimization of a quadratic functional subject to a continuous family of linear inequality constraints. *SIAM J. Control, 11*, 64-79.

Wahba, G. (1978). Improper priors, spline smoothing and the problem of guarding against model errors in regression. *J. R. Statist. Soc. B, 40*, 364-372.

Wahba, G. (1980a). Spline bases, regularization, and generalized cross-validation for solving approximation problems with large quantites of noisy data. pp. 905-912. In *Approximation Theory III,* W. Cheney (ed.), Academic Press.

Wahba, G. (1980b). "Ill posed problems: Numerical and statistical methods for mildly, moderately, and severely ill posed problems with noisy data." University of Wisconsin-Madison Statistics Department Technical Report No. 595 (To appear in the Proceedings of the International Conference on Ill Posed Problems, M.Z. Nashed, ed.)

Wahba, G. (1981). Data-based optimal smoothing of orthogonal series density estimates. *Ann. Statist., 9*, 146-156.

Wahba, G. (1982). Constrained regularization for ill posed linear operator equations, with applications in meteorology and medicine. pp. 383-418. In *Statistical Decision Theory and Related Topics III, Vol. 2*, S.S. Gupta and J.O. Berger (ed.), Academic Press.

Wahba, G. (1983a). Bayesian "confidence intervals" for the cross-validated smoothing spline. *J. Roy. Stat. Soc. B., 45*, 133-150.

Wahba, G. (1983b). "A comparison of GCV and GML for choosing the smoothing parameter in the generalized spline smoothing problem." University of Wisconsin-Madison Statistics Dept. TR No. 712.

Wahba, G. (1983c). "Cross validated spline methods for the estimation of multivariate functions from data on functionals." Dept. of Statistics, University of Wisconsin-Madison Technical Report No. 722, (to appear, Proc. 50th Anniversary Conference of the Iowa State University Statistical Consulting Laboratory), H.A. David, ed.

Wahba, G., (1983d). Surface fitting with scattered, noisy data on Euclidean d-spaces and on the sphere, to appear, Rocky Mountain J. Math,. 14, 1 (1983), 281-299.

Wahba, G., and Wendelberger, J. (1980). Some new mathematical methods for variational objective analysis using splines and cross-validation. *Monthly Weather Review, 108*, 36-57.

Wegman, E. (1984). Optimal non-parametric function estimation. *J. Stat. Plan. Inf.* 9, 375-388.

Wegman, E. and Wright, I. (1983). Splines in statistics. *J. A. S. A., 78*, 351-366.

Wendelberger, J. (1981). "The computation of Laplacian smoothing splines with examples." University of Wisconsin-Madison TR No. 648.

Wendelberger, J. (1982). *Smoothing noisy data with multidimensional splines and generalized cross-validation.* University of Wisconsin-Madison Statistics Dept. PhD. Thesis.

Data Analysis in Three and Four Dimensions with Nonparametric Density Estimation

David W. Scott

Department of Mathematical Sciences
Rice University
Houston, TX

ABSTRACT

Nonparametric density estimation, which has proven to be a powerful and versatile data analytic tool for univariate and bivariate data, can provide even greater analytical and graphical impact for data in 3 and 4 dimensions.

1 INTRODUCTION

In this chapter, we consider the role of nonparametric density estimation in three challenging problems in modern data analysis: handling non-Gaussian data, coping with many variables, and accommodating very large data sets. Analyzing data sets with even one of these characteristics presents practical problems. Yet it is increasingly common to face all three characteristics simultaneously with important problems in signal detection, classification, and image processing. For data in one and two dimensions, the kernel method of density estimation has proven to be a versatile data analytic technique for discovering non-Gaussian features and depicting such features graphically in data sets of moderate size. Bivariate examples may be found in Scott *et al* (1978; 1980) and Silverman (1981), which contains a useful survey. Significant features uncovered in these analyses included multiple modes and asymmetries, as well as outliers and heavy tails. Because nonparametric density estimates are consistent for a large class of true densities, confirmatory analyses may be undertaken. This has been done for discrimination in small medical data set with two and more variables by Habbema, Hermans, and van den Broek (1974); see also Hand (1982).

Multivariate kernel density estimation was first discussed by Cacoullos (1966). Most multivariate kernel applications use the product kernel form given by Epanechnikov (1969):

$$\hat{f}(y) = \frac{1}{n} \sum_{i=1}^{n} \left[\prod_{k=1}^{d} \frac{1}{h_k} K \left(\frac{y^{(k)} - x_i^{(k)}}{h_k} \right) \right],$$

where $y \in R^d$, $x_i^{(k)}$ is the value of the kth variable for the ith case, and K is a univariate probability density called the kernel. The constants $h_1, \ldots, h_d$ are called smoothing parameters and control the bias and variance of the estimator. Some authors assume $h_1 = h_2 = \ldots = h_d$ for theoretical simplicity, but this is not generally a good idea for real applications. There has been substantial progress in our understanding of how to choose these smoothing parameters for actual data analysis, but we shall not address that important research here. Smoothing parameters in all our examples were chosen subjectively by the author.

We face two substantial obstacles when we consider applying density estimation methods in a computer graphics workstation environment to large data sets with more than two variables: computation and representation. Although the theory of multivariate kernel methods is well-developed, the estimators suffer from the curse of dimensionality, as pointed out by Huber (1983). Hence we may require enormous sample sizes, and the computational burden quickly becomes prohibitive as the dimension increases. Furthermore, the density function of d variables is a surface in R^{d+1} so that representational difficulties set in for trivariate data. We expect continued rapid progress in the speed of computational engines, but the problem of representation will always remain. In section 3, we address these two practical problems, focusing on the important special cases of trivariate and quadrivariate data. These dimensions are close to our physical reality, and the extra geometrical degrees of freedom should prove as useful as that observed with bivariate versus univariate data. For data in very high dimensions, we expect to project into 3 or 4 dimensions, rather than 1 or 2. Section 4 contains several examples that deal primarily with exploratory clustering of simulated and real data. We begin by discussing some important work in multivariate data analysis.

2 MULTIVARIATE DATA ANALYSIS

The transition in statistics from the rather free-form data analysis of Karl Pearson to the fully parametric approach of R. A. Fisher was due in part to severe limitations on feasible computation and the prohibitive expense of maintaining computational resources such as Pearson had at his disposal. Based largely on Gaussian assumptions, multivariate parametric models were and continue to be powerful tools for data analysis. These parametric procedures can fail in ways that are difficult to diagnose but, when appropriate, can provide easily interpreted results for data sets of virtually any size or dimension. Non-Gaussian data will have highly non-linear interactions among variables, and the corresponding difficulties in interpretation often raise more questions than they answer. Since it is difficult to adequately parameterize non-Gaussian data, we naturally rely heavily on graphical presentation to summarize and emphasize important features in the data. Over the past decade, Tukey and colleagues have developed tools that are model-free and highly graphical for examining such data (Tukey 1977; Tukey and Tukey 1981). Tukey (1984) has pointed out that data analysis encompasses a great deal more than even statistics, but includes a willingness to look for appearances. Tukey and Tukey (1981) describe the presence of non-Gaussian features as an opportunity, while the challenge is visualizing such data in $\geqslant 3$ dimensions, given our familiarity with a 3-D world and the limitations of 2-D displays. We believe density estimation provides excellent tools for this goal and complements other methodology.

For exploring one-dimensional data, the histogram or stem-and-leaf plot is the primary tool. For bivariate data, the scatter diagram is the most commonly used graphical tool. For $\geqslant 3$ variables, all pairwise plots and other variants on scatter diagrams have been introduced and proven to be useful exploratory tools. For example, scatter diagrams have been enhanced by windowing, linked by various cues, generalized by using glyphs to represent higher dimensional values, and extended by using kinematic capabilities of computer workstations to provide true 3-D capabilities (Friedman *et al.*1982, Tukey and Tukey 1981). Our emphasis on very large data sets excludes other very interesting multivariate data representers (Wang 1978) such as Chernoff faces, which emphasize individual data points. For data rich environments, individual data points become relatively unimportant.

Scatter diagrams have well-known limitations. For large multivariate data sets, scatter diagrams emphasize the edges of the data cloud. Thus it is easy to spot outliers or well-separated clusters. But it is quite difficult with scatter diagrams to glean density information and to recognize moderate features in the middle of a data cloud, because of "too much ink" (Tufte 1983) and overstriking. One interesting idea is to use intensity or glyphs (Carr and Nicholson 1984) to deal with overstriking, but these histogram-like figures can be improved by smoothing; see section 3.1. Nevertheless we strongly advocate the use of these enhanced scatter daigrams—as a first step. In many cases it is natural to further smooth these diagrams in order to sharpen our view of the data. We believe scatter diagrams point to the density function (Scott and Thompson 1983).

An interesting task that arises when looking for appearances in high dimensional data is finding the dimension of a feature. By the dimension of a feature, we mean the minimum dimension of an orthogonal subspace in which the feature is discernible, but not present or correctly perceivable in any smaller subspace. Thus when we use parametric tools or sophisticated color-linked scatter diagrams of 14 variable data, we do so not because we believe there exists or we hope to find a 14-dimensional feature, but rather for ease of interpretation. Usually a subspace of at most several dimensions can be formed that adequately represents the structure in the data. Finding this subspace may not be easy for non-Gaussian data. Nonlinear transformations clearly can be helpful in further reducing the dimension of a feature. It is easy for the mathematicians to construct data with features of arbitrary dimension, for example, mixture densities, but it is interesting to speculate whether there exist real data sets for which it will prove necessary to perform data analysis in a properly chosen subspace of dimension greater than four or five. For these reasons, we again emphasize that 3 and 4-D data are noteworthy and deserve special attention.

3 COMPUTATIONAL AND REPRESENTATIONAL PROBLEMS IN DENSITY ESTIMATION

In this section we describe our approach to the two major problems faced in attempting to use kernel density estimation on data in three and four dimensions. The first problem involves computational speed and the second involves effective representation.

3.1 The Averaged Shifted Histogram

For very large data sets in 3 and 4-D, the tradeoff between computational speed and statistical efficiency is acute. Clearly a histogram is the most computationally efficient density estimator. This is especially true for very large data sets, because the data are compressed to bin counts, which can be used to quickly evaluate the histogram at any point. On the other hand, a kernel estimator requires a pass through the entire data set to evaluate the density at a point. Hence the histogram is doubly effective in that there is significant data compression, and this data compression phase acts as a preprocessor, reducing the lengthy evaluation portion to a table lookup. In many practical situations, reading data from mass storage is the slowest part. However, a kernel estimator is substantially more efficient statistically than a histogram, assuming the true density function is sufficiently smooth. This is of greater importance in 3 and 4 dimensions. In addition, the continuity of a kernel estimator can be put to good use in representing the density, as we shall discuss shortly. Yet some experiments with evaluating a kernel estimator of trivariate samples with 20,000 points over a 30^3 mesh indicated minimum requirements of several hours of CPU on a VAX 11-780, hardly adequate for an interactive graphics workstation environment.

We discovered a compromise density estimator, called the averaged shifted histogram (ASH), that achieves efficiency in both domains (Scott 1983, 1985b). The basic construction is easily described in one dimension and extendible to several dimensions. Consider an equally-spaced histogram with bin width h and suppose zero is a bin edge. Call this histogram g_0. Pick an integer m. Construct histograms $\{g_i, i = 1, \ldots, m-1\}$ using the same data and bin width as g_0, but with a bin edge at $x = i/m$. Each of these m histograms is an equally good choice. Consider their pointwise average:

$$f_m(x) = \frac{1}{m} \sum_{i=0}^{m-1} g_i(x).$$

Letting $\delta = h/m$ and $\nu_k = \{\#$ points in the interval $I_k = (k\delta, (k+1)\delta]\}$, it is easy to see that

$$f_m(x) = \frac{1}{mn\delta} \sum_{i=1-m}^{m-1} \left(1 - \frac{|i|}{m}\right) \nu_{k+i} \quad x \in I_k.$$

In practice, we use the continuous piecewise linear function interpolating $f_m(x)$ at the midpoints of adjacent bins, that is, a frequency polygon of the ASH (FP-ASH). The estimator only requires the data in binned form. In Figure A1, we show g_0 (same as f_1) for the chondrite data, twenty-two samples of the percentage of silica, analyzed by Good and Gaskins (1980) and others, using a bin width of $h = 2\%$. The histogram suggests three modes, but the evidence does not seem very strong. In Figures A2-A4, we display f_2, f_4, and f_8. Clearly the ASH not only smoothes the histogram but also sharpens our view of these data and increases the weight of evidence in favor of trimodality. Substantially larger bin widths are required to suppress the 3 sample modes. This greatly improved efficiency in our use of the data is important in several dimensions. For moderate values of m (3-5), the FP-ASH is indistinguishable statistically and graphically from a triangle kernel estimator (consider f_m as $m \rightarrow \infty$). A detailed comparison of the integrated mean squared error and the various estimators may be found in Scott (1985b). A multivariate ASH, formed by shifting the histogram bins m_i times along the ith axis, approximates a product triangle kernel estimator. The d-dimensional FP-ASH is formed by linearly interpolating certain $d + 1$ adjacent ASH mid-bin values. This, then, is our solution to the problem of computational speed. We note that other kernels may be approximated in f_m and that adaptive meshes may be considered in g_i.

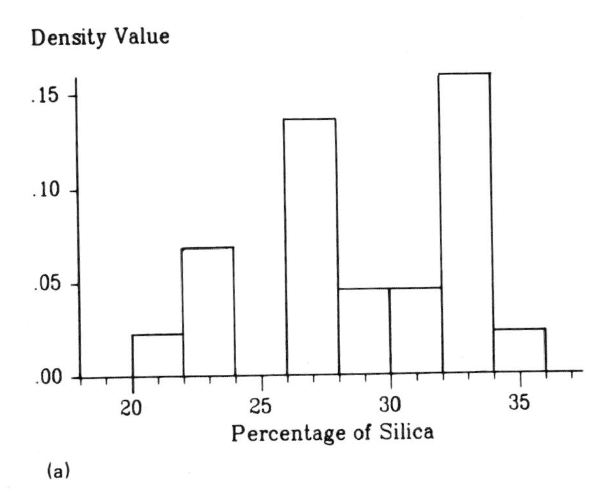

Fig. A1 — Histogram of 22 chondrite data points with bin width of 2%.

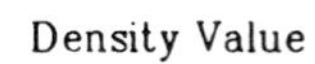

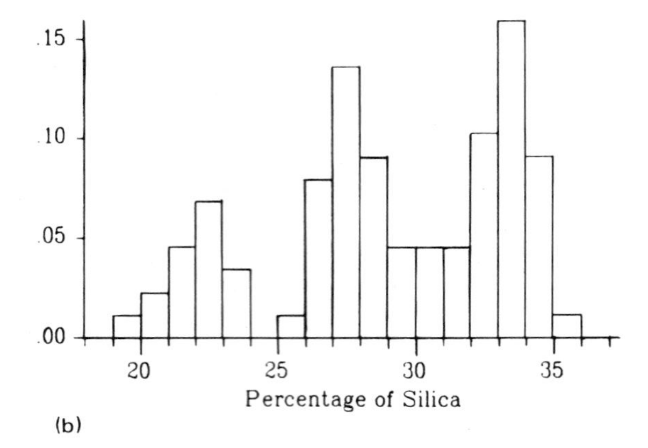

Fig. A2 — Average of 2 shifted histograms, f_2.

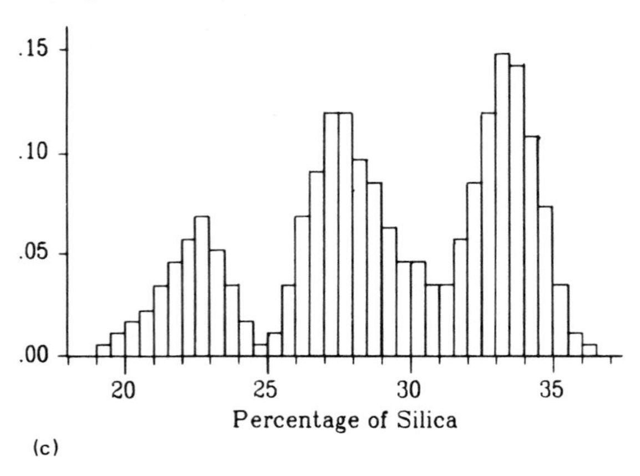

Fig. A3 — ASH f_4.

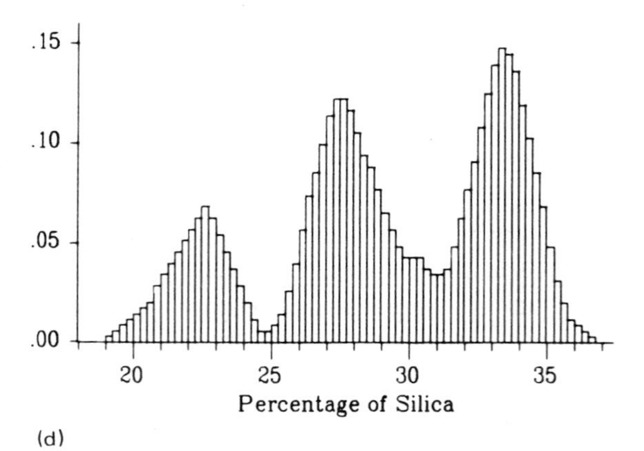

Fig. A4 — ASH f_8.

3.2 Representation of Multivariate Density Functions

3.2.1 Trivariate Data

For univariate and bivariate data, the estimated density functions are easily drawn, in perspective for bivariate data. In general, for d-dimensional data, we need $d + 1$ dimensions to plot the entire density function, which is clearly impossible for $d > 2$. On the other hand, the contour or level set representation of a density function requires only d dimensions, the same as for scatter diagrams. We have chosen to represent a trivariate density function by simultaneously displaying several contours, distinguishing each contour level by using a color cue. The contour representation is not as complete as the perspective representation, giving only a fraction, say 1/2 or 3/4 of the extra dimension. Extra detail is available if individual contours are shown sequentially in real time, starting with the mode and continuously expanding to a very low contour level. However we have found that several contours are usually sufficient for static display. Specifically an α-level contour set is given by the set

$$C_\alpha = \{(x,y,z): f_m\ (x,y,z) = \alpha\ f_m\ (\text{mode})\},$$

where f_m refers to the continuous piecewise linear FP-ASH. C_α is a surface in R^3. We refer to a contour by its α percentage value. For Gaussian data, these surfaces are concentric ellipsoids. Rather than using hidden line routines or shading algorithms, we have chosen to display a CAD-CAM-like shell such as that shown in Figure S1. A FP-ASH contour is intersected by a series of planes orthogonal to an axis, and the intersections displayed. In fact, we chose to

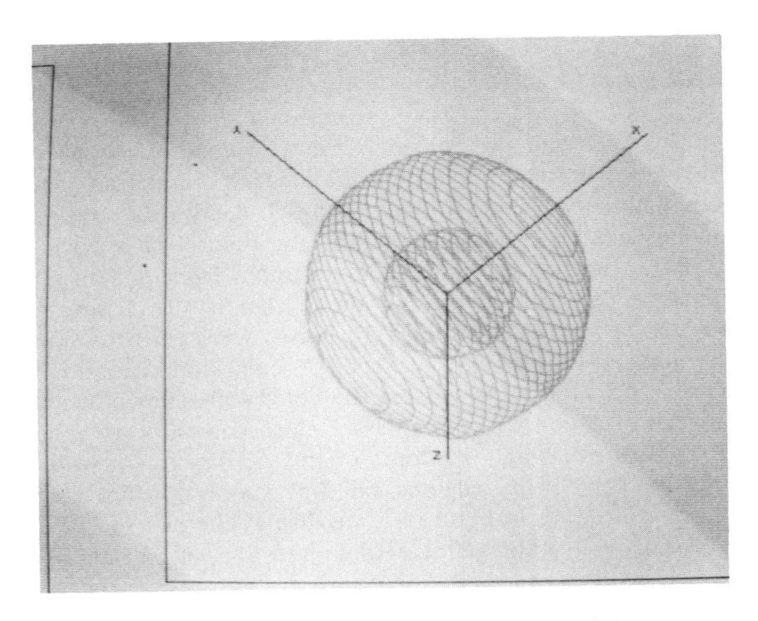

Fig. S1 — 10% contours of FP-ASH of simulated data.

intersect the axis in multiples of $\delta_i = h_i/m_i$, so that the contours are similar to those of a two-dimensional frequency polygon determined by the 2-D array of bin values. Contouring algorithms for this choice are extremely fast and well-known in crystallography work (Diamond 1982). With a little experience (especially seeing them drawn and displaying shadows to give a depth cue), the density function is easily examined for interesting features. For very large data sets, it is interesting to do this side by side with a scatter diagram; see Scott (1985a). Fwu *et al* (1981) used an onion peel analogy to display portions of individual contours for a nearest neighbor estimator.

3.2.2 Quadrivariate Data

For 4-D data, we represent the density function $f_m\ (x,y,z,t)$, which is a surface in five dimensions, by considering the doubly indexed sets

$$C_{\alpha,t} = \{(x,y,z)\colon f_m\ (x,y,z,t) = \alpha\ f_m\ (\text{mode})\}.$$

These sets, which we may visualize as corresponding to *space* + *time*, will be empty for many values of t and α. For several values of α, these sets are displayed continuously as t increases $t_{\min}$ to $t_{\max}$. For independent Gaussian data, this view is of expanding then contracting nested spheres. The choice of variable for the time or reference frame is not critical, although certain choices may be easier to "see." As before color can be used to distinguish several α-levels. We have sometimes found a thunderstorm analogy to be helpful. A trivariate function is represented by a snapshot of a thundercloud, while a quadrivariate function is represented by time-lapse photography of a thunderstom from beginning to end.

4 EXAMPLES

4.1 Simulated Data [1]

Before examining real data, we consider some simulations in order to become familiar with the density contour representation. In simulations with trivariate Gaussian data, estimated FP-ASH's (not shown) clearly indicated the spherical symmetry in the data, closely approximating the theoretical spherical contours; see Scott (1984). Instead we present an example from a density that has a higher dimensional feature, namely, a mode that is not a single point, but rather the surface of a sphere of radius 3. Points were perturbed from the surface by generating $N(3,.6^2)$ radii values. A color-coded scatter diagram of these data is shown in Figure C5, using the color scale shown to indicate value in the orthogonal direction. Red points are closest in the orthogonal direction and blue points furthest away. Contours of this density for fixed α-levels arise in pairs, since higher-density regions fall not simply within a contour, but between pairs of contours. In Figure S1, we display the pair of 10% contours of an FP-ASH based upon 5,000 points. In Figure S2, we display 50% contours.

[1] See Plate 18 for Figure C5.

Notice how the outer contour has shrunk and the inner contour has expanded. Of course, a finite sample estimator will always have modes at points, so that the contours break apart at higher levels, for example, at $\alpha = 80\%$ as shown in Figure S3. However, the important feature in the data is clearly displayed in this estimate.

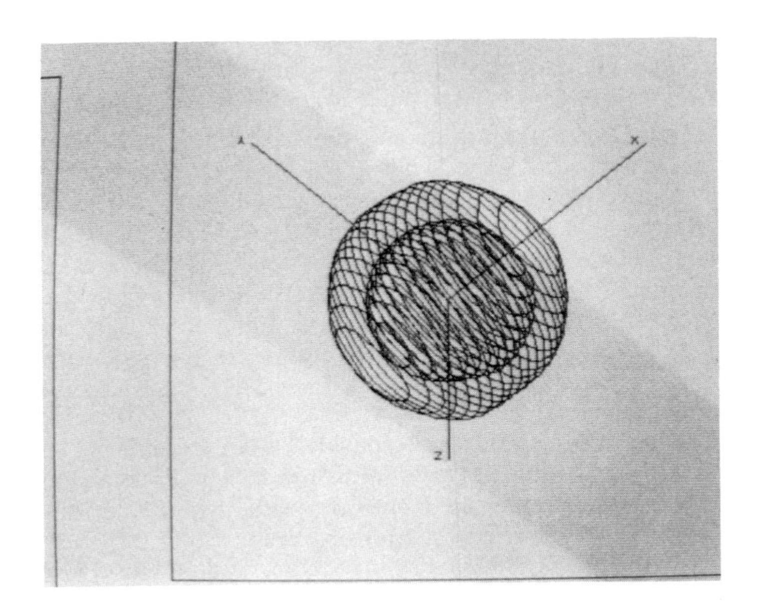

Fig. S2 — 50% contours of FP-ASH of simulated data.

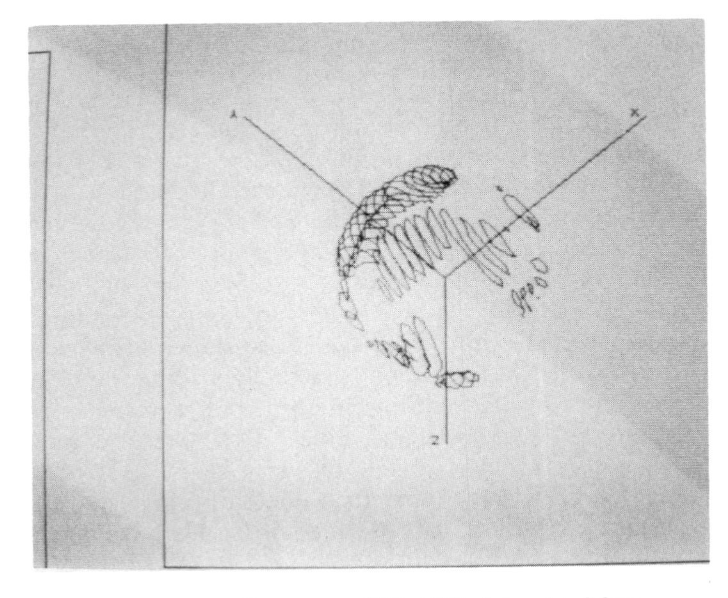

Fig. S3 — 80% contours of FP-ASH of simulated data.

4.2 Graphical Clustering of Real Data

Our final examples are of data that would be appropriate for discrimination since they are labeled, but which we shall treat as unlabeled for exploratory graphical clustering testing. Modes in the density estimates indicate the presence of clusters in the data.

4.2.1 NASA Remote Sensing Data [2]

We initially designed our algorithms and plots to handle large remote sensing data sets obtained from NASA's Landsat *IV* satellite. A passive four-channel multispectral scanner (MSS) on the Landsat satellites measures the intensity (0-255) of reflected light in four spectral bands, two of which are visible (.5-.6, .6-.7, .7-.8, .8-1.1 μm). The satellite's ground-level resolution is 80 m and the corresponding smallest picture element (pixel) is 1.1 acres. Each pixel is imaged every 17 or 18 days. The data have been organized into sampling units or LACIE (Large Area Corp Inventory Experiment) segments of 117 by 196 pixels, covering an area of 5 by 6 nautical miles. The data have been used to identify crops and estimate yields.

The particular data set we shall consider is an agricultural scene during 1977 in North Dakota, segment 1663, furnished by Dr. Richard P. Heydorn at NASA's Johnson Space Center in Houston. Although theoretically new data are available every few weeks, clouds may obscure a scene. Five useful acquisitions were obtained, on days 121, 138, 156, 174, and 211 of 1977. Thus the raw data are 20 dimensional, in addition to the 2 spatial coordinates. Each acquisition was transformed to two commonly used principal components, which measure the brightness and greenness of a pixel. A plot of the greenness of an individual pixel over time is called a greenness profile. A particularly useful nonlinear growth model describing the greenness profile has been developed for pixels in argricultural use by Badhwar (1980). The model extracts three pieces of information from the sample greenness profile on a pixel-by-pixel basis: (x) the time of maximum greenness; (y) a ripening period or half-power width of the greenness profile; and (z) the level of the maximum greenness. In Figure C1, color enhanced images of the Badhwar parameters for segment 1663 are shown. Histograms of the 22,932 values for each parameter are shown for the range of values 0-255. The color scale shown was used to construct the false color images by linearly transforming an interval of these values, with red representing the maximum and blue the minimum. Field structure is clearly observable in each image. Also shown is a false color image of the ground truth, which was obtained by actually visiting the region. Yellow represents (3694) pixels planted entirely in corn, red represents (3811) spring wheat pixels, green (892) barley pixels, blue (731) soybeans, magenta (732) harvested barley, and cyan (1177) pixels of sugar beets, flax, or oats. White represents 10,256 pixels in which more than one crop was planted. Remaining pixels are also white. While spatial information is clearly of great value, we

---

[2] See Plates 16 and 17 for Figures C1 to C4.

shall consider only the trivariate Badhwar data cloud for this segment. In Figure C2, a color-coded scatter diagram is shown. We notice a large body of data, a number of outliers, and also a smaller narrow cloud of points partially hidden behind the main cloud of points. Most of the outliers represent singularities in the fit of the Badhwar model. Those values were set to 0 or 249. The smaller cloud of points primarily represents many acres of sugar beets, which are havested in a different manner than sunflower and grain crops. In Figure C3, we display the $\alpha = 1\%$ contour of the FP-ASH ($m_x = m_y = m_z = 2$). The scale and orientation are the same as in Figure C2. Notice how much of the detail in the scatter diagram is well outside this contour, which contains more than 95% of the data. The density at the 1% contour is actually bimodal, a feature that is observable in the scatter diagram. Only 3,956 of the 22,932 pixels are displayed in the 400 by 400 pixel display area, as the result of overstriking. Thus the scatter diagram mainly displays the surface of the data cloud and outliers. Expanding the scale with the same orientation and looking inside the big data cloud, we find some more interesting features (clusters), as shown by the 10%, 20%, 40%, 60%, and 80% contours in Figure C4, color-coded using the same color scale. Two modes are seen here, sunflower on the left and small grains on the right. Closer examination reveals 2 bumps in small grain region. Clearly sunflower pixels are well-discriminated by the Badhwar model, but the discrimination power is less for the small grains. We have used estimated ASH's in a Bayesian discriminator for sunflower, spring wheat, and barley crops (Scott and Jee 1984). The performance is surprisingly better than a Gaussian Bayesian classifier, in spite of the fact that the individual data are nearly Gaussian.

4.2.2 Fisher's Iris Data [3]

In this section, we show how we represent the estimated density of 4-D data. Without kinematic display, our ability to do this is limited. We show a "frame" from our movie of Fisher's iris data, where we again treat the 150 data points as unlabeled. The variables were linearly scaled onto the interval [−2,2], and the sepal width was placed on the t or time reference axis. Sepal length, petal width, and petal length were placed on the x, y, and z axes, respectively. The time axis was divided into 30 equally spaced steps steps for the movie; in Figure C6, we show the 5% and 20% contours for $t = 17$. In this frame, the trimodal feature of the density is revealed, reflecting the three varieties of iris, *setosa, versicolor,* and *virginica*, with 50 points each in the set. Of course, these data do not represent a properly weighted sample of these varieties. Other investigators have used 4-D glyph methods to show graphically that the iris *versicolor* and *virginica* varieties do not separate in the 4 dimensional data space (Nicholson and Littlefield 1983). We have shown that while the data for these two species do not separate in 4-D, the estimated density function is multimodal, revealing the presence of the two species. All other choices for the t-variable also reveal this fact, but not necessarily in one time frame. It is

[3]See Plate 18 for Figure C6.

interesting to note this success in 4-D with only 150 points. We have also investigated other 4-D data sets: the particle physics data (Friedman and Tukey 1974) and some aspen and spruce data from Minnesota's Superior National Forest collected by Bob MacDonald of the Johnson Space Center; see Scott and Thompson (1983) and Scott and Jee (1984), respectively. We remark that *space + time* describes exactly certain geological data sets collected over time.

5 FINAL REMARKS

We have emphasized the graphical side of nonparametric density estimation for multivariate data analysis and hinted at other uses. The graphics tools for representing density functions work well in a color graphics workstation environment and complement existing multivariate exploratory methods. If the graphical representation suggests that assumptions for a standard parametric model may not be valid, then the estimated density may be used to continue in the confirmatory or production stage. Thus these procedures can be used for diagnostic purposes in parametric analysis.

For data much beyond four dimensions, we generally must be content to view certain selected projections. It is difficult to imagine, for example, that we could find true 7-D features in data by purely graphical means without the guidance of a mathematical model. The ASH representation may be extended to 5-D data by introducing a bivariate reference frame. It should be possible to devise artificial intelligence algorithms to guide movement in this reference frame. Higher dimensional representation is theoretically possible by constructing nested trivariate reference frames, although we suspect 5-D is near the boundary of feasibility. Other combinations of stimuli could be introduced. Principal components, which maximizes a variance criterion, can provide a useful subspace for viewing. An alternative criterion is a projection that results in an "interesting" ASH. Huber (1983) has shown that Friedman and Tukey's original 1974 projection pursuit criterion produces interesting views for a univariate kernel estimate with uniform kernel. Other projection pursuit criteria have been considered.

Smoothing, noise reduction, and signal extraction are important in all areas of statistics. Running medians applied to a noisy raw time series can yield a surprising "smooth" (Tukey 1977). Density estimates smooth raw data displayed in scatter diagrams, although they provide an indirect view of the data. An important advantage of our smooth density representation is that it does not fundamentally change as the sample size increases. Thus it is possible to extend experiences gained with small and moderate sample sizes to very large data sets. Many statistical procedures, such as interpreting chi-squared tests, differ dramatically for large data sets. But for larger and larger data sets requiring sophisticated analysis, we believe that density-based methods in 3 and 4 dimensions will be both efficient and effective.

ACKNOWLEDGMENTS

I would like to thank Dick Heydorn, George Terrell, and Jim Thompson for their many contributions towards this work. This research was supported by

NASA/JSC, ARO, and ONR under grants NCC 9-10, DAAG-29-82-K0014, and N00014-85-C0100, respectively.

REFERENCES

1. Badhwar, G.G. (1980), "Crop emergence data determination from spectral data," *Photogram. Eng. Remote Sens.*, 46:369-377.

2. Cacoullos, T. (1966), "Estimation of a multivariate density," *Annals of the Institute of Statistical Mathematics*, 18, 179-189.

3. Carr, D.B. and Nicholson, W.L. (1984), "Graphical tools for analysis of multiple two- and three-dimensional scatterplots," *Proceedings of the Fifth Annual National Computer Graphics Association Conference*, Volume II, pp. 743-752.

4. Diamond, R. (1982), "Two contouring algorithms," in *Computational Crystallography*, David Sayre, Editor, Clarendon Press, Oxford, pp. 266-272.

5. Epanechnikov, V.A. (1969), "Nonparametric estimators of a multivariate probability density," *Theory of Probability and its Applications, 14, 153-158.*

6. Friedman, J.H., McDonald, J.A., and Stuetzle, W. (1982), "An introduction to real time graphical techniques for analyzing data," *Proceedings 3rd Annual Conference of National Computer Graphics Association.*

7. Friedman, J.H. and J.W. Tukey (1974), A projection pursuit algorithm for exploratory data analysis, *IEEE Trans. Comp.*, C-23, 881-890.

8. Fwu, C., Tapia, R.A., and Thompson, J.R. (1981), "The nonparametric estimation of probability densities in ballistics research," *Proceedings of the Twenty-Sixth Conference on the Design of Experiments in Army Research Development and Testing,* pp. 309-326.

9. Good, I.J. and Gaskins, R.A., (1980), "Density estimation and bump-hunting by the penalized likelihood methods exemplified by scattering and meteorite data," *J. American Statistical Association*, 75:42-73.

10. Hand, D.J. (1982), *Kernel Discrimination Analysis,* John Wiley & Sons, Chichester.

11. Habbema, J.D.F., Hermans, J., and van den Broek, K., A.T. (1974), "A stepwise discrimination analysis program using density estimation," *Compstat 74*, G. Bruckman, Ed., Vienna: Physica-Verlag, 101-110.

12. Huber, P.J. (1983), "Projection pursuit," Technical report MSRI 009-83, Berkeley.

13. Nicholson, W.L. and Littlefield, R.J. (1983), "Interactive color graphics for multivariate data," *Computer Science and Statistics: Proc 14th Symposium on the Interface,* K.W. Heiner, R.S. Sacher, and J.W. Wilkinson, eds., Springer-Verlag, New York.

14. Scott, D.W. (1983), "Nonparametric probability density estimation for data analysis in several dimensions," *Proceedings of the Twenty-Eighth Conference on the Design of Experiments in Army Research Development and Testing*, pp. 387-397.

15. Scott, D.W. (1984), "Multivariate density function representation," *Proceedings of the Fifth Annual National Computer Graphics Association Conference,* Volume II, pp. 794-801.

16. Scott, D.W. (1985a), "Frequency polygons: theory and application," To appear in *J. American Statistical Association.*

17. Scott, D.W. (1985b), "Averaged shifted histograms: effective non-parametric estimators in several dimensions," to appear in *The Annals of Statistics.*

18. Scott, D.W., Gorry, G.A., Hoffmann, R.G., Barboriak, J.J., and Gotto, A.M. (1980), "A new approach for evaluating risk factors in coronary artery disease; a study of lipid concentrations and severity of disease in 1847 males," *Circulation* 62:477-484.

19. Scott, D.W., Gotto, A.M., Cole, J.S., and Gorry, G.A. (1978), "Plasma lipids as collateral risk factors in coronary artery disease: a study of 371 males with chest pain," *Journal of Chronic Diseases* 31:337-345.

20. Scott, D.W. and Jee, R. (1984), "Nonparameteric analysis of Minnesota spruce and aspen tree data and Landsat data," *Proceedings of the Second Annual Symposium on Mathematical Pattern Recognition and Image Analysis,* NASA publication, pp. 27-49.

21. Scott, D.W. and Thompson, J.R. (1983), "Probability density estimation in higher dimensions," *Proceedings of the 15th Symposium on the Interface of Computer Science and Statistics,* J.E. Gentle, Ed., North-Holland, Amsterdam, pp. 173-179.

22. Siverman, B.W. (1981), "Density estimation for univariate and bivariate data," in Barnett, V. (ed.), *Interpreting Multivariate Data*, New York: John Wiley & Sons.

23. Tufte, E.R. (1983), *The Visual Display of Quantitative Information*, Cheshire, CT: Graphics Press.

24. Tukey, J.W. (1977), *Exploratory Data Analysis*, Reading MA: Addison-Wesley.

25. Tukey, J.W. (1984), "Data analysis: history and prospects," in *Statistics: An Approach*, H.A. David and H.T. David, eds., pp. 183-202, Iowa State University Press, Ames.

26. Tukey, P.A. and Tukey, J.W. (1981), "Graphical display of data sets in 3 or more dimensions," in Barnett, V. (ed.), *Interpreting Multivariate Data*, New York: John Wiley & Sons.

27. Wang, Peter C.C., ed. (1978), *Graphical Representation of Multivariate Data*, Academic Press, New York.

Dimensionality Reduction in Density Estimation

Thomas W. Sager

Department of General Business
University of Texas
Austin, TX

ABSTRACT

All truly nonparametric density estimators suffer from embarrassingly poor error properties in higher dimensions. Even with optimal choices, the behavior of an error measure like mean-integrated-squared-error is $O(n^{-4/(d+4)})$ where d is the dimension. Enormous data sets and fantastic computational power would be required to obtain reasonable density estimates even for modest values of d. This situation stands in marked contrast to such standard statistical techniques as regression, which has good error properties essentially independent of the dimension and which is applied routinely to high-dimensional data. The key to the success of regression is its skillful exploitation of presumed linear relationships in the data. In density estimation, the statistician's presumed knowledge of something of the structure of the density may prove similarly exploitable. In fact, it is shown that if the *isopleths* (or contours) of the density can be specified or even modelled approximately, then the high-dimensional density estimation problem can be collapsed into a one-dimensional density estimation problem, with the error properties of the latter.

1 INTRODUCTION

Few statisticians would challenge the assertion that our tools for multivariate density estimation are not entirely satisfactory. But it is not widely appreciated just how turgid are the waters we row. High dimensional density estimation is, in fact, mired in a slough of impracticality from which not even the modern Computer can easily effect a rescue. This point deserves more attention than it has received, so in section 3 I proffer some observations on our predicament. However, recognition of our plight merely sounds the alarm. All paths to salvation seen to recognize - implicitly or explicitly - that conventional multivariate density estimators remain asymptotically sensitive to too large a class of densities. The statistician frequently has more knowledge than the nonparametrician admits but less than the parametrician claims. Thus the appropriate class of candidate densities may really lie between the parametric and nonparametric dichotomy. From this point of view, it is therefore natural that inference on such a class should utilize techniques which embody features of both approaches to inference. In section 4 I discuss a hyrbrid parametric/nonparametric technique for exploiting a certain kind of knowledge about the density: the general form of its contours, or isopleths. When this knowledge is available, some of the theoretical and computational problems of section 3 may be ameliorated. In particular, the deterioration of estimator performance with increasing dimension can be arrested. Section 5 concludes with some numerical and graphical examples. Because an understanding of basic density estimation techniques facilitates appreciation of the discussion in Section 4, we begin with a brief overview of methods.

2 UNIVARIATE DENSITY ESTIMATION

Until the 1950's, the histogram was the sole statistical method for estimating a density function nonparametrically. To apply this method, the real line would be partitioned into a number of intervals ("windows") of (usually) equal width. Suppose $\{a_i\}$ denotes the window boundaries, with $\ldots < a_{-1} < a_0 < a_1 < a_2 < \ldots$. Then the histogram density estimate for an x in the window (a_i, a_{i+1}) would be based only on the data visible through that window. Moreover, the density estimate would be the same anywhere in the window, namely, the proportion of the data in the window divided by the length of the window, or $[F_n(a_{i+1}) - F_n(a_i)]/(a_{i+1} - a_i)$ where F_n denotes the empiric distribution function of the data. The idea is that this ratio should be a good estimate of $[F(a_{i+1}) - F(a_i)]/(a_{i+1} - a_i) \sim F'(z) = f(z)$ for some z in the window, where F is the cumulative distribution function and $f()$ is the probability density function. Note that $f(z)$ represents the average value of the density over the window. If f did not vary too much over the window, then $f(z)$ might be a reasonable proxy for the desired $f(x)$.

The histogram estimator is characterized by stuck rectangular windows. That is, the window locations are fixed and the estimate assigns equal weight to all data in the window but zero weight to all data outside the window. In the 1950's, Fix and Hodges (1951) and Rosenblatt (1956) pioneered new estimators which allowed for sliding windows and nonrectangular windows. The general theory for the new approach was set forth in a seminal paper by Parzen (1962) and the technique was christened the kernel method. The idea behind the kernel method is that those data values closest to x contain the most information about $f(x)$ and therefore should be weighted most in estimating $f(x)$. Thus rectangular windows, with their 0-1 weights, should be inferior. In fact, they are, although *sliding* rectangular windows are not bad. The major disadvantage of the histogram is its stuck windows: Consider estimating $f(x)$ for an x near the boundary a_i of a histogram window. Data just outside the window on the other side of a_i contribute nothing to the estimate of $f(x)$ although such data intuitively may contain considerable information about $f(x)$. In estimating $f(x)$, it is more desirable to slide the window along the real line until it is centered over x. Thus we see best those data closest to x, with our vision becoming more obscured as we gaze further away from x. Because the histogram windows are stuck, the estimates of f near window boundaries suffer bias effects which cannot be alleviated even by asymptotically increasing the number of windows and reducing their width. Although by this procedure the histogram can be made to converge to f, its convergence rate remains inferior to the kernel and other methods.

1) The kernel estimator

The kernel method begins with a function K, called the kernel, which satisfies certain regularity conditions (see Parzen (1962), e.g.). The role of K is to assign weights to the data depending on their distance from x. The width of the kernel window, and hence the rate at which the weights tail off, is controlled by specifying a window width parameter h_n. The kernel estimator is $f_n(x) = (nh_n)^{-1} \Sigma_{i=1}^{n} K((x-x_i)/h_n)$ where $x_1, \ldots, x_n$ denote the data. Frequently a unimodal density will be used for K. For instance, if K is the $N(0,\sigma^2)$ density, then specifying σ controls the window width. It is now recognized that all the major density estimators can be expressed as generalized kernel estimators.

(2) The nearest neighbor estimator

The nearest neighbor estimator is $f_n(x) = 2\ r_n\ n^{-1}/(Y_{k+r_n}\ Y_{k-r_n})$ where Y_{k+r_n} is the r_n-th smallest datum greater than x and Y_{k-r_n} is the r_n-th largest datum less than x. Note that $f_n(x) = [F_n(Y_{k+r_n}) - F_n(Y_{k-r_n})]/(Y_{k+r_n} - Y_{k-r_n})$ so that the nearest neighbor estimator is a kind of histogram, but with the sliding boundaries Y_{k-r_n} and Y_{k+r_n} determined by the data. r_n plays the role of h_n in controlling window size, and the kernel K is rectangular and data dependent. Loftsgaarden and Quesenberry (1965) defined this estimator.

(3) The trigonometric series estimator

There is a class of estimators based on expanding f in a series using orthogonal functions and estimating the coefficients of those functions from the data. As an example, the estimator based on the trigonometric series $\{\cos(\pi kx);\ k=1, 2, \ldots\}$ is $f_n(x) = \Sigma_{k=1}^{q(n)} a_k \cos(\pi kx)$ where $a_k = n^{-1} \Sigma_{i=1}^{n} \cos(\pi kx_i)$ and $q(n)$ is the truncation point for the series expansion. $q(n)$ plays the role for orthogonal series estimates that the window width parameter h_n does for kernel estimates. By setting $K(x) = \pi[\sin(y/2)/\ (y/2)\ ]^2/2$ in the kernel estimator, the above version of the trigonometric series estimator is obtained. Kronmal and Tarter (1968) is a representative paper.

(4) The maximum penalized likelihood estimator

Unrestrained maximum likelihood estimation of continuous densities makes no sense. The likelihood $\Pi\ f(x_i)$ can be made arbitrarily large by choosing f with sufficiently tall spikes at the data. But by imposing suitable constraints, it is possible to obtain useful constrained maximum likelihood estimates. One such approach is to penalize the roughness of the density estimate.

A spiky density, for example, is "rough". Rough densities have large first and second derivatives. Therefore, maximizing $\Sigma \log f(x_i) - \alpha \int [f'(t)]^2 dt - \beta \int [f''(t)]^2 dt$ for an appropriate choice of $\alpha \geqslant 0$ and $\beta > 0$ becomes attractive. More generally, one maximizes $\Sigma \log f(x_i) - \Phi(f)$ in some appropriate space of functions. The rougher the density, the larger will be the penalty $\Phi(f)$ subtracted from the likelihood. The solution is the maximum penalized likelihood density estimator. An IMSL computer routine is available for calculating it. Under appropriate conditions, the solution is a polynomial spline. Good and Gaskins (1971) introduced this technique.

(5) The isotonic estimator

A second approach to maximum likelihood estimation of densities is to impose the constraint of unimodality. If $\Pi_{i=i}^{n} f(x_i)$ is maximized subject to f being unimodal, the solution is a step function with steps at some of the data, increasing to the left of the (estimated) mode and decreasing to the right. This is a form of kernel estimate with a rectangular kernel whose window width is variable and highly data dependent. Representative of a number of papers in this area is Wegman (1970).

(6) The spline estimator

A spline estimate is essentially a smoothed histogram. A piecewise polynomial of low degree, the spline preserves all the information present in the histogram on which it is (at least implicitly) based. In a generalization of the method, if the "histogram" is just the original dataset and the spline estimate is built up by positioning a deltaspline at each data point, then the resulting estimate is a kernel estimate with deltaspline kernel. Boneva, et al (1971) introduced this technique.

There are hundreds of articles in the literature of density estimation (e.g., see bibliography by Wertz and Schneider (1979)). Wegman (1983) and Fryer (1977) are good introductory overviews. Tapia and Thompson (1978) is a sophisticated mathematical monograph with emphasis on maximum penalized likelihood methods. Hand (1982) contains a comprehensive treatment of the kernel method. Kronmal and Tarter (1976) is an elementary overview with emphasis on trigonometric series. Wertz (1978) is a compilation of theorems without proofs.

3 PRACTICAL DIFFICULTIES WITH MULTIVARIATE DENSITY ESTIMATION

The literature on multivariate density estimation is much less extensive than the vast body of work on the univariate problem. For the most part, such multivariate estimators as have been studied are relatively straightforward analogues of univariate estimators. Thus multivariate kernels, multivariate nearest neighbors, multivariate series, multivariate penalized likelihoods replace univariate kernels, nearest neighbors, series, splines, and likelihoods in the analysis. These "hand-me-up" methods work poorly if the dimension exceeds 2 or 3, but it is not widely appreciated how quickly the situation becomes desperate.

Table 1 Sample size yielding equivalent error for estimating mean as error in estimating density with sample of 500, as a function of dimension d

d =	1	2	3	4	5	6	7	8	9
Comparable sample size for mean	144.3	63.0	34.9	22.4	15.8	12.0	9.6	7.9	6.8
d =	10	15	20	25	30	35	40	45	50
Comparable sample size for mean	5.9	3.7	2.8	2.4	2.1	1.9	1.8	1.7	1.6

The impracticality of analogue estimators came sharply into focus on a recent U.S. Navy project. The objective was to estimate the densities for a number of datasets, each containing about 500 independent observations and 10, 20 or more dimensions. It was felt that many of the densities might be normal, or close to normal, but no great confidence was invested in this belief. Some, in particular, were believed to be non-normal. Pre-analysis quickly established that all the standard, analogue nonparametric multivariate density estimators (NMDE's) suffered fatal flaws which rendered then useless for the required objective. The flaws are both theoretical and practical and apply to other multivariate density estimation settings, as well.

For example, there is the problem of deterioration of error with increasing dimension ("the curse of dimensionality" (Hand, 1982)). A common measure of the closeness of a density estimator to the true density is mean integrated square error (MISE), defined as $E\{\int(f_n - f)^2\}$. The MISE of all analogue NMDE's deteriorates as the dimension d increases. (Published claims to the contrary — e.g., Schwarz (1967) — do not use a full nonparametric class for the eligible densities. Recall that all the estimators in section 2 are species of kernel estimators, so a kind of duality applies. In imposing restrictions on the density, Schwarz shares the *spirit* of the solution I propose, but his restrictions are difficult to interpret, difficult to verify, and thus uncertain of application.) Similar deterioration applies to other error measures. A typical example is the behavior of the multivariate kernel estimator (Cacoullos, 1966). Its MISE is proportional to $n^{-4/(d+4)}$. Compared with, say, estimation of the mean, density estimators do not seem to exploit efficiently the information in datasets. The mean squared error of $\bar{x}$ is proportional to n^{-1}, independent of d. If we ignore proportionality factors, then for a sample of n, the kernel estimator yields approximately as much information about the density as a sample of $n^{4/(d+4)}$ yields about the mean. For $n = 500$, Table I shows how this equivalent sample size deteriorates with d.

The values in the table are distressingly low for a nominally large sample of 500. It may be argued that things are not as bad as they seem. The $n^{-4/(d+4)}$ rate is only asymptotic and no one is sure for what n the rate becomes a good approximation. Moreover, the constants of proportionality (omitted in the above), which depend on the true density and the particular kernel func-

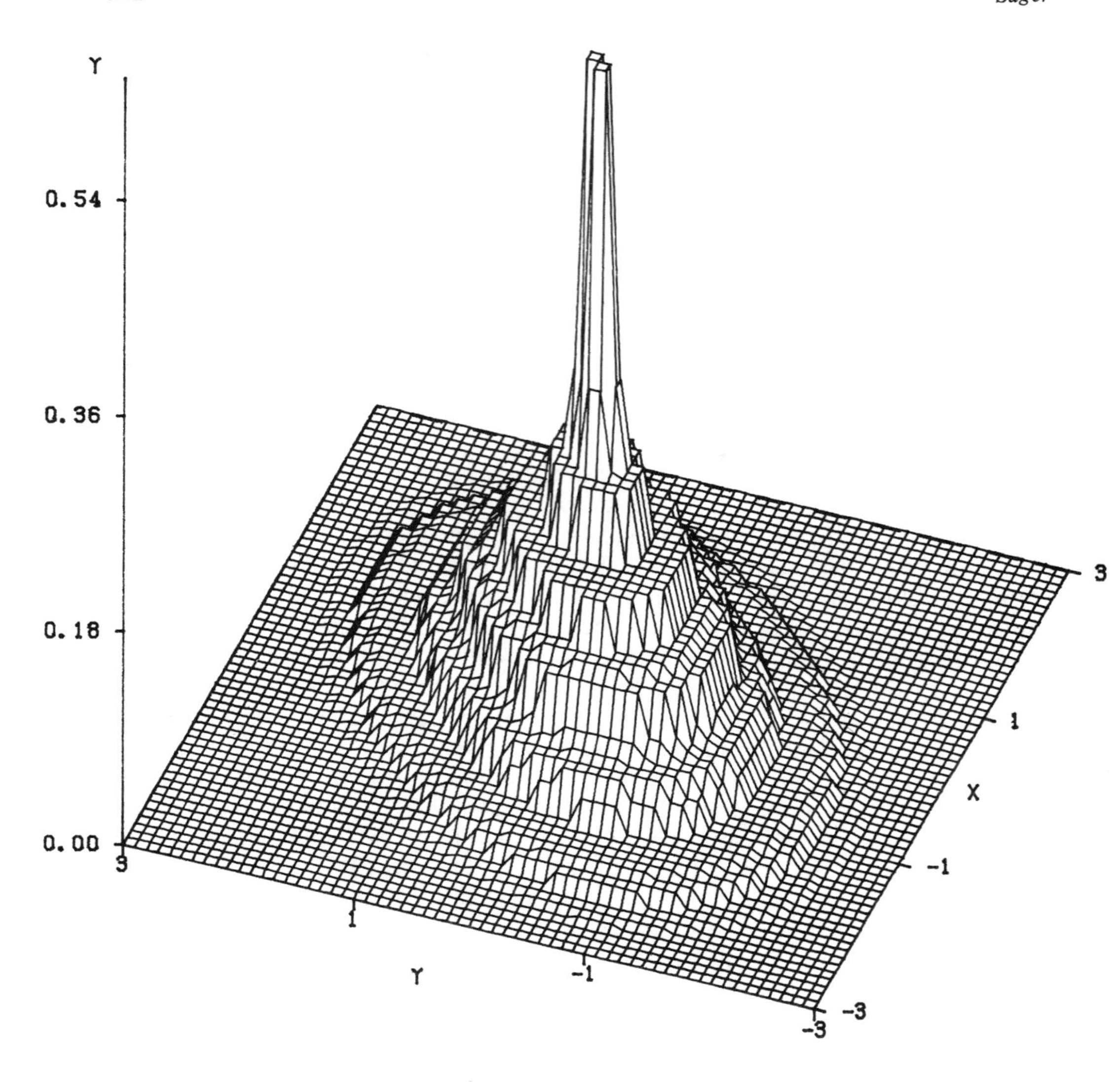

Figure 1 Estimated bivariate density. Elliptical isopleths assumed. True density is normal (0,0,1,1,.5).

tion, could have a meaningful ameliorative effect. Theoretical studies for finite n can establish approximate bounds for MISE, with the best case arising from a fortuitous choice of the optimal kernel. Since the optimal kernel depends on the true density, in practise it will not be known. Moreover, most attention has focused on the case $d = 1$. The situation of $d > 1$ remains unsettled, although it appears that Table I will not need substantial revision.

Many people believe that computers, in conjunction with techniques like Monte Carlo, bootstrapping, and jackknife, will shortly alleviate all our problems. I think this is unlikely. Consider, for example, a hypothetical Monte

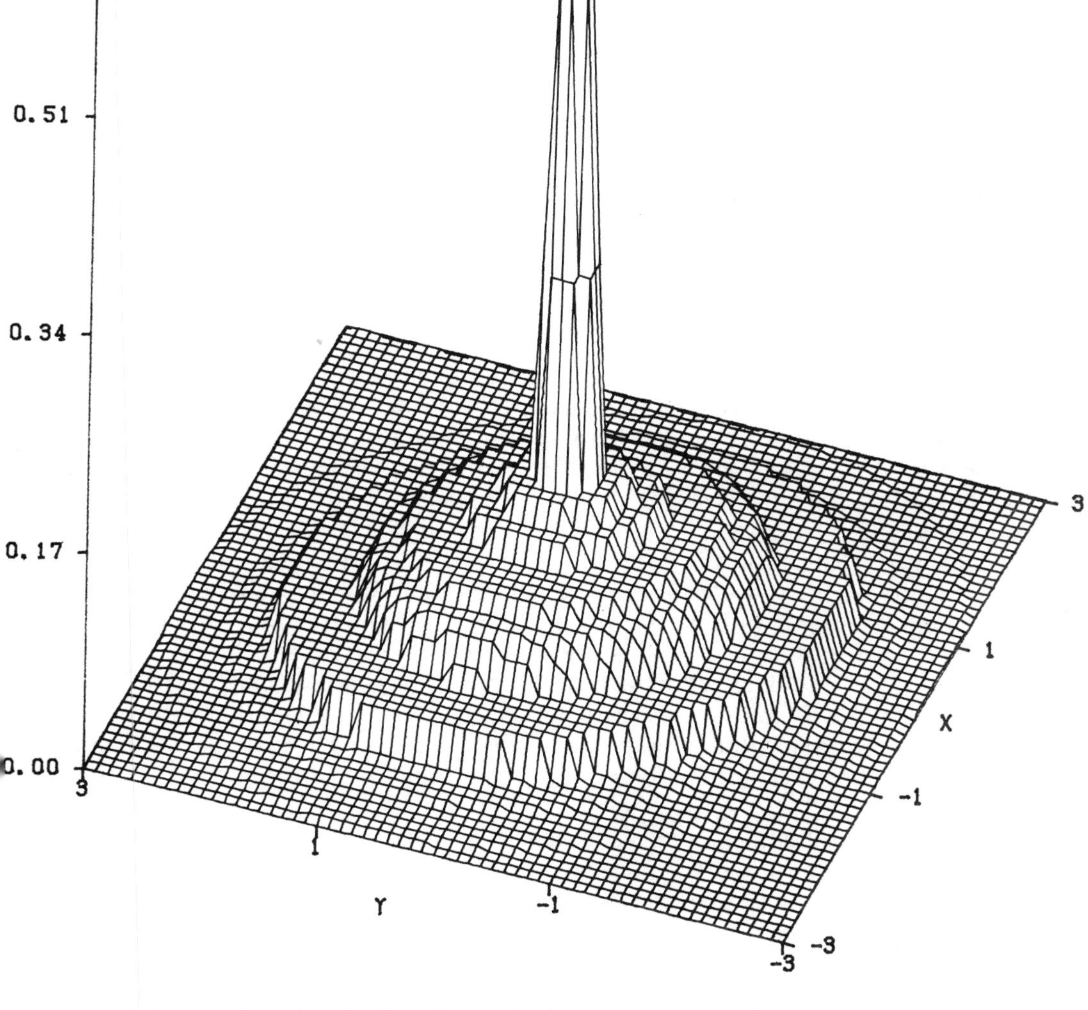

Figure 2 **Estimated bivariate density. Elliptical isopleths assumed. True density is normal (0,0,1,1,0).**

Carlo study of MISE of an NMDE. To calculate MISE requires numerical integration. Typically, a grid is laid down in the space of the independent variables and the integral is approximated by Riemann sums over the grid. If as few as 30 gridmarks are deemed sufficient for each dimension (cf. Figures 1 and 2, which use 60) then every MISE calculation would require 30^d Riemann sums. It is instructive to estimate how much computer time would be required to calculate this. The Cray 1, for example, is one of the fastest modern computers. It can perform between 10^8 and 10^9 elementary operations per second. If the Cray could compute 10^8 Riemann sums per second (and note that each sum first requires computation of the estimated density at the gridpoint), then

Table 2 Estimated Cray computation time for a single MISE as a function of dimension

d =	6	7	8	9	10	20
Time =	7.3 sec	3.7 min	1.8 hrs	2.3 days	68.3 days	10^{14} yrs

Table II shows the approximate computation time for a single MISE. For similar reasons, the estimate could not be stored - it would require 30^d words of memory to contain it. The estimate would have to be recomputed for each application.

Regression applications with 500 observations and 10 or 20 variables are routine. The solutions are well-behaved and major computational difficulties do not arise. But the models are parametric. The problem with NMDE's seems to be that they are *too* nonparametric. They retain sensitivity to too many alternatives, most of which the practitioner would consider pathologies. With NMDE's, the density is really the parameter. The density (and thus, the parameter) can be changed a lot in a small region without having much impact on the overall distribution of probability. For an NMDE to respond promptly to local perturbations in the density, data points must be laden with exploitable information about the density's behavior in every direction. When $d = 1$, a data point need only inform on the density to the right and left. When $d = 2$, a data point must cover the density's eccentricities in a 360° arc. As d increases, it is intuitively clear that the directions to guard quickly overwhelm a point's ability to provide useful information. Consequently, either the NMDE's performance deteriorates or many more data are required. If points could be relieved of some of this burden, NMDE performance could be improved.

In many situations, the statistician may not need an unrestricted nonparametric class of densities. He may possess exploitable knowledge about the true density that, by sharpening the admissible class of densities and controlling the amount of permissible local perturbation, significantly reduces the informational burden on the data. For example, in the Navy project mentioned previously, it was believed that most of the distributions were unimodal and had elliptical contours, without necessarily being normal. Thus an NMDE constructed to exploit this knowledge should possess more reasonable error properties than a general purpose NMDE designed to be asymptotically sensitive to all nonparametric alternatives. The approach taken in the Navy project began by estimating the location and orientation of density contours through the mean vector and the covariance matrix. In this, the estimator anticipated finding a normal distribution. But then a density value was estimated for each contour by isotonic regression [see Sager (1982)]. Departures from normality were thereby permitted. The resulting estimator has an MISE proportional to $n^{-2/3}$ (not dependent on the dimension) and calculates very rapidly on a computer (even a micro!). Even better convergence could be obtained by preliminary grouping. This estimator is representative of the class of estimators I wish to discuss. They are hybrids, partially parametric and partially nonparametric, incorporating aspects of both approaches to inference. They profitably reduce the complexity of the admissible class of densities by eliminating irrelevant or unlikely alternatives while retaining its essential nonparametric flavor.

4 THE IOSPLETH DENSITY ESTIMATOR

Throughout this section $f(x)$ denotes a d-dimensional probability density function, $d = 1, 2, \ldots$. In some problems, a statistician may not have enough knowledge to specify a parametric form for f but also will not be so ignorant as to require a completely nonparametric class. In such cases he may be able to express his knowledge in the form of assertions about the shape of the density. Such assertions may be descriptions of the contours of the density. For example, he may believe that the density is similar to that of a multinormal: unimodal with elliptical contours. Or he may feel that the density is bimodal with circular contours near each mode. We shall call such contours isopleths.

Definition. An isopleth of f is a set of the form $I_c = \{x;\ f(x) = c\}$, $o < c < \infty$. A modal region of f is a set of the form $M_c = \{x;\ f(x) \geqslant c\}$.

To eliminate pathologies, we shall suppose that each modal region is bounded. This is necessarily the case if f is uniformly continuous. The boundary of a modal region is an isopleth, which is most simply thought of as a curve. However, an isopleth may be a region of positive area if f has "flat spots". We shall rule this out by assuming that the Lebesgue measure of all isopleths is zero.

If f has elliptical contours, then f depends on x only through the value of $g(x) = (x - \theta)' A (x - \theta)$ *for some* θ *and positive definite matrix* A. *Thus* f is a composition $f = h(g)$ for some univariate function h. Whenever f is written in this way, we shall call g an *isopleth form* for f, and h the *transfer density* for g. Observe that h is a legitimate density function on the space of the values of g. Note also that g is not unique: if g is an isopleth form, then so is $a \cdot g$, for constant $a \neq 0$. Occasionally, the statistician may be able to specify g completely (up to a). In the case of an elliptical isopleth form, this entails providing values for θ and A. More usually, the parameters of the form (e.g., θ and A) would have to be estimated. To avoid encumbering the development, we shall suppose for the moment that g is specified completely. We shall also suppose that g is *maximal* in the sense that if g_1 is any isopleth form, then each isopleth of g_1 is contained in an isopleth of g. A maximal isopleth form always exists. Moreover, since it is the isopleths of g (rather then the values of g) which are primary, we may always redefine a maximal isopleth form to be monotone, in the sense that the value of g on an isopleth is greater than the value of g on any interior isopleth.

Now let $X_1, \ldots, X_n$ be i.i.d. random variables distributed according to f. Let $X_{(j)}$ be the X_j which yields the jth smallest value in $\{g(X_i);\ 1 \leqslant i \leqslant n\}$ and let $C_j = g(X_{(j)})$. It is natural to refer to $X_{(1)}, \ldots, X_{(n)}$ as *order statistics*: If $i < j$ then $C_i < C_j$, and $X_{(i)}$ lies higher up on f than does $X_{(j)}$. Moreover, the corresponding modal regions are nested: $M_{C_1} \subset \ldots \subset M_{C_n}$. And any sub-collection of $F(M_{c_1})$, $F(M_{c_2}) - F(M_{c_1})$, ..., $F(M_{c_n}) - F(M_{c_{n-1}})$ follow a joint Dirichlet distribution, so the areas (or volumes) between modal regions play the role of spacings for order statistics. The essential role played by an isopleth form is thus seen to be that of imposing a natural order in d-dimensional euclidean space. It now becomes clear how to estimate a multivariate density with specified isopleth form: simply apply any univariate density estimator directly to the univariate order imposed by the imposed by the isopleths. Any x_o for which an estimated $f(x_o)$ is sought can be thought of as an equivalence class $I_{C_o} = \{ x; g(x) = g(x_0)\}$, where $C_0 = g(x_0)$. The true density is constant on this equivalence class I_{C_0}; therefore, the estimate should also be constant on the isopleth. Data points near the isopleth convey the same information about the density regardless of whether or not the points are near each other in a euclidean sense.

The examples which follow discuss the adaptations necessary to implement this idea with the univariate estimators of section 2.

Examples

(1) Kernel estimator.

To adapt the kernel estimator to the multivariate setting, choose a univariate kernel K and a window parameter h_n. Then the adapted isopleth kernel estimate is $f_n(x_0) = (nh_n)^{-1} \Sigma_{i=1}^{n}=K((\lambda(M_{C_0}) - \lambda(M_{C_i}))/h_n)$ where $\lambda(M)$ denotes the area (volume) of the modal region M. Note that the *volume* between modal regions is the multivariate analogue of univariate *distance* between points. So the univariate volume $\lambda(M_{C_0}) - \lambda(M_{C_i})$ is used in the kernel K in place of the multivariate distance $x_0 - X_i$. Note also that the volumes of modal regions will be easy to calculate for many isopleth forms. For example, with elliptical contours of the form $g(x) = (x - \theta)' A (x - \theta)$, the volume of M_C is $(C\pi)^{d/2}$ (det $A)^{-1/2}/\Gamma(d/2 + 1)$ (Cramer, 1946).

(2) Nearest neighbor estimator.

$f_n(x_0) = 2\, r_n\, n^{-1} / [\lambda(M_{C_{k+r_n}}) - \lambda(M_{C_{k-r_n}})]$. Here, $M_{C_{k+r_n}} - M_{C_{k-r_n}}$ is the "spacing" between the r_n −th closest data point "less than" x_0 and the r_n −th closest data point "greater than" x_0, as determined by their isopleths.

(3) Trigonometric series estimator.

$f_n(x_0) = \Sigma_{k=1}^{g(n)} a_k \cos(\pi k\, \lambda(M_{C_0}))$ where $a_k = n^{-1} \Sigma_{i=1}^{n} \cos(\pi k \lambda(M_{C_i}))$ and M_{C_0} is the modal region on the boundary of which x_0 lies.

(4) Maximum penalized likelihood estimator.

Maximize $\Sigma \log f(\lambda(M_{C_i})) - \Phi(f)$ for an appropriate Φ function. The solution is a *univariate* f. Then the density estimate is $f(x_0) = f(\lambda(M_{C_0}))$.

(5) Isotonic estimator.

Maximize $\Pi_{i=1}^{n} f(\lambda(M_{C_i}))$ subject to f being isotonic and evaluate $f(x_0) = f(\lambda(M_{C_0}))$.

(6) Spline estimator.

Begin with the "histogram" of the $\lambda(M_{C_i})$'s that the particular spline method requires. Then apply the spline to the resultant "histogram" and evaluate $f(x_0) = f(\lambda(M_{C_0}))$.

It should now be apparent that the isopleth technique is a dimensionality reduction approach to density estimation. The inferential problem is transferred from the multivariate f to the univariate h. Since $f(x) = h(g(x))$, the transfer density h is nonnegative and integrates to 1; so it is a probability density function. Moreover, h is monotonically decreasing on $[0, \infty)$: if $g(x_1) < g(x_2)$ then $M_{g(x_1)} \subset M_{g(x_a)}$, so $f(x_1) > f(x_2)$. Except for the isotonic estimator, none of the estimators of h discussed in this section need be monotonic. Therefore, they may be improved by computing their isotonic regression (cf. Barlow., et al (1972)).

Given the above framework, it is not surprising that NMDE's based on isopleth forms should inherit the properties of their univariate forebears.

Theorem. Let $f = h(g)$ be a multivariate density function having a known isopleth form g. Let $\hat{h}_n$ be a univariate density estimator for h based on $g(X_1), \ldots, g(X_n)$. For each x, let the multivariate density estimate of f be $f_n(x) = \hat{h}_n(g(x))$. Then

1) f_n inherits any almost sure convergence rate of $\hat{h}_n$: if $a_n(\hat{h}_n - h) \to 0$ *a.s.*, then $a_n(f_n - f) \to 0$ *a.s.*

2) f_n inherits any distributional limit of $\hat{h}_n$: if $a_n(\hat{h}_n - h) \to W$ in law, then $a_n(f_n - f) \to W$ in law.

3) f_n inherits any L_p limit of $\hat{h}_n$: if $a_n||\hat{h}_n - h||_p \to 0$ and the support of f is bounded, then $a_n||f_n - f||_p \to 0$.

4) f_n inherits any mean L_p limit of $\hat{h}_n$ (in particular, MISE): if $a_n E(||\hat{h}_n - h||) \to 0$ and the support of f is bounded, then $a_n E(||f_n - f||_p) \to 0$.

Proofs: For 1), choose x, let $c = g(x)$ and apply the hypothesis. For 2), again let $c = g(x)$ and apply the hypothesis to the characteristic function $E\{\exp(i\, a_n(f_n(X) - f(X))t)\}$. For 3), $a_n||f_n - f||_p$ is approximated by the Lebesque-Stieltges sum $a_n\{\Sigma|\hat{h}_n(c_i) - h(c_i)|^p[\lambda(M_{c_{i+1}}) - \lambda(M_{c_i})]\}^{1/p}$ for a partition $c_1 < \ldots < c_n$. Since the support of f is bounded, this sum is bounded above by constant $\times$ $a_n\{\Sigma|\hat{h}_n(c_i) - h(c_i)\ |^p\}^{1/p}$, which is an approximation for constant $\times$ $a_n||\hat{h}_n - h||_p$. For 4), take expectations in 3).

Actually, propositions 3) and 4) hold more generally. It is not necessary to bound the support of f, but the analysis is more delicate and will not be pursued here.

Every multivariate density has an isopleth form g and a transfer density h. But the isopleth form may not be simple and it may not be known. If the form is too complicated, the problem may be outside the scope of this paper, which is limited to the moderately complicated case between the simplicity of parametric forms and the complexity of the completely nonparametric. Therefore, if these techniques are to be used, the isopleth form should be relatively simple. But the form may still be unknown. In that case, the form must be estimated. With elliptical isopleths and a predisposition toward normality, as in the Navy project previously alluded to, it is natural to estimate the mean vector and covariance matrix. If the isopleth form is a polynomial of known order but unknown coefficients, a least-squares or method of moments approach any be used to estimate coefficients. If the order of the polynomial is unknown, then techniques analogous to stepwise regression might be employed. For other than polynomial forms, the estimation techniques multiply prodigiously. The proper investigation of such methods and their properties lies beyond the scope of this paper. For a review of methods of estimating isopleths and related features of the density, see Sager (1983).

5 EXAMPLES

Figures 1 and 2 are illustrations of the isotonic regression isopleth density estimator for $d = 2$. For Figure 1, 100 i.i.d. N(0,1) pairs of variables with correlation coefficient zero were generated by computer. The mean vector and covariance matrix were estimated from the data, giving rise to an (estimated) elliptical isopleth form. The areas of all 100 ellipses on which the data lay were calculated. These areas then became the "data" to which the isotonic density estimator was applied. This estimator groups the data into contiguous intervals, on each of which the density estimate is constant and decreasing as the intervals move away from the mode. These constants become the density estimates and are represented in the figure by the heights of the concentric plateaus surrounding the mode. The univariate isotonic density estimate has a known and correctible tendency to peak too extremely near the mode. In Figure 1, that tendency is clearly visible as the too tall central peak. In order to show the estimate in its pure form, no attempt was made to correct for peaking. Nor were the data preliminarily grouped — a procedure known to improve the convergence rate.

Figure 2 was obtained in the same way as Figure 1, but the computer generated data were 100 i.i.d. N(0,1) pairs with a correlation coefficient of .50. Including estimating the covariance matrix and calculating areas, the computation of the density estimate for Figures 1 and 2 required a fraction of a second on a small CDC Cyber machine. The graphics for each figure took about 20 seconds on an IBM 370/158.

REFERENCES

Barlow, R.E., Bartholomew, D.J., Bremner, J.M., Brunk, H.D., (1972), *Statistical Inference Under Order Restrictions*, Wiley, New York.

Boneva, L., Kendall, D., and Stefanov, I., (1971), Spline transformations: Three new diagnostic aids for the statistical data-analyst. (With discussion), *J. Roy. Statist. Soc. B*, 33, 1-71.

Cacoullos, T., (1966), Estimation of a multivariate density, *Ann. Inst. Statist. Math.*, 18, 178-189.

Cramer, H., (1946), *Mathematical Methods of Statistics*, Princeton University Press, Princeton, New Jersey.

Fix, E. and Hodges, J. Jr., (1951), Discriminatory analysis, nonparametric discrimination: consistency properties, *Report No. 4, Project No. 21-49-004*, USAF School of Aviation Medicine, Randolph Field, Texas.

Fryer, M.J., (1977), A review of some nonparametric methods of density estimation, *J. Inst. Maths. Applics.*, 18, 371-80.

Good, I.J. and Gaskins, R., (1971), Nonparametric roughness penalties for probability densities, *Biometrika*, 58, 255-77.

Hand, D.J., (1982), *Kernel Discriminant Analysis*, Wiley, New York.

Kronmal, R. and Tarter, M., (1968). The estimation of probability densities and cumulatives by Fourier series methods, *J. Amer. Statist. Assoc.*, 63, 925-952.

Kronmal, R. and Tarter, M., (1976). An introduction to the implementation and theory of nonparametric density estimation, *Amer. Statistician*, 30, 105-112.

Loftsgaarden, D.O. and Quesenberry, C.P., (1965). A nonparametric estimate of a multivariate density function, *Ann. Math. Statist.*, 36, 1049-1051.

Parzen, E., (1962). On estimation of a probability density function and mode, *Ann. Math. Statist.*, 33, 1065-1076.

Rosenblatt, M., (1956). Remarks on some non-parametric estimators of a density function, *Ann. Math. Statist.*, 27, 832-837.

Sager, T.W., (1982). Nonparametric maximum likelihood estimation of spatial patterns, *Ann. Statist.*, 10, 1125-36.

Sager, T.W., (1983). Estimating modes and isopleths, *Commun. Statist. Theor. Methods*, 12, 529-557.

Schwarz, S.C., (1967). Estimation of probability density by an orthogonal series, *Ann. Math. Statist.*, 38, 1261-65.

Tapia, R.A. and Thompson, J.R., (1978). *Nonparametric Probability Density Estimation*, Baltimore and London: The Johns Hopkins University Press.

Wegman, E.J., (1970). Maximum likelihood estimation of a unimodal density function, *Ann. Math. Statist.*, 41, 457-471.

Wegman, E.J., (1983). Density estimation, *Encyclopedia of Statistical Sciences*, (Vol. 2), Eds. Kotz & Johnson.

Wertz, W., (1978). *Statistical Density Estimation: A Survey*, Vandenhoeck and Ruprecht, Gottingen.

Wertz, W. and Schneider, B., (1979). Statistical density estimation: a bibliography, *Inter. Statist. Review*, 47, 155-175.

Volumetric 3-D Displays and Spatial Perception[1]

David J. Getty and A.W.F. Huggins

BBN Laboratories
Cambridge, MA

ABSTRACT

We perceive and comprehend a visual world of three spatial dimensions. However, current computer-generated graphical displays generally fail to utilize the full potential of visual perception in that they create and present only flat 2-D images. We describe here methods for extending computer-generated displays to provide three-dimensional images, and focus on the recent development of a practical, true volumetric display. We then summarize the results of several perceptual experiments, conducted with this volumetric display, designed to futher our understanding of the perception of orientation and direction in a displayed three-dimensional volume. The accuracy of perceived orientation and direction is an important issue in considering possible applications for 3-D displays.

1 INTRODUCTION

The human visual system has evolved in a three-dimensional world, and so it is perhaps no surprise that we in fact *perceive* three spatial dimensions. This is, however, a remarkable achievement for the visual system given that each of our eyes receives only a two-dimensional projection of a viewed scene. Moreover, the visual system is exquisitely tuned to detect visual structure and relationships of many different kinds—about equally well in all three dimensions. For this reason it seems advantageous to explore and exploit these three-dimensional perceptual capabilities in designing and using graphical displays.

[1] This work was supported by the Office of Naval Research under contract N00014-80-C-0750.

We have two objectives in this paper. First, we discuss briefly the major methods for generating three-dimensional displays, and then describe one type in more detail namely, volumetric displays. Second, we describe some results we have obtained using SpaceGraph, a particular volumetric display. This research concerns, first, the perception of orientation of objects in space, and, second, the perception of direction in space.

2 GENERATION OF 3-D DISPLAYS

Before discussing the generation of 3-D displays, it is important to observe that the visual system is highly opportunistic. There exist many different visual cues that can potentially provide information about the three-dimensional structure of a viewed scene, and the visual system will take advantage of all that are made available. Most of these cues are monocular—that is, they can be extracted from the image in one eye alone. These include linear perspective, texture gradients, the size of familiar objects, the interposition of objects, shadows and shading, and motion parallax. Only one cue, stereopsis, requires differential information from the two eyes. Of these cues, stereopsis and motion parallax play the most important roles, and it is these two that provide the principal methods for generating computer-driven 3-D displays.

Stereopsis is a cue based on the slightly differing views of a scene received by our two eyes as a result of their horizontal separation. Because of this separation, the visual angle subtended between any two objects located at different depths will necessarily be different in the two eyes. With both eyes fixated on object F, shown in Figure 1, the angle subtended from the fovea by another object A, located at a different depth, will be larger in the left eye than

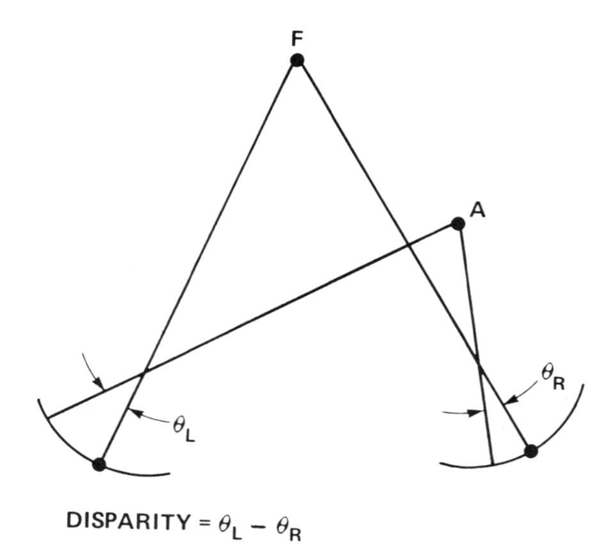

Figure 1 Binocular disparity generated between the two eyes when viewing objects at different depths.

in the right. This difference in subtended angle is called binocular or retinal disparity, and is a sufficient condition for the perception of depth.

Two of the three methods for generating 3-D displays depend upon stereopsis for the creation of depth. The first class of displays, *stereo-pair displays*, are familiar in the form of childhood stereoscopes and the occasional 3-D movie that comes along every few years. These examples, along with computer-generated 3-D displays, all have in common that, for each scene, a pair of images are constructed from two horizontally-separated vantage points. These two disparate views are delivered independently, one to each of the viewer's two eyes. A number of different techniques for image delivery have been used, including prisms and lenses (as in the stereoscope), red-and-green filter glasses or cross-polarizing glasses (as in the 3-D movies), and alternating shutter glasses. The primary difficulties with stereo-pair methods are that they are often unwieldly and uncomfortable for the viewer, that they require careful image registration to avoid viewer fatigue, and that they present only a single point of view on the scene determined by the system generating the images rather than by the observer.

The second class of stereopsis-based displays are the *volumetric* or *space-filling* displays. As suggested by the name, these displays are based on a *single* real or virtual image which quite literally fills a three-dimensional volume of space. Examples are displays produced by rapid rotation of a flat, dense matrix of LEDs through a volume, displays produced by holograms, and displays produced by oscillating movement of a flexible, variable-focus mirror. We return to discuss a realization of the oscillating mirror display (SpaceGraph) in the section that follows.

While stereo-pair and volumetric displays depend upon stereopsis to communicate depth information, a third class of display is possible, one that relies on the presence of *motion parallax*. One may think of motion parallax as the temporal analogue of binocular disparity, where two different moments in time are substituted for the differing vantage points of the two eyes. If the viewer is moving through the scene, of if there are objects moving within the scene, then two images of the scene separated by a short interval of time will generate two different viewpoints. The visual system is able to reconstruct the three-dimensional structure of objects in motion relative to the observer from the continuously varying patterns of movement in the image received by either eye. The sufficiency of motion cues to convey structure has been demonstrated in an increasing body of psychological studies (Dover, Lappin & Perfetto, 1984; Ullman, 1979; Braunstein, 1976; Johansson, 1975; Gibson & Gibson, 1957; Wallach & O'Connell, 1953). The primary limitation of this method of display to provide depth information is the need to keep the objects being viewed in motion—typically in rotation.

3 SPACEGRAPH

SpaceGraph is a particular realization of the space-filling or volumetric display. As discussed above, it differs from stereo-pair displays in that the image viewed by the observer is truly three-dimensional: the luminous points of which the image is composed actually exist at different depths from the observer. This

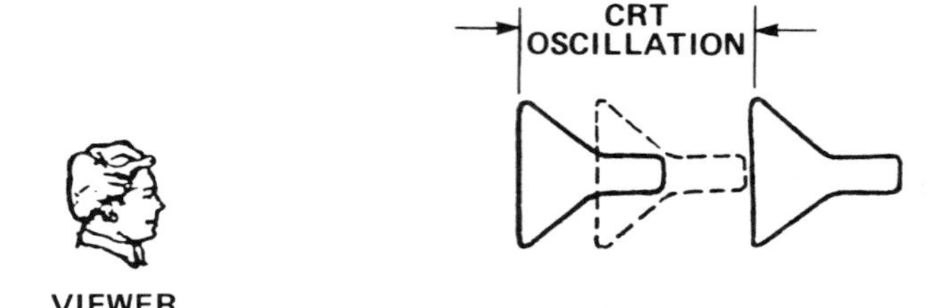

Figure 2 The conceptual principle for a volumetric three-dimensional display: points are displayed at the appropriate moment in time on the face of an oscillating CRT.

contrasts with stereoscopic displays which recreate, with two flat displays, what the observer's left and right eyes *would* see if they were looking at a three-dimensional scene.

The optical principle which underlies SpaceGraph is shown in Figure 2: if the observer could be made to perceive that a flat CRT display was moving back and forth in depth, then a controlling computer could display and refresh a three-dimensional image in the volume swept out by the oscillating phosphor screen. However, moving a real CRT through a useful distance, and at a useful repetition rate, would be a formidable task. The difficulty is side-stepped by using a mirror, which creates a world of *virtual* objects behind the mirror which are visually indistinguishable from real objects.

In particular, we can arrange to have the observer view the face of a stationary CRT by looking at its reflection in a rigid mirror as shown in Figure 3. By moving the mirror it is possible to move the weightless virtual CRT. The observer cannot distinguish the appearance of the moving *virtual* CRT from a moving *real* CRT. Furthermore, a movement of the mirror by 1 unit in depth results in a movement of 2 units by the virtual CRT. However, this limited 2-to-1 amplification in movement created by simple translation of the mirror is not sufficient for any practical device.

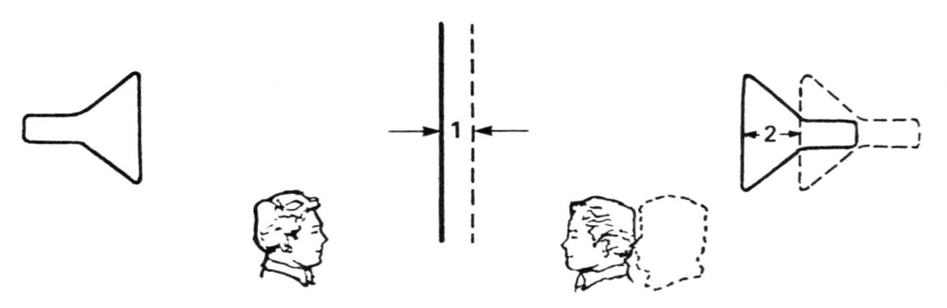

Figure 3 A realizable volumetric display using an oscillating mirror and a stationary CRT. The observer sees a virtual image of the CRT face in the moving mirror.

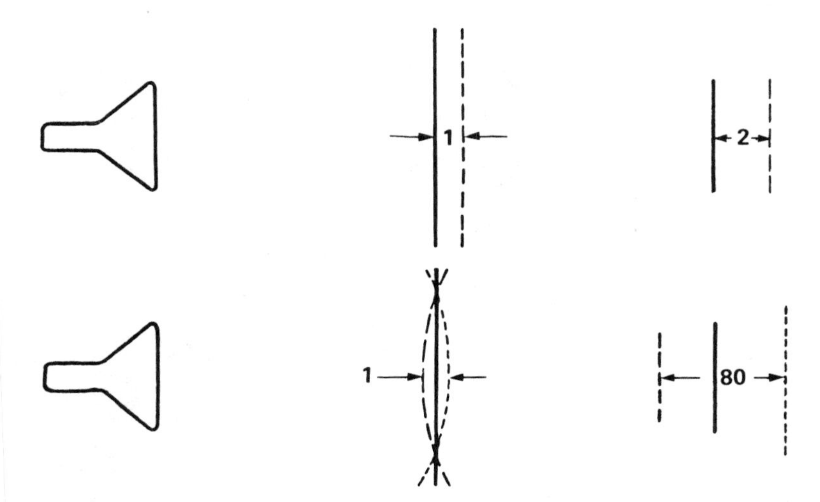

Figure 4 A translating mirror provides only a 2-to-1 amplification of depth in the viewed virtual volume: a mirror flexing in the shape of a spherical lens provides up to an 80-to-1 amplification of depth.

There is, however, a better way to create large movements of the virtual CRT, and this is by using a flexible rather than a rigid mirror, and by causing it to flex rather than translate in an oscillatory manner, as shown in Figure 4. By flexing the mirror, we cause it to act as a spherical lens whose focal length changes continuously as the mirror cycles successively through flat, concave, flat, and then convex shapes. With this method, it is possible to gain an 80-fold amplification of movement.

This concept has been realized in SpaceGraph by mounting a silvered acrylic plastic plate, shown at the left in Fig. 5, on the front of a low-frequency loudspeaker. When the loudspeaker is excited by a 30 Hertz sine wave, the mirror vibrates, approximately spherically, at 30 cycles per second. As it does so, the virtual image of the face of the CRT, which appears to the observer to be behind the mirror, sweeps cyclically through a depth of about 30 cm. If a point on the face of the CRT is momentarily brightened at the same instant in every mirror cycle, the observer will see a luminous point suspended at a specific depth in the dark void behind the mirror. The depth of the point can be varied by changing the instant in the phase of the mirror cycle at which the CRT beam is intensified. Thus the depth dimension is specified by the *time* in each mirror cycle when a point is displayed. The lateral and vertical positions of the point in the display volume correspond to the lateral and vertical positions of the point, respectively, on the CRT.

The mirror acts like a window through which the observer sees whatever is displayed in the volume. An image appears within the volume as a 3-dimensional distribution of luminous points. Since the points have no physical substance, they are transparent in the sense that two, one behind the other, are seen to be brighter than one.

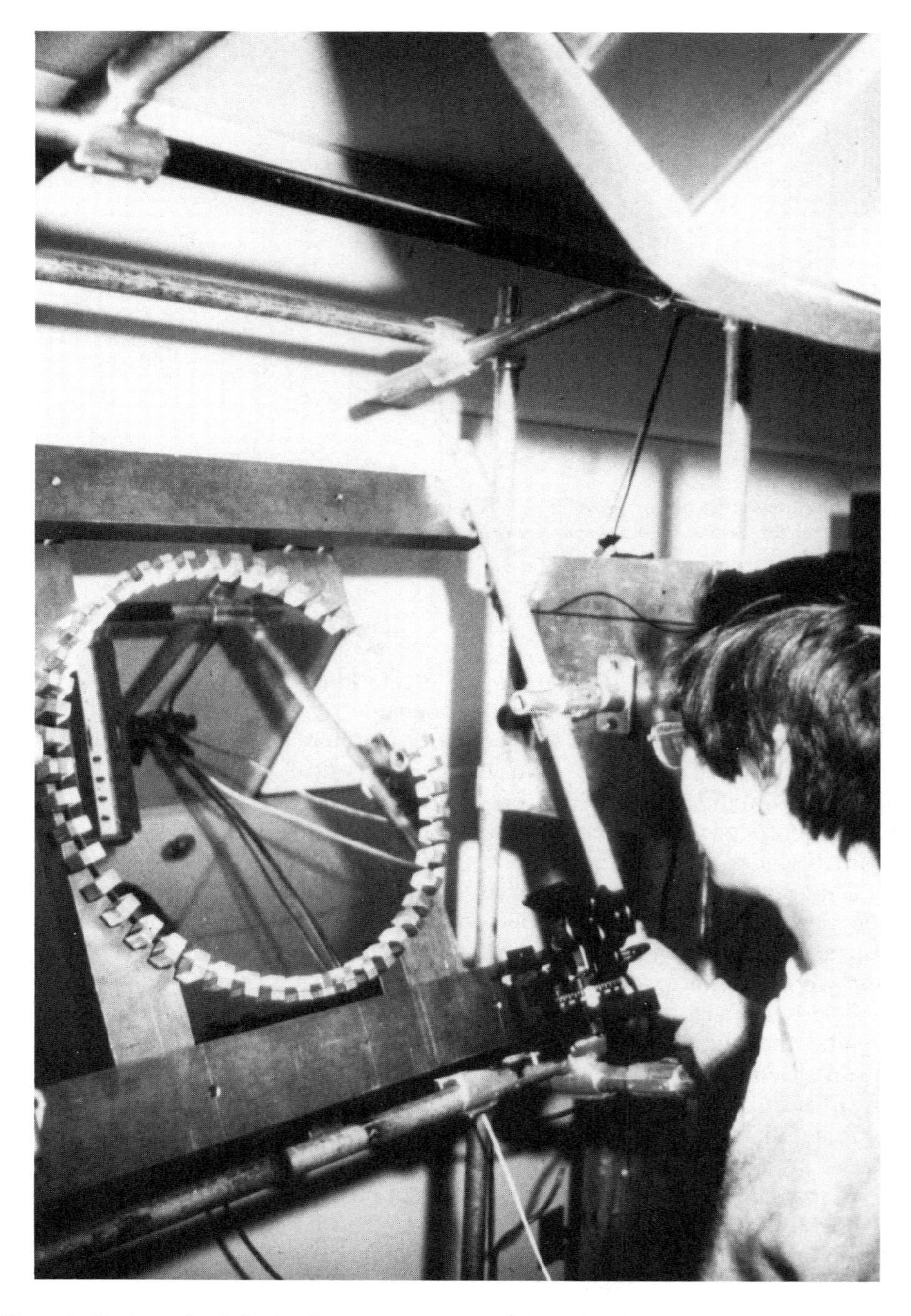

Figure 5 Photograph of the development prototype of SpaceGraph. The observer views the face of the CRT (part of which is visible at the upper right, above the observer) in the circular flexible mirror (shown at the left). The volumetric image of the displayed object appears to the observer to be behind the mirror.

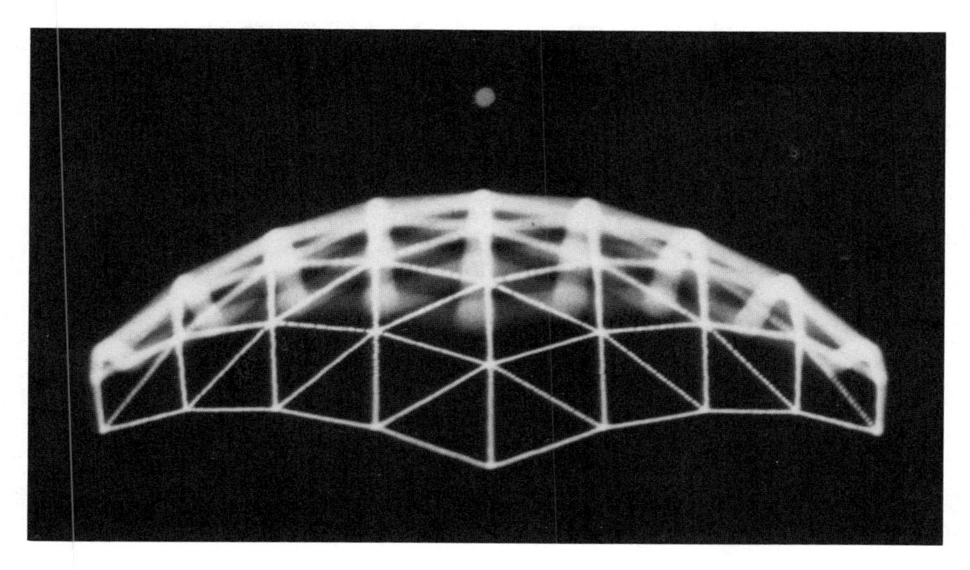

Figure 6 Photograph of a three-dimensional geodesic dome, taken directly from the SpaceGraph display. The progressive loss of camera focus for more distant parts of the dome suggests the true three-dimensionality of the displayed image.

Images are built of points and linear arrays of points. In the prototype model on which we have done our research, about 3000 points are available for drawing an image, which is sufficient for about 300 cm of line. This is enough for a fairly complex image. Figure 6 is a photograph of a geodesic dome taken directly from the display. The camera focus was set for the front edge of the dome, and the progressive loss of focus for the more distant parts of the dome provides some indication of the depth present in the image. Of course, it is impossible in any two-dimensional photograph to convey the immediacy and conviction of the three-dimensionality of images viewed first-hand in the display.

4 PROPERTIES OF SPACEGRAPH IMAGES

SpaceGraph images have several properties that make them unique. Since the points comprising the image are truly at different depths from the observer, a displayed object or scene shows perspective cues identical with that of a physical 3-D object. Furthermore, the binocular parallax effect arising from eye separation, and the movement parallax effects that occur when the observer moves his head, occur as when viewing a natural scene. The observer is free to move his head laterally or vertically, or even rotate it about the line of sight, and indeed the 3-D percept is enhanced by doing so. Movements by the observer literally change the point of view of the image, just as they would when viewing a tangible, physical object. Moving one's head upward, it is possible to look down on a displayed object; moving one's head downward, to then look up at it, and so on. The amount of movement possible is constrained only by the requirement that the viewer not lose sight of the CRT face reflected in the mir-

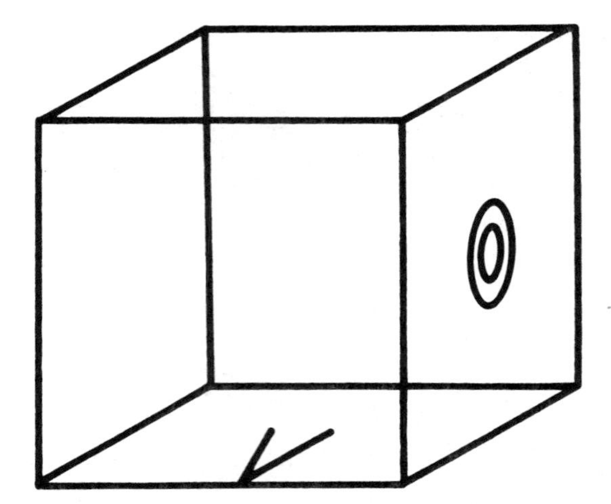

Figure 7 Appearance of the stimulus cube on a typical trial of the Y-axis orientation experiment. Reproduced from Huggins and Getty, 1981.

ror. The ability to move freely to see a displayed image from different vantage points becomes particularly valuable in exploring 3-dimensional data sets of unknown structure. For example, seismologists have made several new findings about the subduction of the Pacific plate underneath the Asian plate by viewing in SpaceGraph a cloud of points representing the 3-D locations of the epicenters of all recent seismic events in a volume of the Earth's crust in the vicinity of Japan.

5 PERCEPTION OF ORIENTATION

Having described volumetric displays, we turn now to some results of perceptual research we conducted using SpaceGraph. The practical use of volumetric displays, in tasks such as the discovery of structure in high-dimensional data sets, requires some basic understanding of our capabilities to perceive structure and relationships within such a display.

One fundamental question concerns the accuracy and speed with which one can perceive the orientation of a three-dimensional object if it is presented at a random rotation in space (Hintzman, O'Dell & Arndt, 1981; Fitts & Seeger, 1953). We addressed this question in a series of experiments in which we used a simple outline cube as the displayed stimulus object, as shown in Figure 7. The procedures and results of these experiments are described more fully in Huggins & Getty (1981, 1982). An orientation cue was drawn on the cube's bottom face, consisting of a capital letter V with its apex almost touching the front edge. One of the other five sides, chosen randomly on each trial of the experiment, was marked with a "button" consisting of two concentric circles. The observer's task was to decide which face was so marked.

The experimental apparatus is diagrammed in Figure 8. Briefly, the experimental procedure was as follows: To begin a trial, the observer pressed a

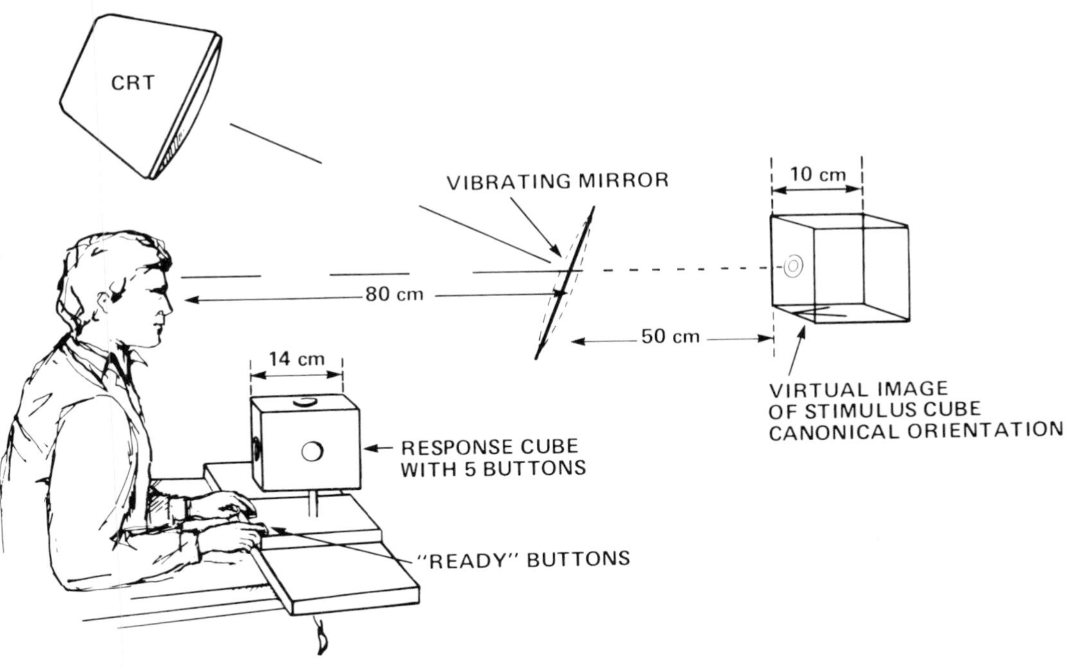

Figure 8 Sketch of an observer seated behind the response cube, viewing the stimulus cube in SpaceGraph. Reproduced from Huggins and Getty, 1981.

"Ready" key. Two seconds later the stimulus cube appeared, with a button showing on one of its five faces, excluding the bottom (which bore the V). The observer held down the Ready key until he had decided *which* face of the stimulus cube was marked, and then pressed the physical button on the corresponding face of a cubical metal response box as quickly as possible. Two times were recorded for each trial: the "reaction" time, from the presentation of the stimulus cube to the release of the Ready key, and the "movement" time from the release of the Ready key to the depression of the response key. We will discuss only reaction time here. It is the measure of principal interest, being the time required by an observer to determine how the cube was oriented and to decide which face of the cube was marked.

In our first experiment, the cube was seen in orientations rotated only about the vertical Y-axis of the cube. It could appear in any one of 24 different orientations, which together comprised a complete rotation of the cube image, about its vertical Y-axis in 15 degree steps. Photographs of the 24 orientations are shown in Figure 9; however, all five buttons appear here in each cube image, for economy of presentation, whereas observers saw only one randomly chosen button on a trial.

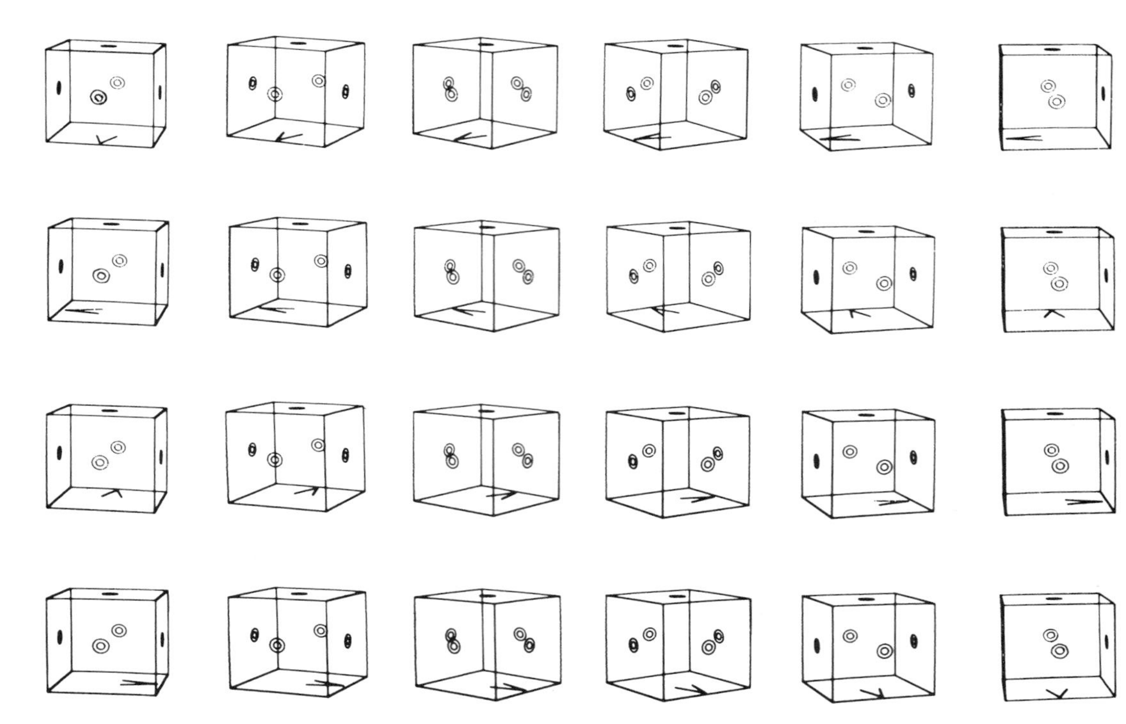

Figure 9 The 24 orientations of the stimulus cube used in the Y-axis orientation experiment. All five stimulus buttons are shown in each view here, for economy of presentation; on each trial, an observer saw only one of the five buttons. Reproduced from Huggins and Getty, 1981.

Figure 10 shows mean reaction time pooled across three observers. Reaction time, in milliseconds, is plotted on the ordinate, while cube orientation, in degrees of rotation, is plotted on the abscissa. Data points for one-quarter revolution in each direction from the head-on view are duplicated at the left and right sides of the plot for the sake of continuity.

The functions shown in Figure 10 fall into three distinct groups. The lowest function, for the TOP face, is flat showing that this decision was not affected by the orientation of the cube. The middle pair of functions, for the NEAR and FAR cube faces, show a generally flat plateau for most orientations, with linearly increasing skirts for the first quarter-rotation in either direction from the head-on orientation. The highest pair of functions, for the LEFT and RIGHT cube faces, exhibit dramatic, roughly linear, increases in decision time for increasing rotation in either direction from head-on, with a ragged peak at one-half revolution. We believe that the three different functions reflect the operation of three different strategies for determining which face of the cube is marked. The strategies differ both in their efficiency, and in the conditions under which they can be applied.

The first of the three is a *Spatial* strategy: Consider the TOP face of the cube in Figure 9. With rotation about the vertical axis, neither the position within the retinal image nor the retinal shape of the TOP key changed as the

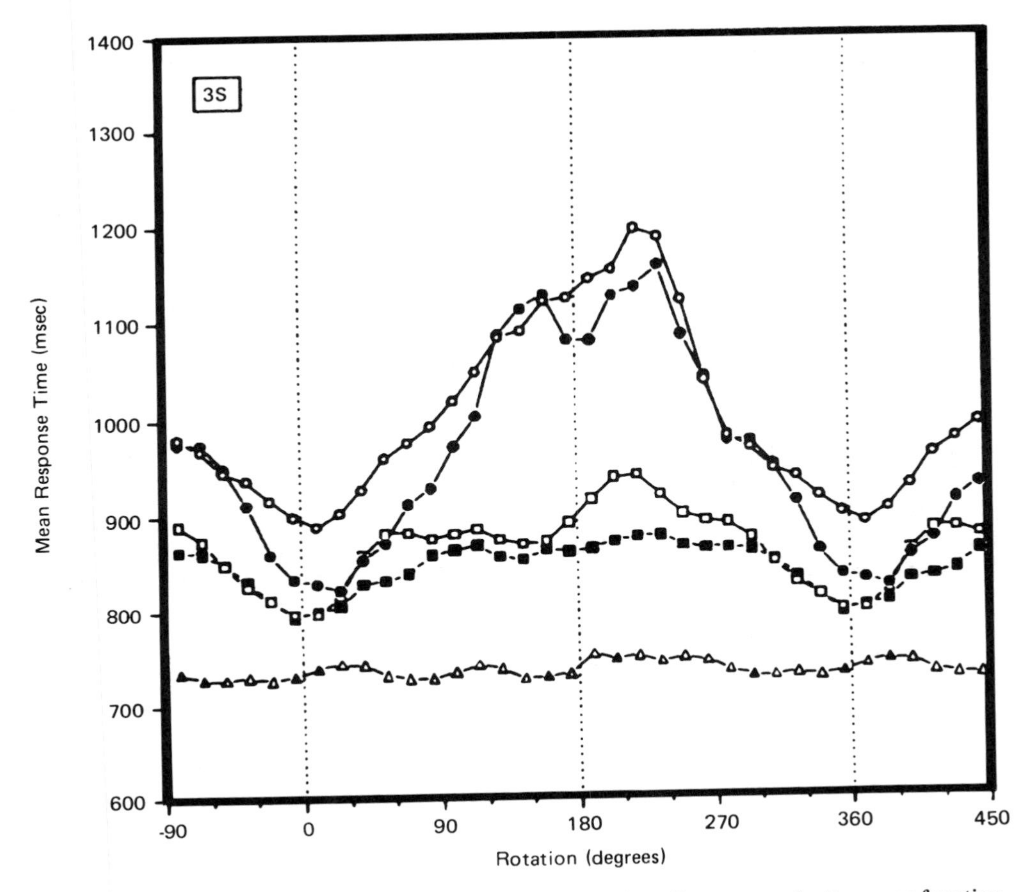

Figure 10 Mean reaction times in the Y-axis task for each of the five response buttons, as a function of the orientation of the stimulus cube. Coding of the functions for the five stimulus cube faces is as follows: left face is given by the open circles, right face by the closed circles, far face by the open squares, near face by the closed squares, and top face by the open triangles. Reproduced from Huggins and Getty, 1981.

orientation of the stimulus cube was altered. Therefore, observers were able to use a highly efficient and compatible spatial mapping strategy for selecting the TOP response: if you see a button near the top of the visual field, then press the top response button.

Now, consider the four keys on the vertical sides of the cube. The *only* information that specifies which face of the cube bears the stimulus key is the position of the V relative to the stimulus key. With respect to the V, the four stimulus keys fall into two groups, with the LEFT and RIGHT keys in one and the NEAR and FAR keys in the other.

Consider first the LEFT and RIGHT keys. The V exhibits lateral symmetry, so that although a stimulus key appearing to one side of the V can be quickly identified as *either* the LEFT or RIGHT key, picking the correct one of the two is difficult. The solution for the observer is to use a *Relational* strategy. Reports from observers suggest that they imagined themselves looking from the apex of the V towards its points, and then deciding whether the displayed key was on the left or the right. This implies a sort of mental rotation, although it is the observer's *body* image that is rotated rather than the stimulus image. The possibility that observers perform some sort of mental rotation gains some credibility from the strong similarity between the shape of our reaction time functions and those obtained by other investigators in studies of mental rotation (Cooper & Shepard, 1973). Those studies have reported results consistent with the notion that human observers are able to mentally rotate an image, that the rotation takes place at a uniform angular velocity, and that during rotation the image passes through a continuum of intermediate rotated states.

With regard to the NEAR and FAR stimulus keys the same Rotational strategy could potentially be applied. That is, the observer could mentally rotate his body image into alignment with the V, and then make a NEAR or FAR response according to the near or far location of the stimulus key. This Rotational strategy would account for the sloping skirts of the NEAR and FAR response functions for orientations near the head-on position. However, the flat plateaus observed in the NEAR and FAR functions over much of the range of orientations probably reflects the use of a third strategy, a *Relational* strategy.

This strategy is possible because the apex of the V always points towards the NEAR key, while the open end always points towards the FAR key. Making use of this invariant relationship would yield reaction times that are relatively independent of cube orientation and, except for orientations near the head-on position, apparently faster than those obtained with the *Relational* strategy.

In other experiments, we have rotated the cube about its X-axis, Z-axis, or randomly about any one of the three major axes. The results are consistent

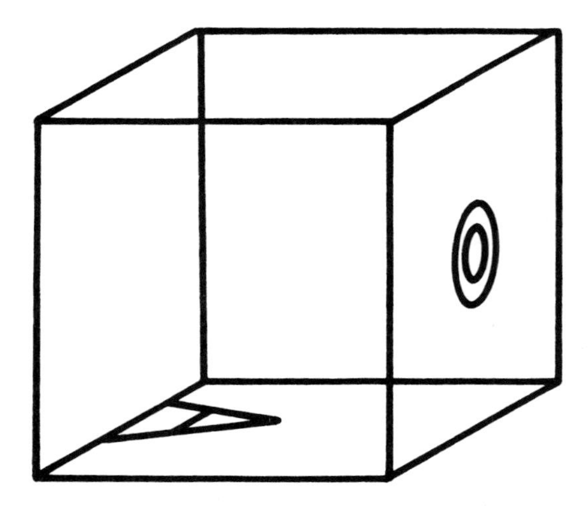

Figure 11 The stimulus cube for the Y-axis experiment in which the orientation cue was a capital letter "A" with its base on the left edge of the bottom face. Reproduced from Huggins and Getty, 1982.

with the use of the strategies just described. We have performed other tests of our understanding by altering the relationship between the displayed symmetry of the orientation cue and pairs of cube faces. In one such experiment, the orientation cue was again a left-right symmetric capital letter drawn on the bottom face of the cube, as shown in Figure 11. The letter was an A, rather than a V, to minimize possible interference from the earlier tasks. It appeared with its base almost touching the *left* edge rather than the near edge of the cube.

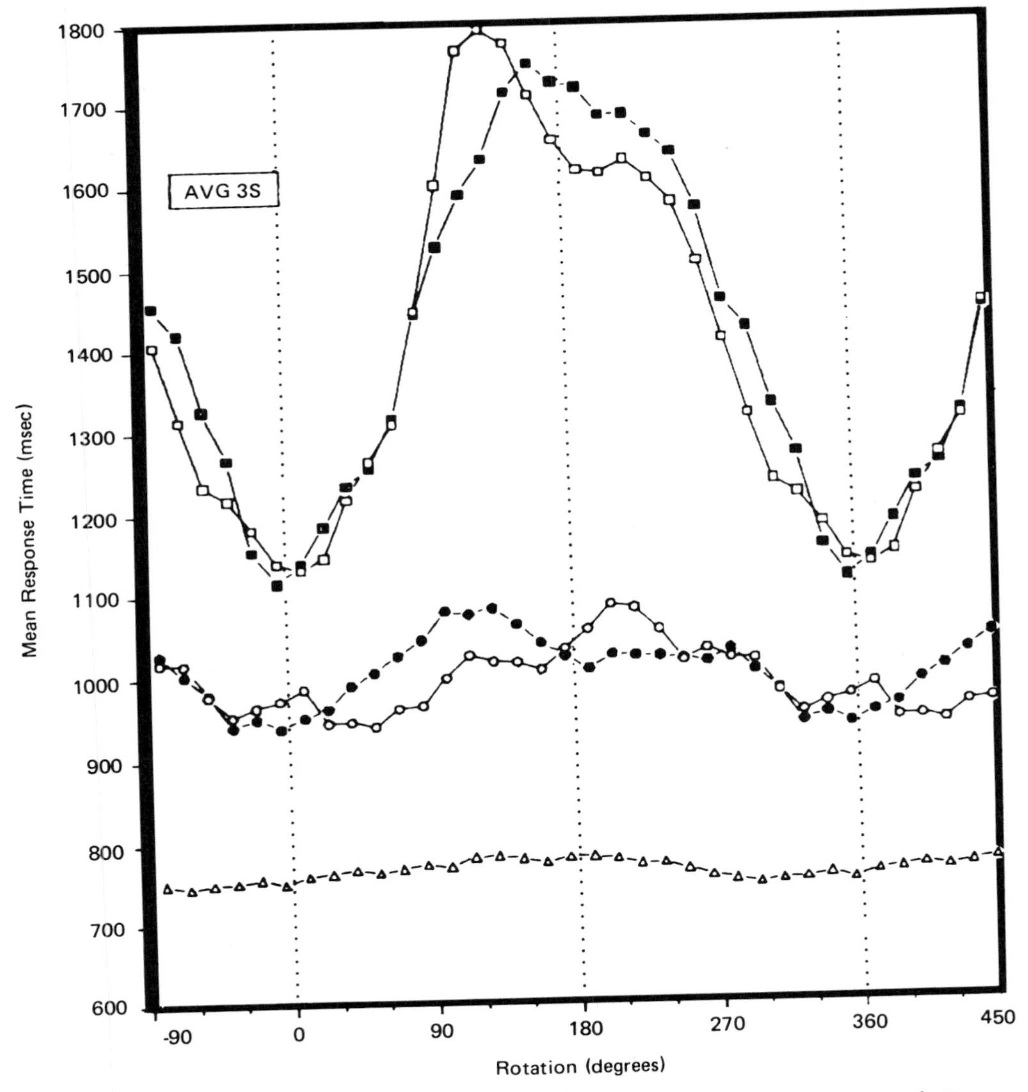

Figure 12 Mean reaction times in the "A" orientation cue task for each of the five response buttons, as a function of the orientation of the stimulus cube. As in Figure 10, the coding of the functions for the five stimulus cube faces is as follows: left face is given by the open circles, right face by the closed circles, far face by the open squares, near face by the closed squares, and the top face by the open triangles. Reproduced from Huggins and Getty, 1982.

The shape and the position of this orientation cue were asymmetric with respect to the LEFT and RIGHT stimulus keys, but symmetric with respect to the NEAR and FAR keys, thus reversing the relationships of the previous experiment.

The results of the experiment, shown in Figure 12, are similar to those we've just described in that there are again three groups of functions. As we would predict, the LEFT/RIGHT responses rather than the NEAR/FAR, now show the relatively flat plateau that we associate with the Relational strategy, and the NEAR/FAR responses, rather than the LEFT/RIGHT, now show the dramatically peaked function we associate with the use of the Rotational strategy. Thus, moving the orientation cue from the front edge of the bottom face to left edge of the bottom face has reversed the types of reaction time function associate with the LEFT/RIGHT and NEAR/FAR pairs of responses.

As one further test, we decided to make the cue asymmetric in shape relative to *both* pairs of cube faces, as depicted in Figure 13. A modified letter V was used, drawn on the bottom face of the stimulus cube with its apex almost touching the front edge as before, but with an additional cross-bar or serif added at the top and to the left. The serif made the V LEFT/RIGHT asymmetric as well as UP/DOWN asymmetric. Under these conditions, we predicted that observers should use a Relational strategy for all four sides of the cube.

The results, shown in Figure 14, are in reasonable agreement with the prediction. The large peaks formerly associated with the LEFT-RIGHT responses and the Rotational strategy have disappeared, although small residual peaks are still present. It appears, then, that asymmetry is important in permitting an observer to utilize the relatively efficient Relational strategy in preference to the slower Rotational strategy.

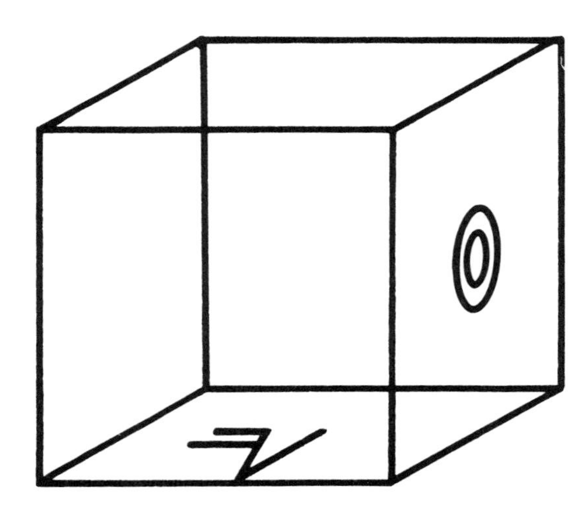

Figure 13 Illustration of the stimulus cube for the Y-axis experiment in which the orientation cue was a capital letter "V" oriented as in the first experiment, but with serifs added to the left arm. Reproduced from Huggins and Getty, 1982.

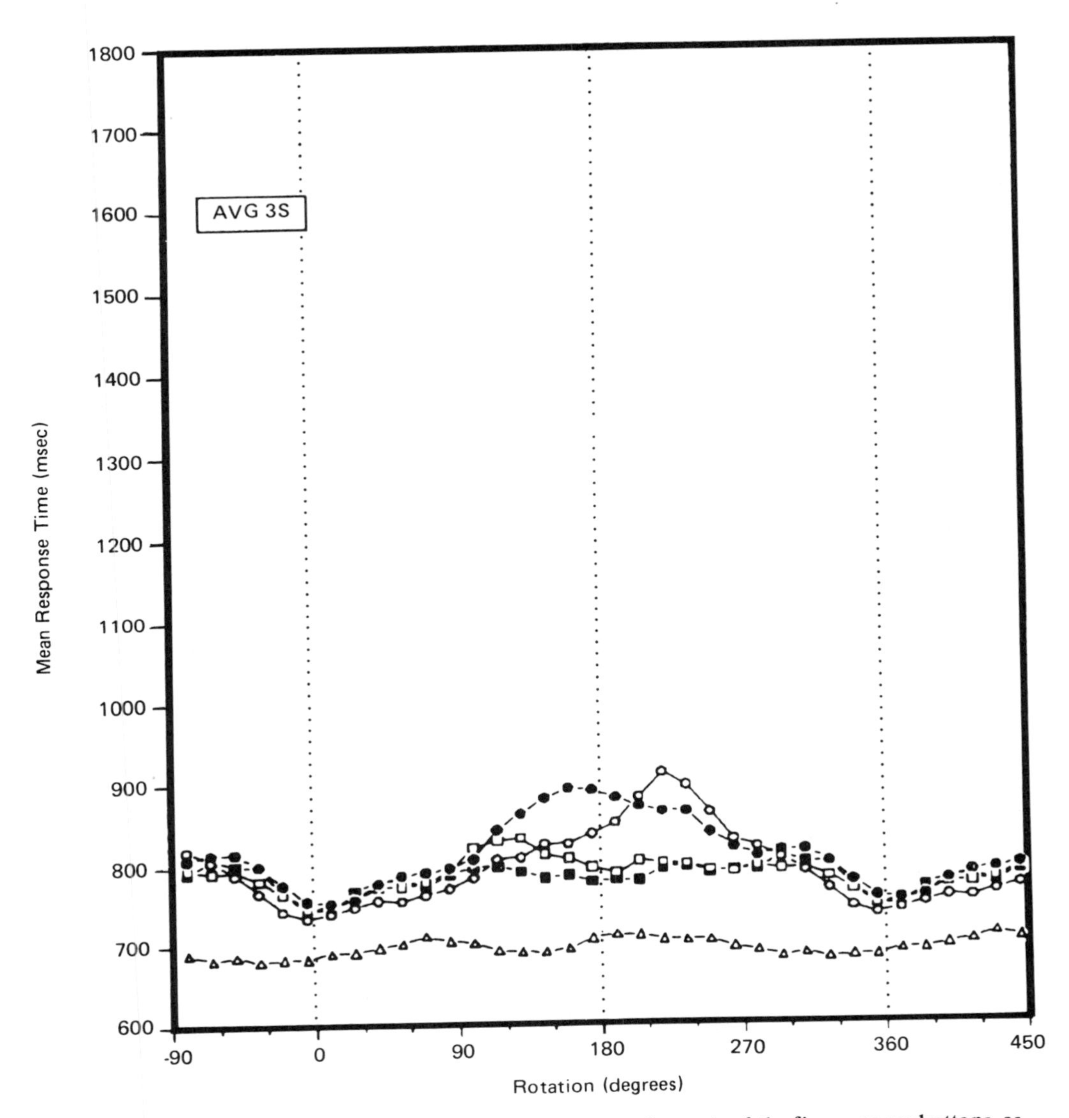

Figure 14 Mean reaction times in the asymmetric "V" task for each of the five response buttons, as a function of the orientation of the stimulus cube. As in Figures 10 and 12, the coding of the functions for the five stimulus cube faces is as follows: left face is given by the open circles, right face by the closed circles, far face by the open squares, near face by the closed squares, and the top face by the open triangles. Reproduced from Huggins and Getty, 1982.

To summarize these results, we have found that observers appear to use at least three strategies in determining orientation of a rotated object: (1) a very fast Spatial strategy, (2) a reasonably fast Relational strategy, and (3) a much slower Rotational strategy. The decision times measured in these studies may be viewed, in a more general sense, as indicators of the perceptual difficulty and cognitive load imposed on the observer by varying types and degrees of object rotation. As such, the results suggest several guidelines for applications in which objects or scatterplots of data are to be rotated in a 3-D display. First the observer should be able to readily locate and identify the axis of rotation. Second, the data should be provided with a reference set of principal axes or directional indicators that rotate with the data. And most importantly, these reference axes or indicators should be distinctive, one from another, and asymmetric. In the case of a displayed object, the object itself should be asymmetric along each of its principal axes. For example, an icon of an airplane in an air-traffic control display should have its bilateral symmetry removed by, say, marking the left wing distinctively from the right.

6 PERCEPTION OF DIRECTION

In a second line of research, we have studied the perception of direction (see Getty and Huggins, 1984, for a fuller discussion). The issue of trajectory and direction perception is clearly important in the context of data exploration. One would like to know, for example, how accurately a direction in space can be perceived and estimated, how accurately the intersection of two trajectories can be predicted (Salthouse & Prill, 1983; Runeson, 1975), how well angles can be judged, or how sensitive one is at detecting parallelism or orthogonality.

In the task that we developed, shown in Figure 15, the observer was presented on each trial with an outline cube, about 18 cm on a side, with a 3

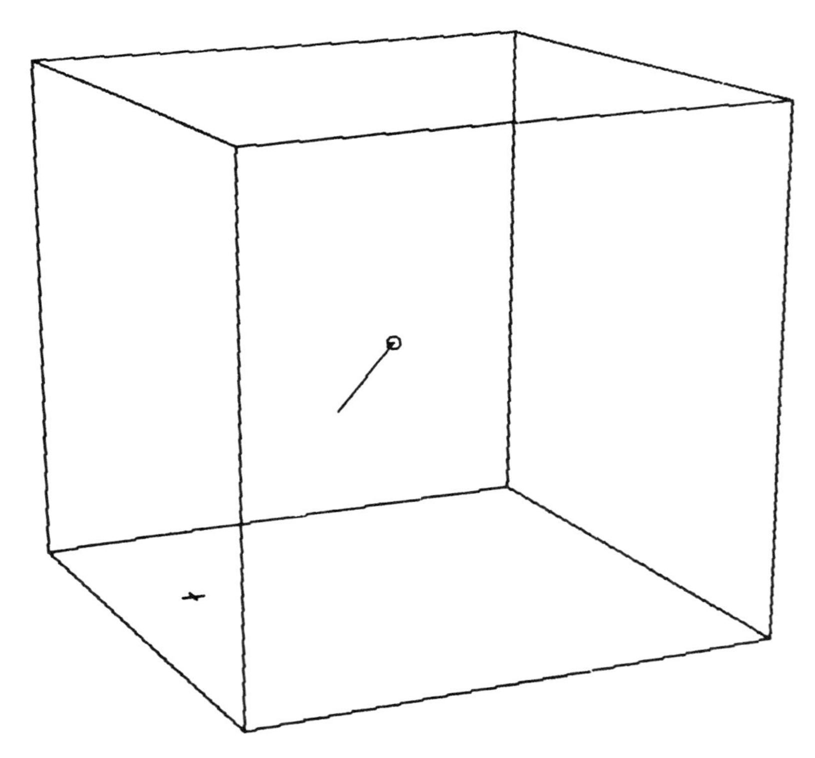

Figure 15 The stimulus cube used in the direction experiment, with a vector anchored at the center of the cube pointing at a target location on the left face, which is marked with a small cross-hair. Reproduced from Getty and Huggins, 1984.

cm long vector pointing from the cube's center towards one of the five sides farthest from the observer. The observer's task was to place a point, using a graphical tablet as a response device, on the cube's surface where the vector would pierce it if extended. The graphics tablet was marked with an outline representing the back five faces of the cube unfolded onto a flat surface, with the back face of the cube in the middle. As the observer moved the stylus within the outline on the tablet, a visible point representing the piercing point—shown in the figure as a small cross—moved correspondingly on the surface of the cube in the display.

The observer viewed the cube from a position back along the depth axis passing through the center of the cube. Figure 16 shows the view in perspective from the observer's vantage point, with an imaginary dotted line connecting the visible vector with it true piercing point on the left side.

There were 80 vectors used in all, 16 pointing towards each of the five faces. Each observer made 20 responses for each of the 80 vectors, in random

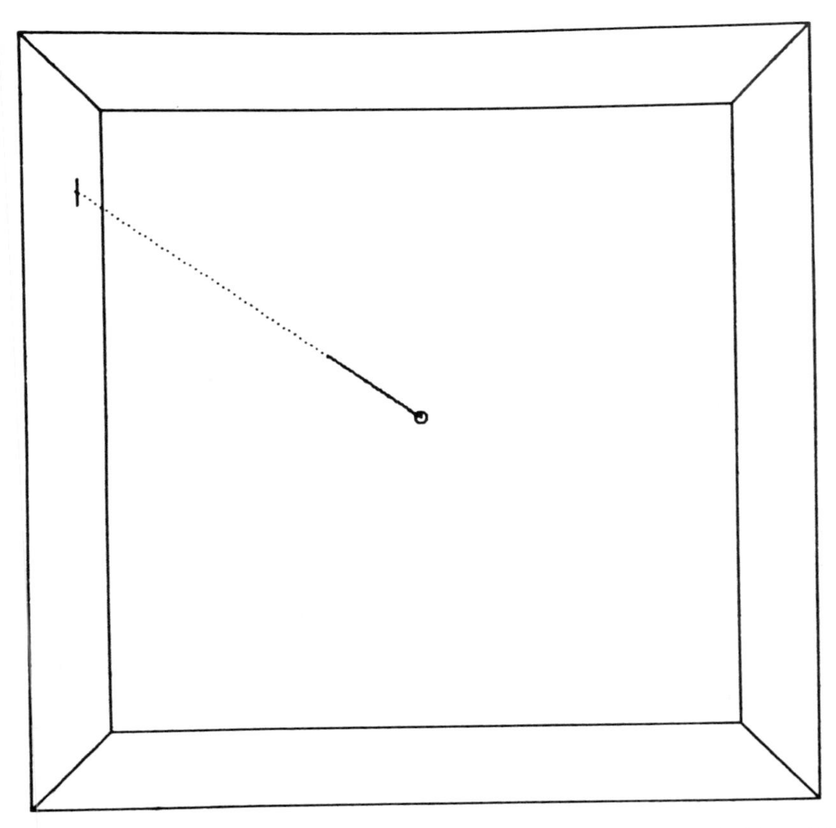

Figure 16 Perspective view of the stimulus cube, vector, and target depicted in Figure 15, as seen from the observer's viewpoint. A dotted line, not actually present in the experiment, projects from the vector to its true intersection with the left face. Reproduced from Getty and Huggins, 1984.

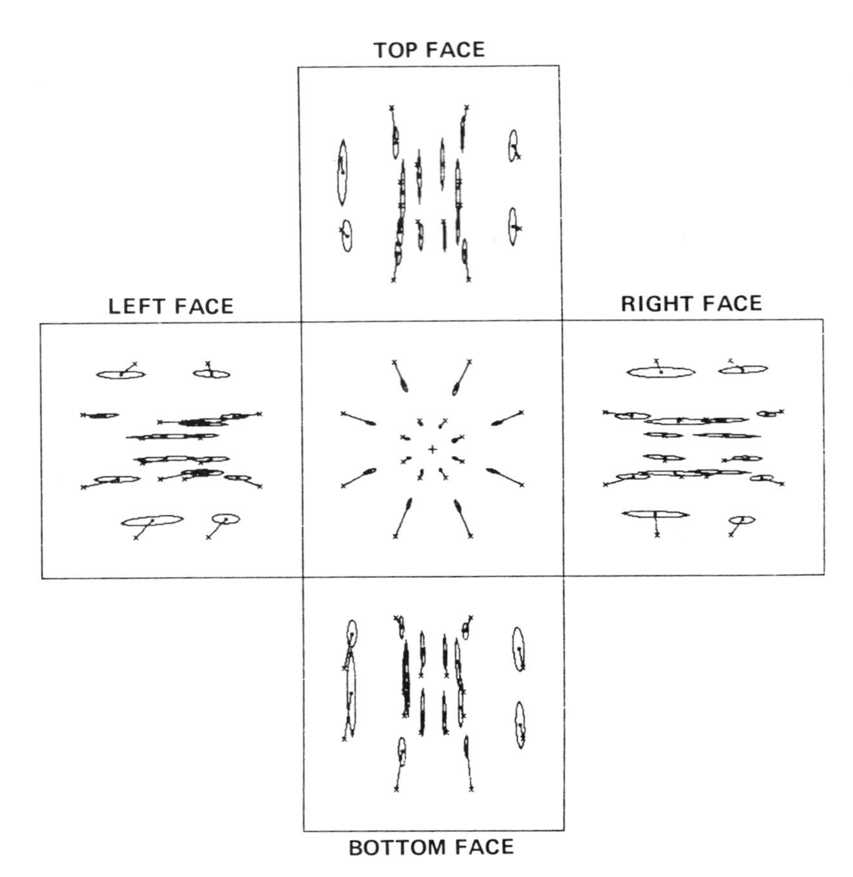

Figure 17 Data for one observer shown on an opened-out cube. The X's mark the true piercing points of the 80 vectors. The dot at the center of each ellipse indicates the mean perceived piercing point for the target connected to it by a straight line. Each ellipse indicates the two-dimensional standard deviation of the responses. Reproduced from Getty and Huggins, 1984.

order. In analyzing the data from this experiment, we found three ways in which our observers' perceptions of direction deviated systematically from the true direction.

Figure 17 shows a summary of the data for one observer, displayed on an opened-out cube with its far face at the center. The small x's on each face represent the true piercing points of the 80 different target vectors, while the dots at the center of the ellipses mark the mean of the observer's perceived piercing points. The connecting linear segments indicate the direction and magnitude of the error. Each ellipse surrounding one of the response-means indicates the distribution of responses, and encloses those responses falling within one standard deviation of the mean. The first of the systematic deviations we found can be easily seen by examining the data for the far face in the center of Figure 17. The response means are all displaced radially inward from the true piercing points towards the center of the face. And, more generally, perceived

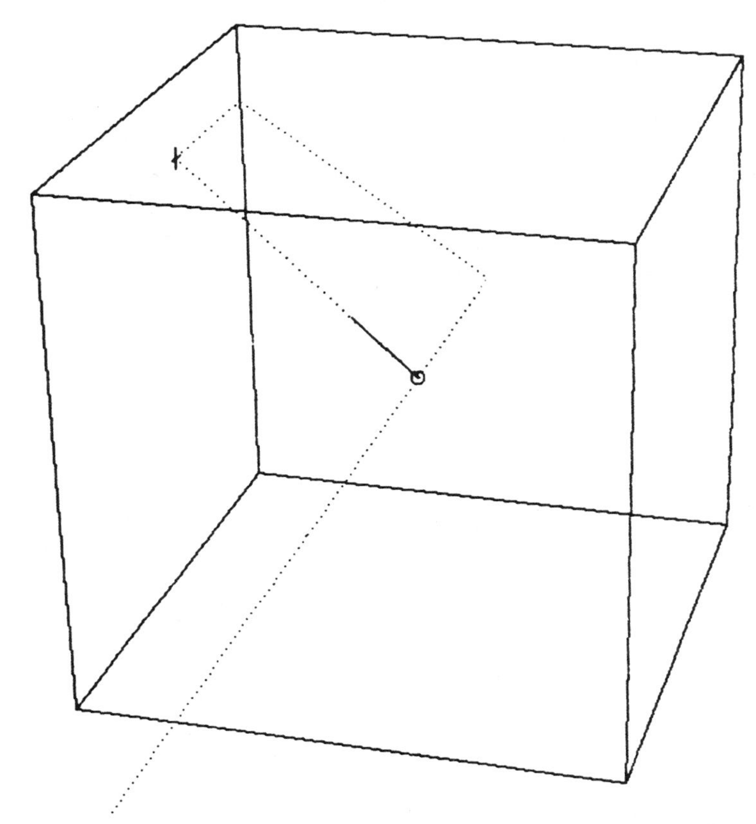

Figure 18 Illustration showing the projection of a vector, and its linear extension to its piercing point, along the far and left sides of the cube, as projected from the observer's viewpoint. Reproduced from Getty and Huggins, 1984.

direction is displaced along the linear extension of a given vector as it would be projected upon the inner surface of the cube from the observer's vantage point.

So, for example, as shown in Figure 18, a vector that would pierce the left face of the cube has a projection, from the observer's viewpoint, that extends radially outward along the far face until it intersects the left face and then continues as a horizontal segment on the left face to the true piercing point. And in looking at the responses for the left face in Figure 17, for example, we see that the ellipses are oriented such that most of the response variability falls along those horizontal projection lines.

The second type of error we found in perceived direction, and perhaps the most important one, is a systematic deviation of the perceived piercing point away from the true point towards the principal vertical or horizontal axes of the space. The effect can be seen for individual targets in Figure 17. For example, looking at the data for the right face, the mean response points for the two targets highest on the face, and those for the two targets lowest on the face, are seen to be displaced towards the horizontal bisector of the face. The full extent of the effect is best seen by looking at response error as a function of the meri-

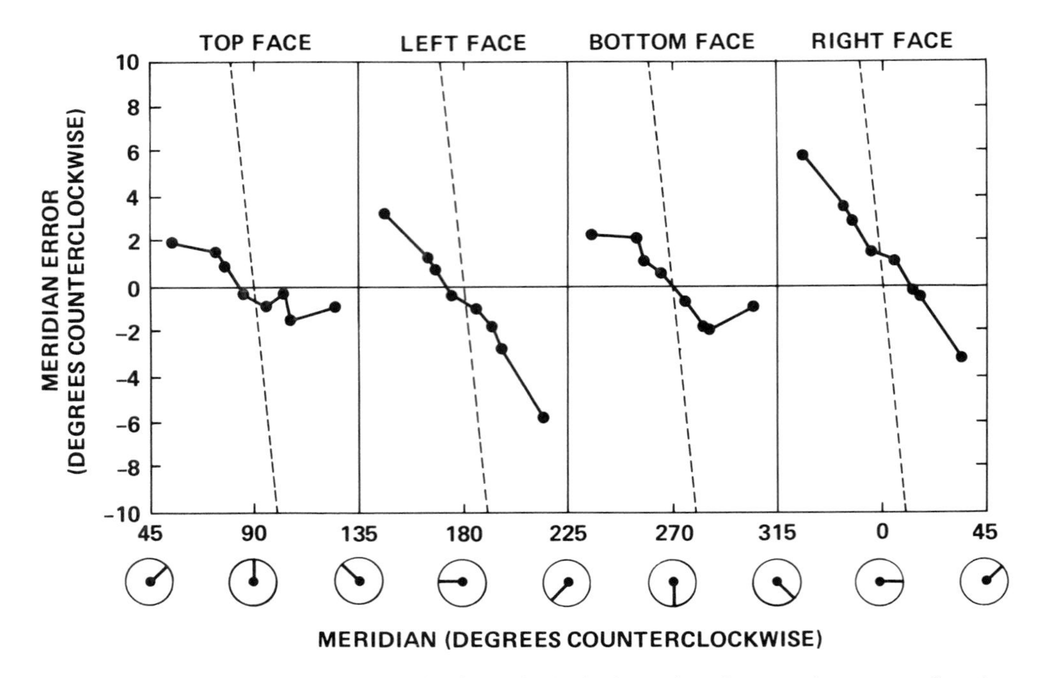

Figure 19 Mean error in meridian angle of perceived piercing points, for one observer, as a function of the meridian angle of the vector. The dashed lines indicate the error that would result if the observer always perceived a vertical vector for the top and bottom faces, or a horizontal vector for the left and right faces. Reproduced from Getty and Huggins, 1984.

dian angle of each vector, which we show for one observer in Figure 19. The meridian angle is defined as the angle between the horizontal axis and the projection of the vector onto the observer's frontal plane. Meridian angle is shown increasing in a counterclockwise direction on the abscissa of the figure—so that vectors pointing at the top face are followed by vectors pointing at the left face, and so on. The meridian directions of the vectors are also shown graphically by the little clock faces distributed along the abscissa. The perceptual error, measured in degrees of meridian angle, is shown on the ordinate.

A first observation is that, for each face, the perceptual error is essentially a linear function of the vector's angle, showing no error when the vector is either horizontal or vertical. This implies that the perceptual error is proportional to the angular deviation of the vector from either the horizontal or vertical positions. Secondly, the direction of the error—shown by the sign of the slope—is always such that the perceived direction is more nearly vertical or horizontal than the true direction. And finally, the magnitude of the effect—as shown by the magnitude of the slope—is greater for the left and right faces than for the top and bottom faces. Some feeling for the magnitude of these effects is provided by the dashed lines which indicate the error that would result if an observer *always* responded with a vertical or horizontal perceived vector, whichever was closer, regardless of the actual vector direction. We have observed analogous effects of error in perception of the depth component

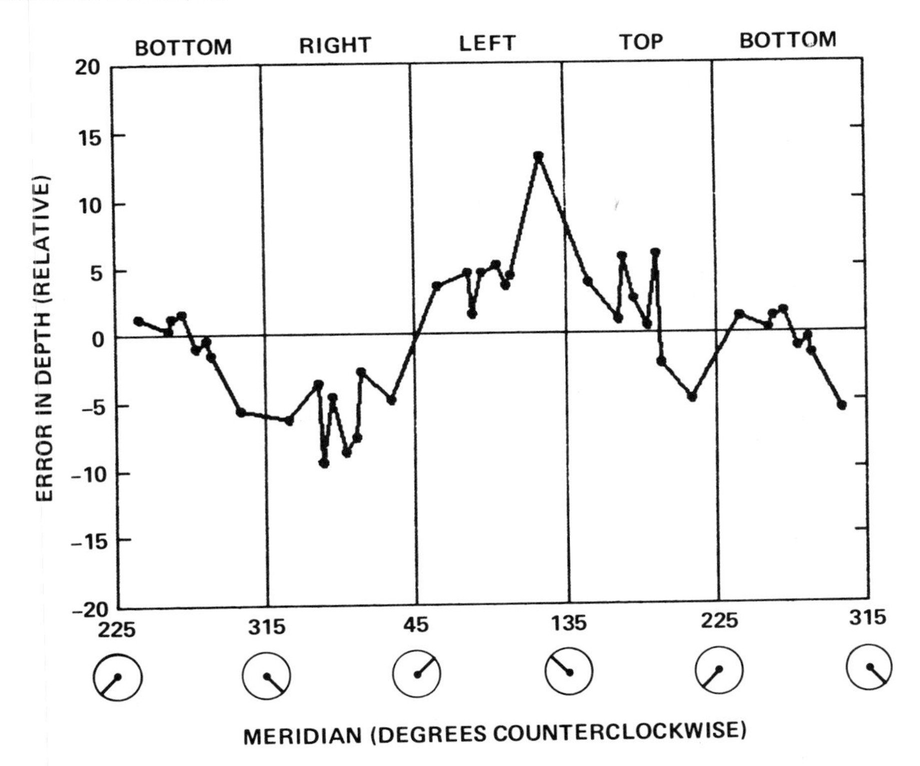

Figure 20 Mean error in depth of perceived piercing points, for one observer, as a function of the meridian angle of the vector. Reproduced from Getty and Huggins, 1984.

of vectors. In short, oblique vectors are generally perceived as being more nearly horizontal or more nearly vertical than they truly are, and this distortion appears to be greater for vectors near the horizontal than for those near the vertical.

Finally, a third, somewhat bizarre, effect we observed was a distortion of the depth component of a vector's direction, as a function of the vector's meridian in the frontal plane. One way to characterize the effect is to say that vectors lying truly in the observer's frontal plane would be perceived as lying in a plane tilted back obliquely in depth. This is revealed in Figure 20 for one observer by a sinusoidal appearing curve relating depth error to meridian angle, with a period of 360 degrees. We believe this effect arises from two additive causes, one of which is a perceptual tilt of the vertical horopter (Tyler & Scott, 1979; Cogan, 1979; Nakayama, 1977). The other is mild, uncorrected astigmatism. Three of our observers, upon examination, have mild, uncorrected astigmatism—as do many of us—and all three show this distortion. The axis and amount of the slant in depth appears to be determined by the axis and amount of astigmatism. Our fourth observer, who has essentially no astigmatism, does not demonstrate this meridian-dependent depth error. In summary, our research on the perception of direction has shown that it is subject to several systematic biases. While all of these effects are relatively small in absolute magnitude, they are quite consistent and capable of affecting fine judgements or discriminations.

7 CONCLUSIONS

In conclusion we suggest that volumetric displays offer several significant advantages over stereo-pair or motion parallax displays, particularly in permitting the viewer to browse naturally and move about the displayed data. This freedom to spontaneously alter one's vantage point on the display is clearly important, for example, in searching for unknown or unsuspected structure in data.

Our research has identified several limitations and biases in the perception of orientation and direction. However, knowing of their existence, most can be either eliminated or corrected for in circumstances where they might be of significance.

REFERENCES

1. Braunstein, M.L. *Depth Perception Through Motion.* New York: Academic Press, 1976.

2. Cogan, A.I. The relationship between the apparent vertical and the vertical horopter. *Vision Research*, 1979, **19**, 655-665.

3. Cooper, L.A. and Shepard, R.N. Chronometric studies of the rotation of mental images. In W.G. Chase (Ed.), *Visual Information Processing.* New York: Academic Press, 1973.

4. Doner, J., Lappin, J.S., and Perfetto, G. Detection of three-dimensional structure in moving optical patterns. *Journal of Experimental Psychology: Human Perception and Performance*, 1984, **10**, 1-11.

5. Fitts, P.M. and Seeger, C.M. S-R compatibility: Spatial characteristics of stimulus and response codes. *Journal of Experimental Psychology*, 1953, **46**, 199-210.

6. Getty, D.J. and Huggins, A.W.F. Display-control compatibility in 3-D displays, 3: Perception of direction. BBN Technical Report 5582, 1984.

7. Gibson, J.J. and Gibson, E.J. Continuous perspective transformations and the perception of rigid motion. *Journal of Experimental Psychology*, 1957, **54**, 129-138.

8. Hintzman, D.L., O'Dell, C.S. and Arndt, D.R. Orientation in cognitive maps. *Cognitive Psychology*, 1981, **13**, 149-206.

9. Huggins, A.W.F. and Getty, D.J. Display-control compatibility in 3-D displays, 1: Effects of orientation. BBN Technical Report 4724, 1981.

10. Huggins, A.W.F. and Getty, D.J. Display-control compatibility in 3-D displays, 2: Effects of cue symmetry. BBN Technical Report 5101, 1982.

11. Johansson, G. Visual motion perception. *Scientific American*, 1975, **232**, 76-88.

12. Nakayama, K. Geometric and physiological aspects of depth perception. *Proceedings of the Society of Photo-Optical Instrumentation Engineers*, 1977, **120**, 2-9.

13. Runeson, S. Visual prediction of collision with natural and nonnatural motion functions. *Perception and Psychophysics*, 1975, **18**, 261-266.

14. Salthouse, T.A. and Prill, K., Analysis of a perceptual skill. *Journal of Experimental Psychology: Human Perception and Performance*, 1983, **9**, 607-621.

15. Tyler, C.W., and Scott, A.B. Binocular vision. In R.E. Records (Ed.), *Physiology of the Human Eye and Visual System*. New York: Harper & Row, 1979.

16. Ullman, S. *The Interpretation of Visual Motion*. Cambridge, MA: MIT Press, 1979.

17. Wallach, H. and O'Connell, D.N. The kinetic depth effect. *Journal of Experimental Psychology*, 1953, **45**, 205-217.

Index